Intelligent and Soft Computing Systems for Green Energy

Scrivener Publishing
100 Cummings Center, Suite 541J
Beverly, MA 01915-6106

Publishers at Scrivener
Martin Scrivener (martin@scrivenerpublishing.com)
Phillip Carmical (pcarmical@scrivenerpublishing.com)

Intelligent and Soft Computing Systems for Green Energy

Edited by
A. Chitra
V. Indragandhi
and
W. Razia Sultana

Scrivener Publishing

WILEY

This edition first published 2023 by John Wiley & Sons, Inc., 111 River Street, Hoboken, NJ 07030, USA and Scrivener Publishing LLC, 100 Cummings Center, Suite 541J, Beverly, MA 01915, USA

For more information about Scrivener publications please visit www.scrivenerpublishing.com.

Wiley Global Headquarters
111 River Street, Hoboken, NJ 07030, USA

For details of our global editorial offices, customer services, and more information about Wiley products visit us at www.wiley.com.

Library of Congress Cataloging-in-Publication Data

ISBN 9781394166374

Front cover images supplied by Pixabay.com
Cover design by Russell Richardson

Set in size of 11pt and Minion Pro by Manila Typesetting Company, Makati, Philippines

Printed in the USA

Contents

Preface

Traditional power grids are being transformed into Smart Grids (SGs) to solve the problems of uni-directional information flow, energy wastage, growing energy demand, reliability and security. Smart grids offer bidirectional energy flow between service providers and consumers, involving power generation, transmission, distribution, and utilization systems. Smart grids employ various devices for the monitoring, analysis and control of the grid, deployed at power plants, distribution centers, and in consumers' premises.

A smart grid requires connectivity, automation, and soft computing techniques for intelligent monitoring. This is achieved with the help of intelligent soft computing systems. This aids SG systems to support various network functions throughout the generation, transmission, distribution, and consumption of energy by incorporating Internet of Things (IoT) devices, such as sensors, actuators and smart meters, as well as by providing the connectivity, automation and tracking for such devices.

The chapters in this book are configured to address the challenges facing intelligent and soft computing techniques deployed in various fields with possible solutions. This book can serve as a resource for industrial professionals, engineers, and researchers working in the domain of intelligent and soft computing systems.

Contributions to this book have been received from various esteemed national and international institutions. We would like to extend our sincere thanks to all the valuable contributors for the wonderful research contributions and time. Our sincere thanks to Vellore Institute of Technology, Vellore for providing all the support to turn this book into a reality. We would like to extend our heartfelt gratitude for the Wiley-Scrivener team for providing endless support.

1

Placement and Sizing of Distributed Generator and Capacitor in a Radial Distribution System Considering Load Growth

G. Manikanta[1], N. Kirn Kumar[2]*, Ashish Mani[1] and V. Indragandhi[3]

[1]*Electrical & Electronics Engineering Department, A.S.E.T, Amity University Uttar Pradesh, Noida, UP, India*

[2]*Department of Electrical & Electronics Engineering, M S Ramaiah Institute of Technology, Bangalore, India*

[3]*School of Electrical Engineering, Vellore Institute of Technology, Vellore, Tamil Nadu, India*

Abstract

Annual load growth in a distribution system is expanding consistently, which results in underprivileged voltage-regulation and increment in power losses. Independent implementation of Distributed generation (DGs) along with capacitors is chosen as alternative techniques to decrease the power loss in the network. Optimal location and capacity of Capacitors along with DGs not only maximize the percentage power loss reduction but also increase the voltage profile, if optimal location and capacity is appropriate. Inappropriate placement and competence of capacitors and DGs leads the system to an increase in power loss. The best location and competence of capacitors and DGs is a difficult nondifferentiable combinatorial optimization problem, which has been applied to solve various engineering optimization problems like improvement in reliability, loadabilty, loss minimization, etc., using various analytical and evolutionary algorithms. In this study economic load growth is modelled with a predetermined yearly load expansion for the base year and next five years. In this work the main contributions are made with placing and sizing the DGs and Capacitors to minimize the power losses for every year. Tabulated results demonstrate that simultaneous implementation of

**Corresponding author*: neelamseeti.kiran@gmail.com

A. Chitra, V. Indragandhi and W. Razia Sultana (eds.) Intelligent and Soft Computing Systems for Green Energy, (1–24) © 2023 Scrivener Publishing LLC

Capacitor and DGs has high reduction in power loss for every year including base year in comparison with independent implementation of DGs and Capacitors. For obtaining the best location and competence of capacitors and DGs an Adaptive Quantum inspired evolutionary Algorithm (AQiEA) is applied successfully. AQiEA uses probabilistic representation with Q-bit and does not require any additional operators. The effectiveness of AQiEA is verified on a standard test bus system, i.e., 85 bus system. Simulated results exhibit that the proposed algorithm has superior performance in comparison to the algorithms in the available literature.

***Keywords*:** Distributed generators, capacitor, load growth, power losses, 85 bus system, AQiEA

1.1 Introduction

The power demand at distribution network keeps on increasing day to day and in some scenarios the generated power is unable to meet the required load demand. Load demand at distributed network is exponentially increasing from day to day, due to industrial, domestic, commercial, municipal, residential and irrigation needs. Load growth in distribution network is a natural phenomenon which results in increased power losses (both active and reactive) and increased voltage drop. Many methods and techniques have been executed in distributed networks in order to decrease the losses. Over the last few decades, DGs and Capacitors in the distributed network are used to reduce the losses. Implementation of DG in the network will reduce losses and also improve the system voltage. DGs are defined as small power generating sources, located nearer to the load centers and size varies from kW to few MW [1]. DGs are used in distribution system due to its ease of implementation, environmental friendly technologies and low maintenance [2]. Different types of DG are available with respect to their modular structure and size. Therefore, their impact on the distribution system varies depending on location and capacity [3]. Compensation of reactive power in the distribution network is generally provided by the capacitors. Installation of Capacitors nearer to load centres reduces the power losses and maintains the voltage profile within permissible limits [4].

Analytical-based methods are easy to implement; their major drawback is implementation of single DG or Capacitor with large size in order to reduce the losses [5–7]. In most of the works, minimization of power loss in distribution system with DG integration, population-based metaheuristics are used as solution strategies. Symbiotic Organism Search [8], Particle Swarm Optimization [9], Simulated Annealing [10], Big-Bang

Big-crunch optimization [11] and Fire Work Algorithm [12] are some of the well-known optimization techniques used to reduce the losses. However, Simulated Annealing [13], Firefly Algorithm [14], Plant Growth Simulation Algorithm [15], Teaching Learning-Based Optimization [16] and Particle Swarm Optimization [17] are some population-based meta-heuristic techniques used for optimal location and capacity of Capacitors to reduce the losses. Some methods have considered only injection of real power (DG) and some other methods considered only injecting the reactive power (Capacitor) into the system, and other methods have considered simultaneous implementation of capacitors and DG [18, 19].

In this study, AQiEA was implemented to find the optimal placement and capacity of Capacitors and DGs in distribution network without violating the limits. The objective of reducing losses was considered by implementing DGs and Capacitor placements both at the same time. The yearly load increment was calculated in advance for continuous successive five years. Effect of load increase was calculated based on with and without inclusion of Capacitors and DGs. In recent times, AQiEA was applied on DG network for optimal location of capacitor was successfully applied [20], optimal DG problem [21–23], DG operation along with the network re arrangement [24], simultaneous implementation of both DG and Capacitor [25], ceramic grinding [26] and constraint handling technique [27].

The rest of the chapter was prepared as follows. The problem formulation section describes the objective to minimize the power loss by implementation of both Capacitor and DGs with predetermined annual load growth. Placement and sizing of Capacitors and DGs is a tough task and also a non-differentiable combinational optimization problem. The problem of optimization of placing and sizing of capacitors was solved by using AQiEA, which is described in the section Algorithm. The efficiency of AQiEA was compared with some other algorithms in the Results and Discussions section. In the final section of the chapter a conclusion is provided.

1.2 Problem Formulation

A load forecast for five years was precalculated based on the past and present load growth, which was used as an objective function for minimization of power losses. This was assumed based on yearly load growth in the DG network and determined by using growth rate plus the initial load in the network [28].

$$P_{Lk}(y) = P_{Lk}(0)(1+g)^y \tag{1.1}$$

$$Q_{Lk}(y) = Q_{Lk}(0)(1+g)^y \tag{1.2}$$

Where $P_{Lk}(y)$ & $Q_{Lk}(y)$ are the real and reactive load in y year, $P_{Lk}(0)$ & $Q_{Lk}(0)$ are the real and reactive load at base year, g is the annual load growth which is assumed as 7.5% and represents number of year (maximum number of years considered for this study is 5).

The primary goal of this research is to reduce power loss for every year by simultaneous implementation of Capacitor and DG. The generalized objective function is given as follows:

$$\text{Min}(F) = \text{Min}(P_{loss}) = \left\{ \sum_{m=1}^{N_b} \left(\frac{P_m^2 + Q_m^2}{V_m^2} \right) R_m \right\} \tag{1.3}$$

Where, P_m & Q_m are real and reactive power injections at m[th] bus. V_m indicates voltage at m[th] bus, R_m & X_m indicates magnitude of resistance and inductance.

Inequality Constraint:

Operation of DG and Capacitor:

$$P_{DG}^{min}(m) \le P_{DG}(m) \le P_{DG}^{max}(m) \tag{1.4}$$

$$Q_{DG}^{min}(m) \le Q_{DG}(m) \le Q_{DG}^{max}(m) \tag{1.5}$$

The power injected by DG into the network should be within the limits. $P_{DG}(m)$ & $Q_{DG}(m)$ are injected real and reactive powers.

Voltage limit:

$$V_m^{min} \le V_m \le V_m^{max} \tag{1.6}$$

The voltage produced at each individual bus in the network has to be in acceptable limits.

Equality Constraint:
Power injection:

$$\sum_{m=1}^{k_1} P_{DG}(m) + \sum_{m=1}^{k_2} Q_{DG}(m) + P_{sub} \leq P_{loss} + P_{load} \tag{1.7}$$

The total power injected by DGs and Capacitor in addition to the substation must be equal or less than its total load demand and losses of the system. Where P_{sub}, P_{loss} and P_{load} represents substation power, power loss and demand in the system.

1.3 Algorithm

Quantum-inspired evolutionary algorithms:
QiEA uses probabilistic representation with Q-bit and it has good characteristic representation of population diversity compared with other representations. It is defined as the smallest unit of information in a quantum computer [29]. It could be represented in two states as state 'a', and state 'b' or the sum of both states. It can be shown in the following equation as

$$|\epsilon = Y_1|a) + Y_2\,|b) \tag{1.8}$$

Here probability amplitudes Y_1 and Y_2 are connected to its corresponding states. $|Y_1|^2$ and $|Y_2|^2$ will provides probability of Q-bits which are to be found in state 'a' or 'b' respectively. A Q-bit individual 'q' with m-bits is given as follows

$$q = \left[q_1\, q_2 \cdots q_m\right] \tag{1.9}$$

Here, $|Y_{1i}|^2 + |Y_{2i}|^2 = 1$, i=1,2,....m

In QiEA, qubit probabilistic nature is widely used for maintaining diversity. Quantum gate operators are used to evolve the solution vectors which are influenced by phase rotation transformation.

Proposed Algorithm:
The limitations associated with EA implementation such as slow convergence, stagnation, etc., are overcome with AQiEA. In AQiEA two qubits

are used, the solution vector's objective function value is stored in the first qubit, while the scaled and ranked objective function value is stored in the second qubit. The objective function value feedback is not used appropriately in EA. The information in the first and second qubits is entangled, and the first qubit influences the second qubit because the first qubit's amplitude controls the objective function value. In the parameter utilized to evolve the first qubit, the second qubit is used as a tuning free adaptive quantum inspired rotation crossover operator/feedback.

A measurement operator is used to generate a solution string from qubit string (Y). In quantum computers, a resultant classical state is observed upon application of measurement operator, which results in collapse of superposition of states. However, in classic computers collapse of states doesn't occur naturally. In order to observe the qubit string, a new string with a random number, whose value varies between 0 and 1 is generated (N_r), which is of same length as that of qubit strings. Hence after measurement operation on the qubit string, a new measured value string (Q_m) is generated, which is of the same length as qubit string. Table 1.1 shows the measured operator on qubit string. The measured value in Q_{mj} is obtained by comparing the generated random number at N_{rj} to square of Y_{1j} at j[th] generation. If N_{rj} is less than the square of Y_{1j}, Q_{mj} is set to square of Y_{1j} otherwise to the square of Y_{2j}.

Minimization of power loss is done by placing and sizing the DG & Capacitor, the solution vector for solving the objective is represented in Figure 1.1.

Table 1.1 Qubit string measured operator.

J	1	2	....	N_p
Y_{1j}^2	0.56	0.17	....	0.41
Y_{2i}^2	0.44	0.83	....	0.59
N_{rj}	0.12	0.24	....	0.07
Q_{mj}	0.56	0.83	....	0.41

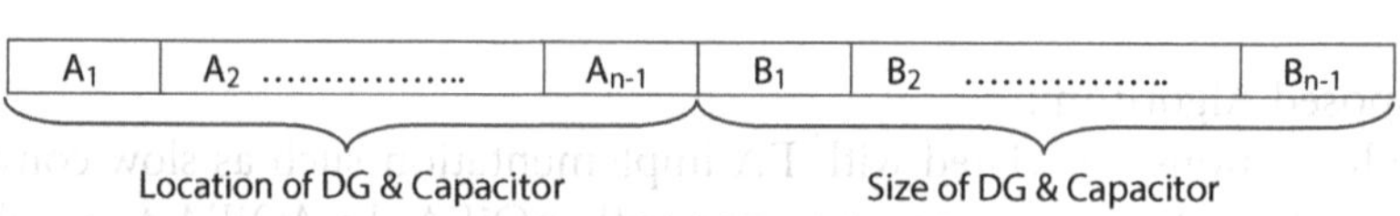

Figure 1.1 Solution vector representation for DG with optimal location and sizes.

From the equation (1.9) the amplitude Y_2 is discarded and computed whenever it is needed. Quantum registers are used to store the qubits. Total No. of qubits used mainly depends on the variables used and qubits considered per quantum register QR_1. This structure is shown below:

$$QR_{1,1} = [Y_{2\,1,1,1}, Y_{2\,1,1,2} \ldots. Y_{2\,1,1,n}]$$

$$\ldots\ldots\ldots\ldots\ldots\ldots\ldots\ldots\ldots\ldots \tag{1.10}$$

$$QR_{1,30} = [Y_{2\,1,30,1}, Y_{2\,1,30,2} \ldots. Y_{2\,1,30,n}]$$

The inspiration of Entanglement and Superposition principles can be represented mathematically as

$$\left|\alpha_{2i}(t)\right\rangle = f_1\left|\alpha_{1i}(t)\right\rangle \tag{1.11}$$

$$\left|\alpha_{1i}(t+1)\right\rangle = f_2\left(\left|\alpha_{2i}(t)\right\rangle, \left|\alpha_{1i}(t)\right\rangle, \left|\alpha_{1j}(t)\right\rangle\right) \tag{1.12}$$

Using the phase rotation transformation and their influence on quantum gate operator, solution vectors are computed by using the quantum operators. In the proposed method of algorithm a parameter tuned based on crossover operator and quantum rotation adaptive derivative are designed. By using second qubit amplitude, the degree of rotation for evolving the first qubit is determined. For this purpose, the following equation is used

$$\left|\alpha_{1i}(t+1)\right\rangle = \left|\alpha_{1i}(t)\right\rangle + f\left(\left|\alpha_{2i}(t)\right\rangle, \left|\alpha_{2j}(t)\right\rangle\right) * \left(\left|\alpha_{1j}(t)\right\rangle - \left|\alpha_{1i}(t)\right\rangle^{\alpha}\right. \tag{1.13}$$

Where $\left|\alpha_{1j}(t)\right\rangle$ and $\left|\alpha_{1i}(t)\right\rangle$ are two solution vectors which are deterministically or generated randomly by adaptive crossover operator. If $\left|\alpha_{1i}(t)\right\rangle$ has lesser rank than $\left|\alpha_{1j}(t)\right\rangle$ then $\left|\alpha_{1i}(t)\right\rangle$ is rotated towards $\left|\alpha_{1j}(t)\right\rangle$. If $\left|\alpha_{1i}(t)\right\rangle$ has a better solution than $\left|\alpha_{1j}(t)\right\rangle$ then $\left|\alpha_{1i}(t)\right\rangle$ is rotated away from $\left|\alpha_{1j}(t)\right\rangle$. Flow chart of AQiEA is shown in Figure 1.2. With the help of three rotation strategy (R-I, R-II, R-III), variation operator converges the search towards a better solution.

Best Strategy towards the Rotation (R-I): In this method of rotation, all the solution vectors are rotated towards the best solution vector. By rotating all the solution vectors towards the best solution, it is expected that better candidate solution will be found for all other vectors.

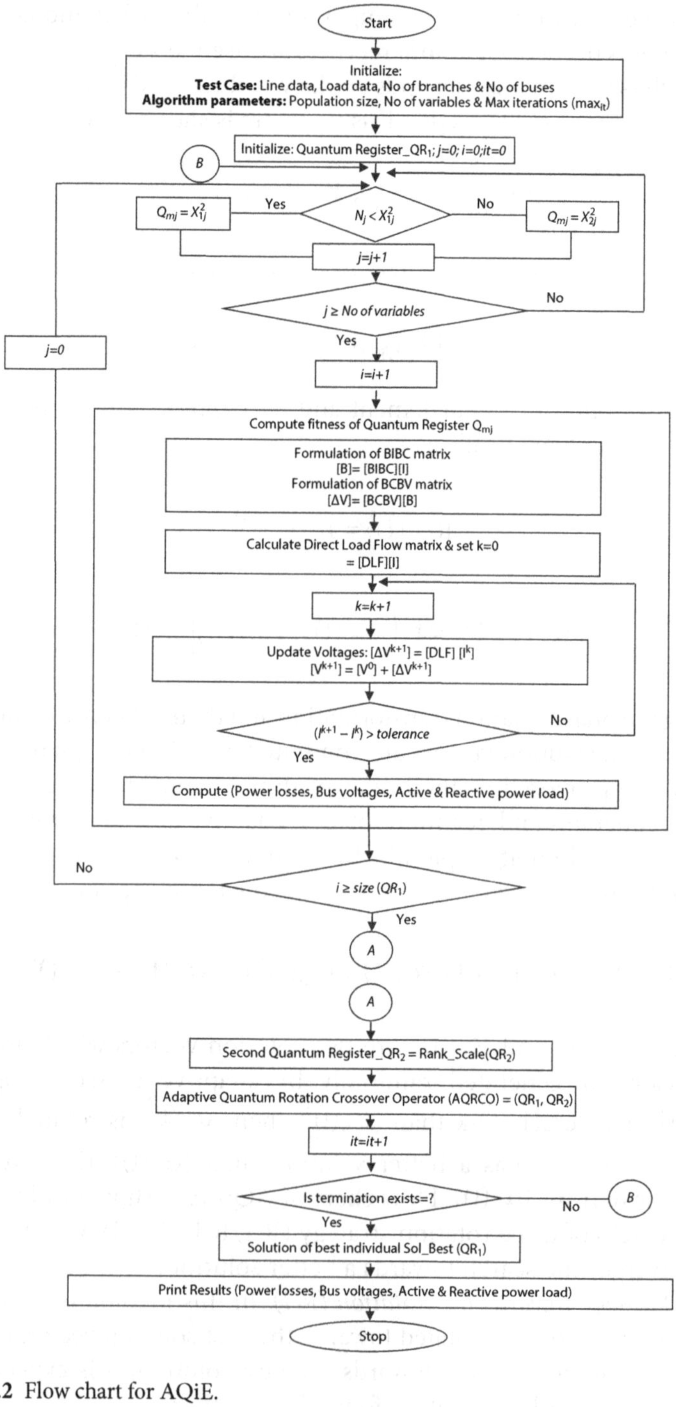

Figure 1.2 Flow chart for AQiE.

Rotation away from the Worse Strategy (R-II): In this method of rotation, the best individual in the population will move away from all other vectors. In the population of individuals, as moving away from worse, the search takes place in all dimensions. This is motivated by the fact that there are better chances of finding a good candidate solution in the vicinity of the best individual.

Better Strategy towards the Rotation (R-III): In this method, two individuals are randomly selected and the individual which has an inferior solution will move towards the better solution in hope of improvement.

The pseudo code of the proposed algorithm along with description is given as follows:

```
    t ← 0
    a. Initialize (QR_1 (t))
    While (! termination_criteria)
{
    b. Q_m = Measurement_operation(QR_1 (t))
    c. f(x)= Compute_fitness (Q_m(t))
    d. QR_2(t)= Rank_Scaled (f(x))
    e. QR_1c= A QRC_(QR_1(t), (QR_2(t))
    f. Tourn_Selection (QR_1(t), f(x))
       t ← t+1
}
```

1.4 Results & Discussions

A medium voltage bus system, i.e., 85 bus network system was considered as a benchmark test bus system to test the effectiveness of AQiEA. Table 1.2 shows the initial data of test bus system such as total real and reactive power demand, Base MVA, etc. Branch and load data for the test bus system is considered from [10].

Simulated results in Table 1.3 show the power losses without DG and Capacitor for five years. The load growth on the system is linearly increasing year by year. The overall power loss (real and reactive) induced in the system is also increasing year by year which results in poor voltage regulation in the system.

Simulated results in Table 1.3 show the determined power loss by not including the capacitor and DG. The total power loss (real and reactive) for base year are 316.12kW and 198.6kVAr with minimum voltage of 0.8713p.u. Similarly for I year total power losses obtained are 372.8kW and

Table 1.2 Initial data for 85 Bus system.

Particular	Value
Total Real Power (MW)	2.571
Total Reactive Power (MVAr)	2.62
Base (MVA)	100
Line Voltage (kV)	11
Buses	1 to 85
Sectionalizing switches	1 to 84
Maximum Allowable Size of DG & Capacitors	1MW & 1MVAr

234.17kVAr with minimum voltage 0.8602. Increment in power loss and poor voltage regulation is observed from base year to I year due to increment in load. Similarly for II year power losses are increased to 440.68kW and 276.76kVAr and for V year high increment in power loss is observed with minimum voltage 0.8019p.u. The total load (real and reactive) on the system for base year is 2.57MW and 2.62MVAr, followed by 2.763MW and 2.81MVAr for I year, followed by 2.97MW and 3.03MVAr for II year, followed by 3.19MW and 3.5MVAr for III year, followed by 3.43MW and 3.5MVAr for IV year and 3.69MW and 3.76MVAr for V year.

The voltage profile of a test bus system for five years including base year is shown in Figure 1.3. It is observed from Figure 1.3 that voltage profile is decreasing from base year to fifth year. Poor voltage regulation is observed in the fifth year which falls below the critical value. DGs and Capacitors are used to improve the voltage profile.

Table 1.3 Power loss without integration of DG & Capacitator for five years.

	Base Year	I-Year	II-Year	III-Year	IV-Year	V-Year
P_{loss} (kW)	*316.1221*	*372.8076*	*440.6897*	*522.3574*	*621.1595*	*741.78*
Q_{loss} (kVAr)	198.6059	234.1780	276.7682	327.9782	389.9129	465.48
Min Voltage (p.u)	0.8712	0.8602	0.8479	0.8342	0.819	0.8019
P_{load} (MW)	2.5703	2.7631	2.9703	3.193	3.4325	3.69
Q_{load} (MVAr)	2.6222	2.8189	3.0303	3.5019	3.5019	3.7645

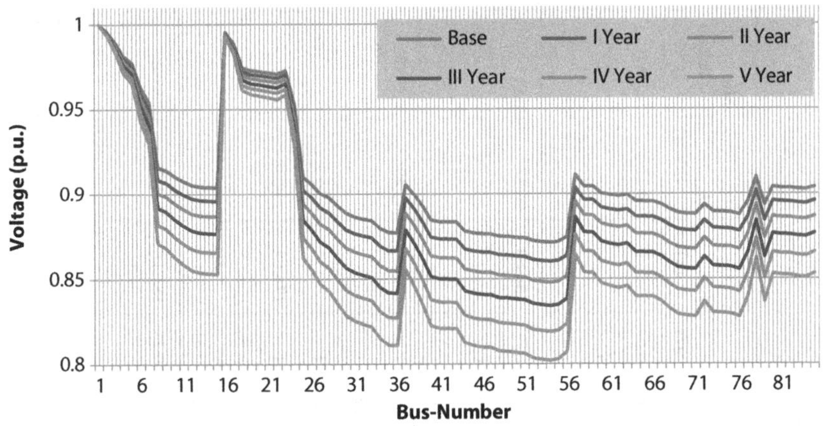

Figure 1.3 85 bus system voltage profile for 5 years.

Placement and capacity play a key role in a distribution system to reduce the power losses. Simulated results in Table 1.4 show the power losses and improvement in voltage by using individual as well as combining both DGs and Capacitors implementation.

It is observed from tabulated results that optimal location and capacity is varied for every year to minimize the losses. With the help of AQiEA the best optimal location and capacity of DG and Capacitor are found. In this study by implementing both two DGs and a single Capacitor is considered for every year. Simulated results show that simultaneous implementation DG and Capacitor had high percentage power loss reduction for every year including base year as compared to the individual implementation of DGs and Capacitors.

Total power loss (active and reactive) produced in the network for base year without DG and Capacitor are 316.12kW and 198.61kVAr, respectively. Independent implementation of DGs and Capacitors with AQiEA reduces the active power loss to 152.75kW and 156.63kW. However, for the same case minimum power loss 67.86kW is obtained with implementation of DG and Capacitor simultaneous. Similarly for other years (I year-V year) power loss is reduced to a minimum value with simultaneous implementation of DG and Capacitor.

Figure 1.4 shows the active power loss reduction with independent implementation of DGs for five years. AQiEA has minimum power loss for all years including base year, i.e., 152.75kW for base year, 174.67kW for I year, 205.06kW for II year, 239.67kW for III year, 278.65kW for IV year and 326.24kW for V year. However, SA has maximum power loss for all years including base year, i.e., 156.85kW for base year, 183.52kW for

Table 1.4 For 85 BUS SYSTEM comparison of AQiEA results with other algorithms.

		SA					PSO					AQiEA				
	Type	Loc	Size	P_{loss} (kW)	Q_{loss} (kVAr)	Min Voltage (p.u)	Loc	Size	P_{loss} (kW)	Q_{loss} (kVAr)	Min Voltage (p.u)	Loc	Size	P_{loss} (kW)	Q_{loss} (kVAr)	Min Voltage (p.u)
Base	***DG***	63	0.83	156.85	96.49	0.939	33	1	155.94	98.87	0.9506	66	0.62	**152.75**	94.52	0.9484
		26	0.82				14	0.45				26	1			
		47	0.34				73	0.56				49	0.46			
	Cap	29	0.44	164.87	101.06	0.9208	64	0.95	157.89	96.76	0.9186	60	0.7	**156.63**	96.37	0.917
		48	0.57				85	0.14				33	0.95			
		11	0.98				32	0.96				85	0.31			
	DG & Cap	32	0.99	72.43	40.73	0.9646	57	0.98	68.86	39.10	0.9538	57	1	**67.87**	38.45	0.9557
		68	0.9				30	0.94				32	1			
		40	1				31	1				32	1			
I Yr	***DG***	66	0.98	183.53	113.14	0.9397	32	0.81	176.22	108.84	0.9452	8	1	**174.68**	108.38	0.953
		32	0.96				9	0.64				34	0.79			
		19	0.28				59	1				71	0.61			
	Cap	45	0.03	187.75	111.79	0.9084	84	0.55	184.36	113.25	0.9095	26	0.78	**181.18**	111.76	0.9142
		62	0.98				64	0.72				59	1			
		31	0.98				29	0.95				33	0.55			

(Continued)

Table 1.4 For 85 BUS SYSTEM comparison of AQiEA results with other algorithms. (*Continued*)

		SA					PSO					AQiEA				
	Type	**Loc**	**Size**	P_{loss} **(kW)**	Q_{loss} **(kVAr)**	**Min Voltage (p.u)**	**Loc**	**Size**	P_{loss} **(kW)**	Q_{loss} **(kVAr)**	**Min Voltage (p.u)**	**Loc**	**Size**	P_{loss} **(kW)**	Q_{loss} **(kVAr)**	**Min Voltage (p.u)**
	DG & Cap	66	0.93	70.19	39.66	0.9619	62	1	66.86	38.59	0.9667	59	1	**64.65**	37.19	0.9664
		27	0.99				32	1				31	0.96			
		32	1				30	0.96				32	1			
II Yr	***DG***	25	0.55	208.7	129.35	0.9373	84	0.5	207.35	127.67	0.9455	31	1	**205.07**	126.87	0.942
		33	0.87				33	1				76	0.59			
		59	0.85				62	0.97				56	1			
	Cap	30	0.97	214.42	131.85	0.9077	67	0.52	211.92	131.46	0.9063	32	1	**210.49**	129.53	0.9063
		11	0.78				47	0.79				79	0.53			
		57	0.97				9	0.96				59	0.98			
	DG & Cap	58	1	105.09	61.84	0.9456	33	1	103.77	61.27	0.952	31	1	**101.66**	60.37	0.9521
		30	1				62	1				62	1			
		39	1				48	1				35	0.99			
III Yr	***DG***	50	0.9	247.13	151.59	0.9375	32	1	243.99	150.58	0.9332	32	1	**239.67**	148.12	0.9466
		59	0.9				84	0.74				66	0.9			
		8	0.87				63	0.9				24	1			

(*Continued*)

Table 1.4 For 85 BUS SYSTEM comparison of AQiEA results with other algorithms. (*Continued*)

		SA					PSO					AQiEA				
	Type	Loc	Size	P_{loss} (kW)	Q_{loss} (kVAr)	Min Voltage (p.u)	Loc	Size	P_{loss} (kW)	Q_{loss} (kVAr)	Min Voltage (p.u)	Loc	Size	P_{loss} (kW)	Q_{loss} (kVAr)	Min Voltage (p.u)
	Cap	33	0.98	251.30	154.49	0.9039	33	1	247.47	153.41	0.8973	63	1	**245.8**	151.78	0.8980
		76	0.98				67	0.49				11	0.76			
		56	0.98				56	1				32	0.99			
	DG & Cap	76	0.83	131.81	80.66	0.9399	66	1	127.30	77.29	0.9438	63	1	**123.37**	75.18	0.9442
		30	0.99				33	1				30	1			
		30	1				35	0.98				33	1			
IV Yr	*DG*	80	0.67	283.95	175.68	0.9235	77	1	281.67	173.68	0.9362	66	1	**278.66**	172.54	0.9253
		33	0.91				33	1				9	0.84			
		59	1				66	1				32	1			
	Cap	59	0.98	293.29	180.64	0.8969	9	1	291.46	179.97	0.8888	33	0.9	**287.30**	177.99	0.8894
		47	0.97				34	0.82				66	1			
		24	0.97				65	1				8	1			
	DG & Cap	29	1	161.81	98.80	0.9305	65	1	158.76	96.46	0.9347	71	1	**156.14**	96.32	0.935
		66	1				31	1				33	1			
		39	1				31	1				32	1			

(*Continued*)

Table 1.4 For 85 BUS SYSTEM comparison of AQiEA results with other algorithms. (*Continued*)

		SA					PSO					AQiEA				
	Type	**Loc**	**Size**	P_{loss} **(kW)**	Q_{loss} **(kVAr)**	**Min Voltage (p.u)**	**Loc**	**Size**	P_{loss} **(kW)**	Q_{loss} **(kVAr)**	**Min Voltage (p.u)**	**Loc**	**Size**	P_{loss} **(kW)**	Q_{loss} **(kVAr)**	**Min Voltage (p.u)**
V Yr	***DG***	63	0.71	333.69	207.75	0.9153	9	1	330.81	204.17	.09154	9	1	**326.24**	202.41	0.9202
		33	0.99				78	1				62	1			
		11	0.98				32	1				33	0.95			
	Cap	63	1	350.24	215.85	0.8841	9	0.86	346.35	213.69	0.8807	33	1	**343.24**	211.83	0.8784
		25	0.99				62	1				63	1			
		51	0.88				48	1				77	1			
	DG & Cap	33	1	202.23	125.22	0.9247	62	1	201.06	124.69	0.9203	33	1	**198.25**	123.45	0.9249
		76	1				48	1				76	1			
		43	1				35	1				47	1			

Cap- Capacitor, $\mathbf{P_{loss}}$- Active Power loss, $\mathbf{Q_{loss}}$-Reactive Power loss, **Min Voltage**, **Yr**-Year, **Loc**- Location of DG & Cap, **Size**-Size of DG & Cap in MW & MVAr.

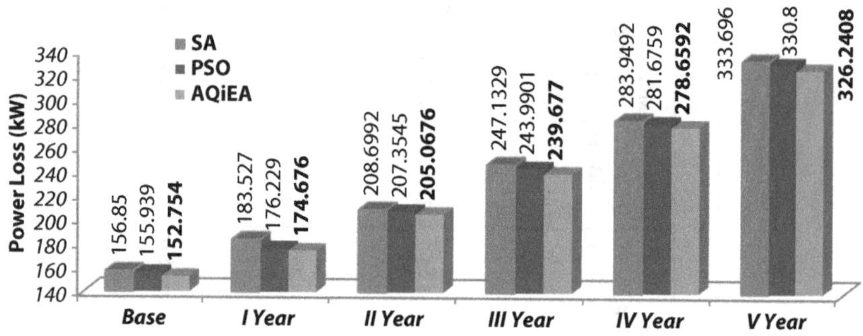

Figure 1.4 Active power losses with DG for five years.

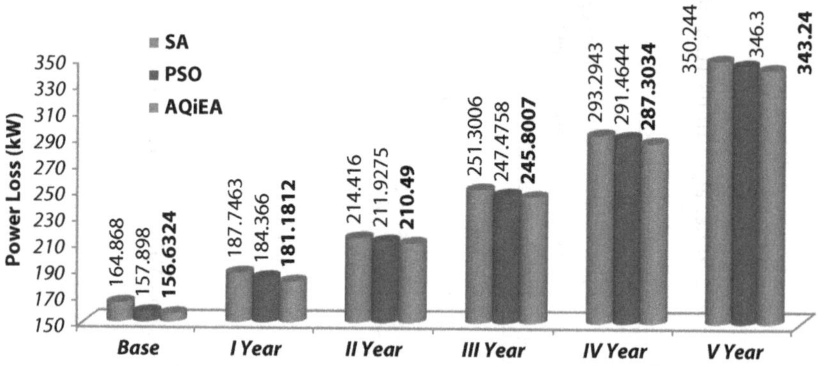

Figure 1.5 Active power losses with capacitor for five years.

I year, 208.69kW for II year, 247.13kW for III year, 283.94kW for IV year and 333.69kW for V year. Similarly, Figure 1.5 shows the active power loss reduction with independent implementation of Capacitors for five years. Minimum power loss is obtained with AQiEA for all years including base year, i.e., 156.63kW for base year, 181.18kW for I year, 210.49kW for II year, 245.8kW for III year, 287.3kW for IV year and 343.24kW for V year. In comparison with SA, PSO has maximum power loss reduction for all years including base year, i.e., 157.89kW for base year, 184.36kW for I year, 211.92kW for II year, 247.47kW for III year, 291.46kW for IV year and 346.3kW for V year. It is observed that optimal allocation of DG in the system has a significant reduction in power loss as compared with optimal allocation of Capacitors.

Figure 1.6 shows the active power loss decrease with simultaneous implementation of DG and capacitor for five years. It is also seen from

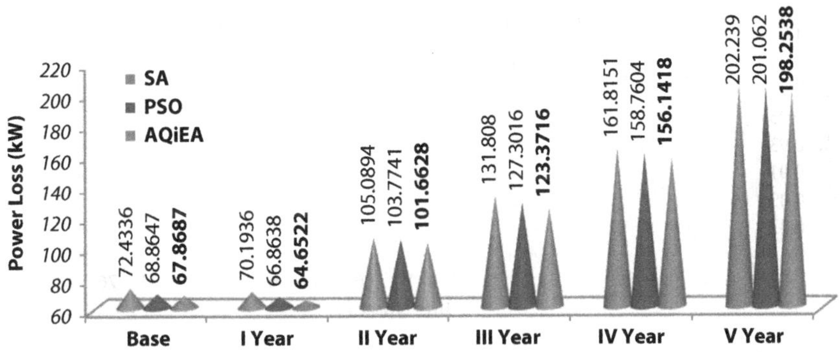

Figure 1.6 Active power losses by placing both DG along with capacitor for five years.

tabulated results that placing both DG along with capacitor with AQiEA has minimum power loss as compared to other algorithms (SA and PSO) that have been reported in the literature. AQiEA has minimum power loss for all years including base year, i.e., 67.86kW for base year, 64.65kW for I year, 101.66kW for II year, 123.37kW for III year, 156.14kW for IV year and 198.25kW for V year. Whereas SA and PSO has power loss of 72.43kW, 68.86kW for base year, 70.19kW, 66.86kW for I year, 105.08kW, 103.77kW for II year, 131.81kW, 127.3kW for III year, 161.81kW, 158.76kW for IV year, 202.2kW, 201.06kW for V year.

Simultaneous implementation of DGs and Capacitor improves the voltage in the system. Figures 1.7–1.12 show the voltage profiles of test bus system for every year including base year. Figure 1.13 shows the power losses without and with DG and Capacitors using AQiEA for five years.

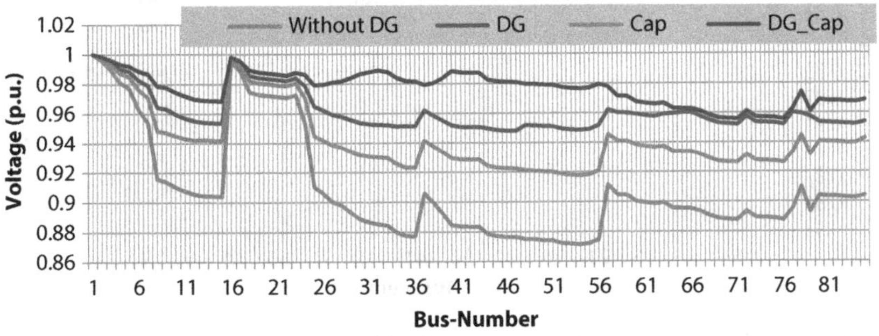

Figure 1.7 85 test bus system voltage profile for base year.

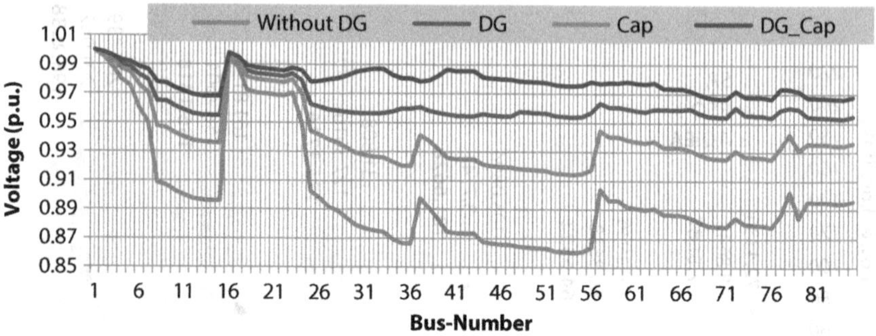

Figure 1.8 85 test bus system voltage profile for I year.

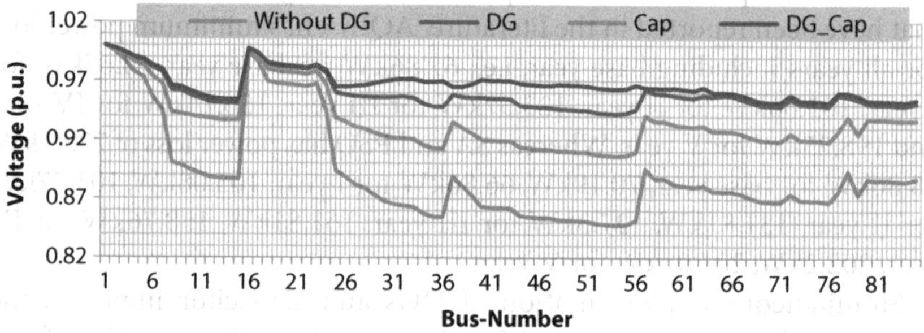

Figure 1.9 85 test bus system voltage profile for II year.

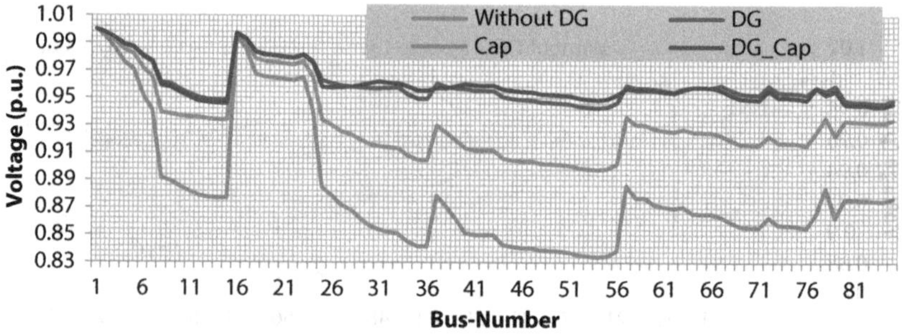

Figure 1.10 85 test bus system voltage profile for III year.

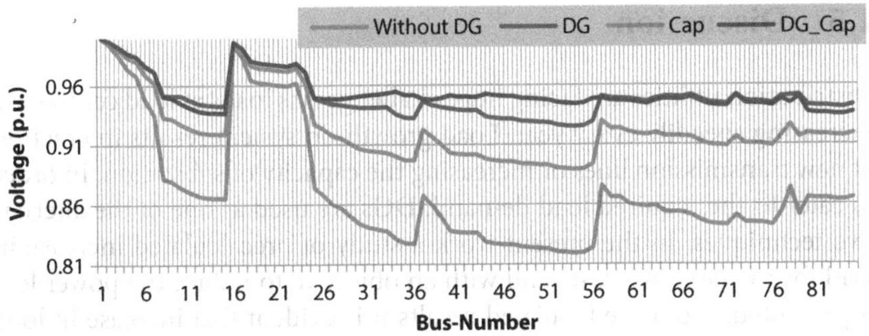

Figure 1.11 85 test bus system voltage profile for IV year.

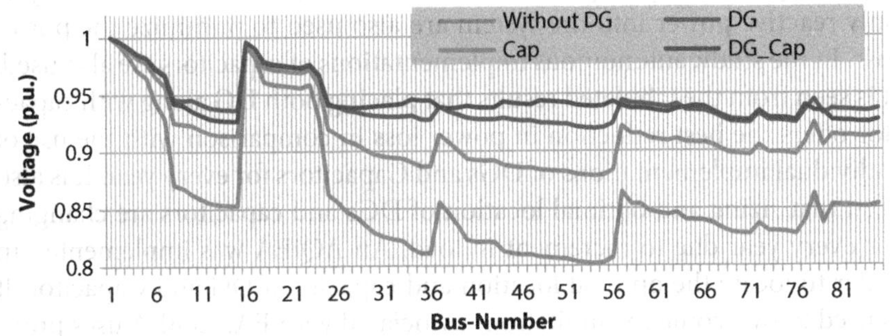

Figure 1.12 85 test bus system voltage profile for V year.

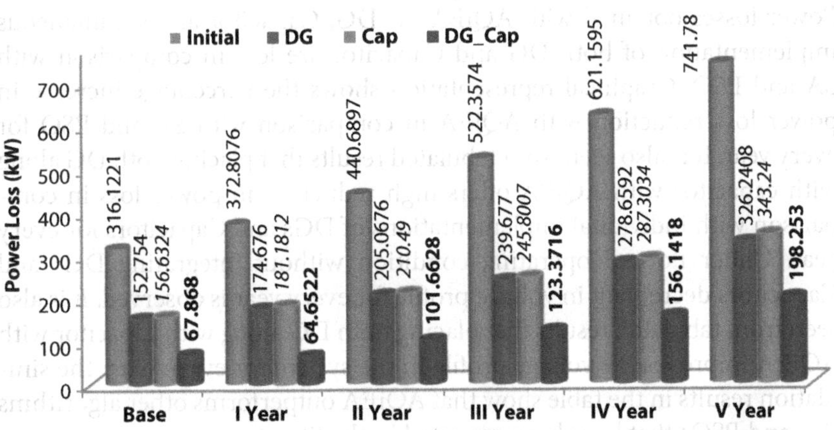

Figure 1.13 Power losses with AQiEA for five years.

1.5 Discussion

Distribution utilities are facing huge challenges, as loads at load centres are increasing speedily every year. Load growth on system results in erection of new transmission lines or increasing the capacity of substation. In order to cater to the required load demand, DGs are used as one of the alternative techniques. In the present work a study of precalculated increase in load for the next five years and with an objective to reduce the power loss is presented. From the tabulated results it is evident that increase in load demand for each year results in an increase in power loss and reduction in voltage. Placement and sizing of DG are two key factors which play a major role in the distribution network system for minimizing the power loss. Inappropriate location and sizing of DG leads to increases in power loss and poor voltage regulation. In some cases, Capacitors which inject only reactive power into the system are also used to minimize the power loss. In the study, autonomous implementations of capacitors are also used. It is seen from the tabulated results that placing both DG along with capacitor offers the best reduction in power loss in comparison with the use of individual implementations of DGs and Capacitors for every year. It is also seen that sizing and optimal location of DGs and capacitors are changing for every year due to increment in load. An AQiEA was implemented in order to locate the optimal location and capacity of DG and Capacitor. It is used to overcome the limitations associated with EA. AQiEA uses probabilistic representation with Q-bit. To improve the convergence rate and minimize premature convergence, AQiEA does not require extra operators such as local search and mutation. Three rotation strategies (R-I, R-II, R-III), variation operator converges the search towards a better solution. Power losses obtained with AQiEA for DG, Capacitor and simultaneous implementation of both DG and Capacitor are low in comparison with SA and PSO. Graphical representation shows the percentage increase in power loss reduction with AQiEA in comparison with SA and PSO for every year. It is also seen from tabulated results that placing both DG along with capacitor with AQiEA offers high reduction in power loss in comparison with individual implementations of DGs and Capacitors for every year. Under normal operating condition without integrating DGs and Capacitors decrement in voltage profile for every year is observed. It is also seen from tabulated results that placing both DG along with capacitor with AQiEA improves the voltage profile in the system for every year. The simulation results in the table show that AQiEA outperforms other algorithms (SA and PSO) that have been reported in the literature.

1.6 Conclusions

In this work, the effect of an increase in load on a distribution network was briefly examined with the use of AQiEA algorithm. In order to cater to the required load demand, DGs and Capacitors are placed at appropriate places in the network so as to reduce the power loss. Multiple numbers of DGs and Capacitors are placed in benchmark test bus system at different nodes optimally. Placement of DGs and its sizing with Capacitors is a difficult multi-variable optimization problem. There have been numerous efforts documented in the literature for solving this critical problem. The best placement with the sizing of DGs and Capacitors is computed with the help of AQiEA. Incorporation of DGs and Capacitors into distribution will decrease the power loss and also improve the voltage profile of the entire distributed network. In this paper simultaneous implementation of Capacitor and DG for an 85 bus radial distribution system is proposed with AQiEA algorithm. The demonstrated result in the table shows that simultaneous implementation of Capacitor and DG for every year including base year gives a better result in comparison with independent implementation of DG and Capacitor for every year. AQiEA was implemented to minimize power losses by locating DGs and Capacitors at an optimal location with the appropriate size. The performance of the proposed algorithm has been compared with the existing techniques proposed by several researchers presented in the literature. The performance shows that the proposed algorithm gives better results as compared with other algorithms.

References

1. T. Ackermann, G. Andersson, and L. Söder, "Distributed generation: a definition," *Electric Power Systems Research*, Vol. 57, pp. 195–204, April 2001.
2. D. Topić, D. Šljivac, M. Gagro: "Influence of Distributed Power Generation from Renewable Energy Sources on Reliability of Distribution Networks," *International Journal on Electrical and Computer Engineering Systems*, Vol. 6, No. 2, 2015.
3. Akorede MF, Hizam H, Aris I, Abkadir MZA. "A review of strategies for optimal placement of distributed generation in power distribution systems". *Res J Appl Sci*; 5(2):137–45, 2010.
4. H. N. Ng, M. M. A. Salama, and A. Y. Chikhani, "Classification of capacitor allocation techniques," *IEEE Trans. Power Del.*, Vol. 15, no. 1, pp. 387–392, Jan. 2000.

5. Tuba Gözel, M. Hakan Hocaoglu, "An analytical method for the sizing and siting of distributed generators in radial systems," *Electric Power Systems Research*, Vol. 79, Issue 6, 2009, pp. 912–918.
6. S. Nojavan, M. Jalali, K. Zare, "Optimal allocation of capacitors in radial/mesh distribution systems using mixed integer nonlinear programming approach", *Int. J. Electric Power Syst.* Res. 107 (2014) 119–124.
7. S. Gopiya Naik, D.K. Khatod, M.P. Sharma, "Optimal allocation of combined DG and capacitor for real power loss minimization in distribution networks," *International Journal of Electrical Power & Energy Systems*, Vol. 53, 2013, pp. 967–973.
8. G. Manikanta, A. Mani, H. P. Singh and D. K. Chaturvedi, "Minimization of power losses in distribution system using symbioitic organism search algorithm", *IEEE PES Asia-Pacific Power and Energy Engineering Conference (APPEEC)*, Bangalore, 2017, pp. 1–6.
9. Manoj Kumawat, Nitin Gupta, Naveen Jain & Ramesh C. Bansal, "Swarm-Intelligence-Based Optimal Planning of Distributed Generators in Distribution Network for Minimizing Energy Loss", *Electric Power Components and Systems*, 45:6, 589–600, 2017.
10. M. Gandomkar, M. Vakilian and M. Ehsan, "A combination of genetic algorithm and simulated annealing for optimal DG allocation in distribution networks", *Canadian Conference on Electrical and Computer Engineering*, 2005, Saskatoon, Sask., 2005, pp. 645–648.
11. Mostafa Sedighizadeh, Masoud Esmaili, Mobin Esmaeili, "Application of the hybrid Big Bang-Big Crunch algorithm to optimal reconfiguration and distributed generation power allocation in distribution systems", *Energy*, Vol. 76, 2014, pp. 920–930.
12. A.Mohamed Imran, M. Kowsalya and D.P. Kothari. "A novel integration technique for optimal network reconfiguration and distributed generation placement in power distribution networks". *International Journal of Electrical Power & Energy Systems*, pp. 461–472, 2014.
13. H.D. Chiang, J.C. Wang, O. Cockings, H.D. Shin, "Optimal capacitor placements in distribution systems: part 1: a new formulation and the overall problem", *IEEE Trans. Power Deliv.* 5 (2) (1990) 634–642.
14. P. Das, S. Banerjee, "Optimal sizing and placement of capacitor in a radial distribution system using loss sensitivity factor and firefly algorithm", *Int. J. Eng. Comput.* (2014) 5346–5352.
15. R.S. Rao, S.V.L. Narasimham, M. Ramakingaraju, "Optimal capacitor placement in a radial distribution system using plant growth simulation algorithm", *Int. J. Electr. Power Energy Syst.* 33 (2011) 1133–1139.
16. S. Sultana, P.K. Roy, "Optimal capacitor placement in radial distribution systems using teaching learning based optimization", *Int. J. Electr. Power Energy Syst.* 54 (2014) 387–398.

17. K. Prakash, M. Sydulu, "Particle swarm optimization based capacitor placement on radial distribution systems", *IEEE Power Engineering Society General Meeting*, 24–28 June 2007, pp. 1–5, 2007.
18. Abolfazl Rahiminejad, Seyed Hossein Hosseinian, Behrooz Vahidi & Shohreh Shahrooyan, "Simultaneous Distributed Generation Placement, Capacitor Placement, and Reconfiguration using a Modified Teaching-Learning-based Optimization Algorithm", *Electric Power Components and Systems*, 44:14, 1631–1644, (2016).
19. Rahiminejad, A., Aranizadeh, A., and Vahidi, B., "Simultaneous distributed generation and capacitor placement and sizing in radial distribution system considering reactive power market," *J. Renew. Sustain. Energy*, Vol. 6, p. 043124, 2014.
20. G. Manikanta, A. Mani, H. P. Singh and D. K. Chaturvedi, "Sitting and sizing of capacitors in distribution system using adaptive quantum inspired evolutionary algorithm," *7th India International Conference on Power Electronics (IICPE)*, Patiala, 2016, pp.1–6.
21. G. Manikanta, A. Mani, H. P. Singh and D. K. Chaturvedi, "Placing distributed generators in distribution system using adaptive quantum inspired evolutionary algorithm," *2016 Second International Conference on Research in Computational Intelligence and Communication Networks (ICRCICN)*, Kolkata, 2016, pp. 157–162.
22. G. Manikanta, Ashish Mani, H.P. Singh, D.K. Chaturvedi, "Adaptive Quantum inspired Evolutionary Algorithm for Optimizing Power Losses by Dynamic Load Allocation on Distributed Generators" in *Serbian Journal of Electrical Engineering*, Vol. 16, No. 3, October 2019, 325–357. DOI: https://doi.org/10.2298/SJEE1903325M
23. G. Manikanta, Ashish Mani, H.P. Singh, D.K. Chaturvedi, "Effect of Voltage Dependent Load Model on Placement and Sizing of Distributed Generator in Large Scale Distribution System". *Majlesi Journal of Electrical Engineering*, 14(4), 97–121. (2020).
24. G. Manikanta, Ashish Mani, H.P. Singh, D.K. Chaturvedi, "Simultaneous application of distributed generator and network reconfiguration for power loss reduction using an adaptive quantum inspired evolutionary algorithm". *International Journal of Energy Technology and Policy* 17 (2), 140–179, 2021.
25. G. Manikanta, Ashish Mani, H.P. Singh, D.K. Chaturvedi "Simultaneous Placement and Sizing of DG and Capacitor to Minimize the Power Losses in Radial Distribution Network" *2nd International Conference on Soft Computing: Theories and Applications* at Bhundelkhand University, Jhansi.
26. Mani and C. Patvardhan. "An Improved Model of Ceramic Grinding Process and its Optimization by Adaptive Quantum-inspired Evolutionary Algorithm," *International Journal of Simulations: Systems Science and Technology*, 1473-8031, 2012.

27. A. Mani and C. Patvardhan, "A novel hybrid constraint handling technique for evolutionary optimization," *2009 IEEE Congress on Evolutionary Computation*, Trondheim, 2009, pp. 2577–2583.
28. Das D. "Maximum loading and cost of energy loss of radial distribution feeders". *Elect Power Energy Syst* 2004;26:307–14.
29. K.H. Han, J.H. Kim, "Quantum-inspired evolutionary algorithm for a class of combinatorial optimization", *IEEE Trans. Evol. Comput.* 6 (Dec (6)) (2002)580–593.

2

Security Issues and Challenges for the IoT-Based Smart Grid

Prathiga*, Kavya K., Nanthitha N., Nithishkumar K., Ritika T. and Vishal T.

Electronics and Instrumentation Engineering, Bannari Amman Institute of Technology, Sathyamangalam, India

Abstract

Power demands of the twenty-first century are growing rapidly because of population increase, and efforts are being made to make the energy grid smarter, with greater attention to the power needs of the purchasers, and to offer stepped-up performance and reliability of energy systems. The Internet of Things (IoT) has emerged as one of the allowing technologies for a smart grid network. As IoT-related gadgets keep growing at a fast pace, one of the most important needs is safety, since having gadgets online puts the smart grid at risk of full-scale assaults. Since a primarily IoT-based smart grid might include tens of thousands of nodes, it represents the most important assault floor for an IoT-centered cyberattack. A cyberattack on a smart grid might have devastating results on the reliability of considerable infrastructure given the likely capability cascade effects of shutting down the energy grid, given that most of the gadgets in our homes, offices, hospitals and trains require energy to run. Once a single element tool is compromised, the entire grid becomes at risk of cyberattacks. Such assaults on energy delivery can result in entire towns grinding to a halt, thereby inflicting large economic and monetary losses. This makes safety a crucial issue to remember before the massive-scale deployment of primarily IoT-based smart grid networks.

***Keywords*:** Cyberspace, cyber security, Internet of Things (IoT), smart grid

2.1 Introduction

Over the last few years, the burgeoning Internet of Things (IoT) has made it viable to attach something and the whole thing to the internet. This has

**Corresponding author*: prathiga.ei19@bitsathy.ac.in

A. Chitra, V. Indragandhi and W. Razia Sultana (eds.) Intelligent and Soft Computing Systems for Green Energy, (25–40) © 2023 Scrivener Publishing LLC

caused a global revolution in how we use a multitude of devices. Using IoT it is now possible to attach mild bulbs, refrigerators, drones, puppy feeders, sensors, smart TVs and virtual set-pinnacle boxes, safety cameras, wearables, automobile structures and scientific gadgets to the internet. Numerous industries, from healthcare to production to utilities, transport, and houses have been converted and are actually smarter than ever. The boom of Machine-to-Machine (M2M) communications over the past decade has furnished a verbal exchange paradigm that has enabled connectivity among gadgets alongside the potential to talk autonomously without human intervention. With time, M2M communication has advanced and is now identified because of the realistic implementation for the adoption of IoT [1, 2] (Figure 2.1).

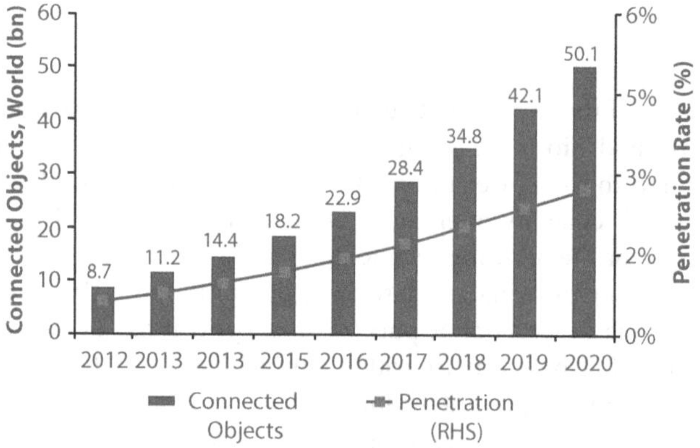

Figure 2.1 Rise in adoption of IoT.

Increasing cell and net usage and social media are among the key factors contributing to the explosive rise in adoption of IoT [3]. However, due to the fact that IoT associated devices are developing at a quick pace, numerous worrying conditions abound. Amongst them is the capacity vulnerability of these net-going-through systems to hackers. Contrary to traditional power grids in which most of the attacks came from physical access to vital infrastructure facilities, the ubiquity of IoT-based totally absolutely smart-grid devices means that the risks are cyber-based infrastructures that can be accessed from anywhere and at any time [3, 6]. As the ubiquity of the IoT era infiltrates further into a smart grid's infrastructure it becomes increasingly prone to cyberattacks. First and foremost, the variety of capacity attack elements in the course of the network is notably huge, and as quickly as a single device is compromised, then the whole grid becomes vulnerable to cyberattacks. Even in instances in which the infrastructure is considered extensively consistent but the communications network is not, then the whole tool is still at risk. The capacity cascade effect of shutting down the power grid makes it a key issue of a cyberattack. Finally, the dire need is to defend the power grid in the most economical way.

This paper is prepared as follows. In segment II we briefly describe IoT, SG and the hyperlink among them. In segment III we observe the advantages collected from the usage of a primarily IoT-based totally smart grid. In segment IV we look into safety problems and demanding situations within the primarily IoT-based totally SG. In segment V we discover different demanding situations of the smart grid in the IoT context. In segment VI we observe what the future holds for primarily IoT-based totally SG, and we present our conclusions in segment VII.

2.2 Usage of IoT in the Smart Grid Context

IoT is considered as the next-step evolution of our cutting-edge grid networks. IoT communications is superior to M2M communications. IoT aims at connecting the devices at a huge scale using IP-based totally completely solutions while letting them engage with every other speakme device over the Internet. As proposed in [7] connectivity is probably the most essential building block of the IoT paradigm.

The National Institute of Standards and Technology (NIST) defines the smart grid as the aggregate of the ultimate century energy grid with the cutting-edge century development in information and verbal exchange technologies [8]. Unlike the traditional energy grid, the smart grid maximizes the energy name for distribution, with growth efficiency, minimizes losses and moreover makes huge-scale renewable energy, which includes solar and wind deployments, a reality. The cutting-edge grid network manages extreme disturbing conditions inclusive of ordinary black-outs, overloading within the direction of peak hours, and company disruptions which may be due to old infrastructure. However, the deployment of some remote sensing gadget capable of measuring, monitoring and speakme information about the grid components makes it extra related and smarter. The received energy records are then used to put in force a self-recovery grid, thereby developing the quantity of self-monitoring and desire making and ultimately developing the overall efficiency [9].

The IoT permits a better and extra related grid via permitting bidirectional information drift and connectivity at some point of the grid infrastructure. Through the IoT, consumers, manufacturers and software program agencies lower human intervention in coping with smart meters, home gateways, smart plugs and unique related domestic device, for that reason ensuring that the grid abilities are optimal in response to the environment.

The Smart Grid (SG) is a smart energy grid and is the most important instance of the IoT network within the near future [10]. The entire energy grid chain, starting from the power energy plant technology to the final strength consumers (houses, constructing, factories, public lighting, electric powered vehicles, smart domestic equipment, etc.), along with the energy transmission and distribution networks, will be equipped with intelligent and two-way conversation abilities to show and manipulate the energy grid remotely. This will be enabled via the usage of smart meters, smart domestic equipment, sensors and actuators. The intention of the SG is to hold a real-time balance amongst power technology and consumption, via allowing monitoring and manipulation over the energy chain, to the two-way talking smart gadgets (smart meters, smart domestic equipment, sensors, actuators, etc.).

The Smart Grid (SG), is considered one of the most important infrastructures, and is seen as one in each of the most important functionality IoT network implementations. Setting up smart grid networks consists of integrating numerous wireless sensors, smart meters, smart domestic equipment, sensors and unique smart gadgets, all of which communicate with each other uniquely over a related network. In this regard, it's crucial to uniquely address each object simply so its far effects are identified within the network. Most smart grids observe the SG reference model as defined via NIST as established in Figure 2.2 below, which shows the form of a smart grid network as defined by NIST.

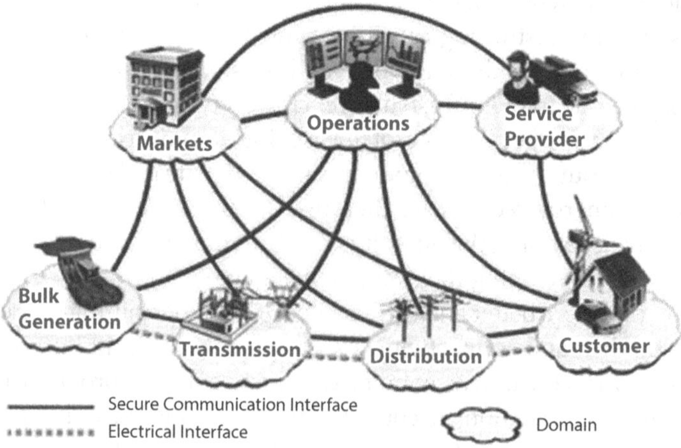

Figure 2.2 Usage of the IoT in smart grid context.

IoT technology plays a crucial role within the smart grid. It permits streamlining the transfer of a large amount of facts over the internet. Moreover, it establishes seamless and effective conversation between most of the sensors, actuators, the smart meters installation at the customer's premises and the software program servers, resulting in power monetary financial savings and charge bargain on top of things of the grid. With an IoT-based absolutely smart grid, software program organizations can leverage the usage of IoT to beautify the development and operation of smart grids as it offers incredible promise of a future enabled through smart devices that enhance overall performance, ease congestion, reduce waste and eliminate human error. Adopting IoT within the smart grid permits a large-scale two-way conversation flow between most of the unique components of the SG. The conversation is made viable due to the massive presence of sensors/actuators and extraordinary smart gadgets alongside the transmission and distribution areas, further to the usage of smart meters and extraordinary smart gadgets at the end-customer side. This permits the tracking of real-time power consumption and phone for the power supply whilst assisting customers to show their non-public usage and modify behaviors [7]. IoT permits devices on the smart grid infrastructure to be sensed and controlled remotely through a scalable conversation network which permits easier integration of most of the physical worldwide grid devices and computer-based absolute control systems essential to advanced overall performance and accuracy, allowing the grid to fulfill the power needs of today and of future generations.

2.3 Advantages of IoT-Based Smart Grid

1. Advanced Metering Infrastructure (AMI).
   Using an IoT smart grid, the superior metering infrastructure can effortlessly be implemented. The AMI is answerable for collecting, analyzing, storing and imparting the metering facts despatched by means of the smart meters to the software company's servers for billing, outage control and call for forecasting. The availability of real-time pricing offers clients and suppliers' precious information to assist them in controlling their electricity needs and supplies, respectively.
2. Improved reliability of the power system.
   An IoT smart grid is a grid that has the cap potential to speedily repair itself (self-healing), in the event of any outside or inner disturbances or threats. It additionally offers

self-recovery of the community, after attacks, disasters, blackouts or disasters of community factors with the aid of using dynamic reconfiguration to repair energy, the cap potential to create microgrids and autonomously powered islands after an energy failure in addition to figuring out reassets of power leakages. This offers higher efficiency, prediction and prognosis of the energy community and for that reason enhances reliability of the energy community.

3. Enhanced functions of SCADA (Supervisory Control and Data Acquisition).
   Due to using a huge quantity of sensors, actuators and smart meters are deployed to reveal the entire strength of the grid infrastructure. They can document statistics periodically, upon request or in reaction to a few occasions to the application even as additionally responding to requests from the application way to their two-manner communication capability. In addition to presenting records at the final mile grid's status, those gadgets may be managed, monitored and managed remotely, therefore presenting stronger features of SCADA.
4. Management of power in the grid.
   It enables bidirectional electric powered go with the drift in which the end-client buys and also can promote any extra strength from the house specifically at some point of peak hours from reassets including from sun or biogas round the house.
5. Demand response.
   Using an IoT smart grid allows dealing with load and call for reaction on a smart grid as a result allowing the effecting of dynamic strength pricing mechanisms. The use of dynamic strength pricing improves the capacity to control height load with the aid of using charging better expenses at some point of height instances to deter intake and decrease expenses at some point of off-peak instances to inspire better intake and make use of the idle capacity.

2.4 Cybersecurity Challenges

There is no doubt that cybersecurity is one of the most important, complicated and demanding situations regarding IoT gadgets. Internet-linked sensors, gadgets, and networks are steady objectives of online probing, espionage, ransom, theft, or even destruction. Since an IoT primarily based totally smart grid includes probably hundreds of thousands of online nodes, spanning extensive geographical regions, it is most at risk of vast cyberattacks. A cyber-assault might consequently bring about devastating results in addition to substantial monetary loss considering that such an assault might bring whole nations to a halt. A recent study on reported assaults in the USA observed that the power infrastructure is the top goal of assaults at 54% at the same time as the range of assaults keeps on growing (Figure 2.3).

Thus, safety is an awesome mission within the deployment and exercise of IoT-based totally definitely smart grid networks. With the demand for IoT products developing, vendors are rushing to supply new products into the market, which ends up in safety being an afterthought and clients being put at risk. What in addition complicates the IoT safety problem is the fact that there can be definitely be no sound way to functional safety of the devices already in the market. Unlike pc structures and specific devices that have the cap potential to reinstall or enhance software program application after the device has been offered to the consumer, most current IoT devices no longer resource that. So, as quickly as the devices have been installed, consumers, organizations and manufacturers overlook them altogether and given the likelihood of a vulnerability later on, getting the vulnerabilities steady is the real mission.

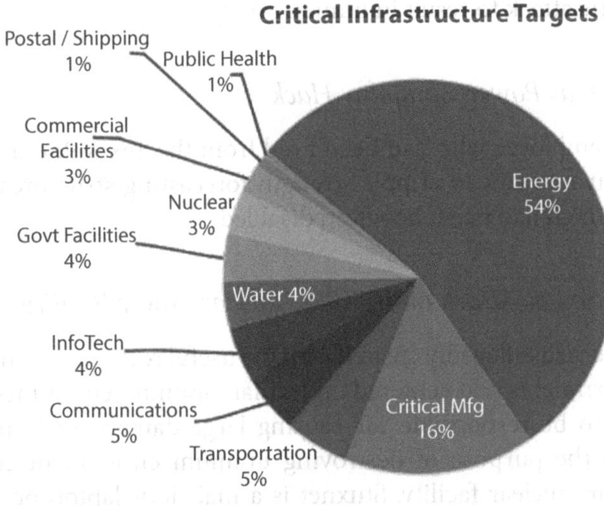

Figure 2.3 Infrastructure targets.

With the developing volume and sophistication of cyberattacks, relentless attention is needed to shield sensitive private and business agency information, further to preserve national safety. Widespread adoption of an approach that will ensure that vital services of most countries are effectively protected in cyberspace networks is essential. At the present time, cyberspace is a totally attractive and strategic medium of attack. In fact, cyber battle has been acknowledged as the fifth region of battle after land, sea, air, and space. Unlike the other domains, cyberspace has trends that make it an attractive frontier for launching attacks. The main one being the benefit of getting right of entry due to the global scope of the internet, which in turn lets attackers roam cyberspace freely and attack systems in locations that would otherwise be no longer viable to get right of entry, and because of this making attacks easier to execute and quicker than conventional battle.

2.4.1 Review of Recent Attacks

In order to better recognize the dangers posed by cyberattacks on essential infrastructure, we are going to evaluate some of the most high-profile examples of cyberattacks throughout the world.

2.4.1.1 Tram Hack Lodz, Poland

In 2008 a tram device hack in the town of Lodz, Poland escalated to a point that a dozen passengers were injured, making this the number one cyber-kinetic attack to result in human injury.

2.4.1.2 Texas Power Company Hack

In 2009 an employee who had been fired from the Texas Power Company hacked their network to cripple strength forecasting structures. He used his logins which had been, however, disabled.

2.4.1.3 Stuxnet Attack on Iranian Nuclear Power Facility

In 2009, a pc virus allegedly created with the useful resources of the U.S. and Israeli governments and centered on Iranian uranium enrichment devices is thought to be responsible for causing large damage to Iran's nuclear utility with the purpose of destroying uranium enrichment centrifuges at an Iranian nuclear facility. Stuxnet is a malicious laptop pc virus that targets SCADA systems with the useful resource of the usage of centered on programmable not unusual place experience controllers (PLCs), which allow the automation of electromechanical processes.

2.4.1.4 Houston, Texas, Water Distribution System Attack

In November 2011, the Water Distribution System on the Water and Sewer Department for the City of South Houston, Texas, was hacked.

2.4.1.5 Bowman Avenue Dam Cyberattack

In 2013, the Bowman Avenue Dam in New York was the victim of a cyberattack and the hackers managed to manipulate the floodgates. Investigations showed they could have changed the settings related to water drift or perhaps changed the amount of chemicals applied in water treatment to catastrophic effect. This would possibly have delivered devastating consequences.

2.5 Other Major Challenges Hindering Growth of IoT Network

Implementation of IoT-enabled smart grids comes with its private set of disturbing conditions, and it is the disturbing conditions that open up opportunities for ultra-modern services and products. In order to tap into the whole cappotential of IoT-enabled smart grids, an understanding of the disruptive conditions is required. Here, we study some of the opportunity disturbing conditions in building and deploying a smart grid network and the one-of-a-kind methods being taken internationally to address them.

2.5.1 Standardization Protocols

While some of the IoT devices can communicate with others, no extensive language exists for the IoT. The absence of necessities has resulted in a fractured collection of solutions that are cobbled together rather than a cohesive set of solutions. Device manufacturers want to therefore choose amongst disparate frameworks at the same time as clients must decide if the devices that they want are desirable and compatible with what they already have. What is even more stressful is that most IoT devices are easy to hack as they lack stringent protection protocols and secure encryption mechanisms. This is an assignment as most of the IoT devices in the market in recent times were developed for connectivity rather than cyber protection.

Although there were attempts at standardization, device manufacturing companies are split regarding the manner of standardization. There are some groups, such as the Open Interconnect Consortium, and the AllSeen Alliance, that are developing private frameworks, with policies regarding intellectual property, structure, and bylaws. It is paramount that these efforts be merged into one collective entity to work towards a single framework so as to useful resource the present products and come up with a now no longer rare certification process.

2.5.2 Cognitive Capability

The rise in type of associated topics translates to extra statistics which requires processing. With the type of associated devices having passed the human population in 2017 and on a path to developing to an extra 50 billion devices by 2020, a question arises as to how to navigate this volume of new statistics. The volume of statistics generated by this massive IoT-based completely smart grid infrastructure is large. Transferring, storing, and analyzing such large portions of statistics might require complex statistics analytics software program application capabilities. Since most IoT devices have low computing power, conventional cryptography cannot work because devices have limited memory which can't address the computing and storage requirements of advanced cryptography algorithms.

A solution that has been proposed is to adopt a cognitive method which could effectively address increasingly extra massive inputs, at the same time as generating tremendous output. Cognitive computing lets in the processing of this developing volume of statistics through interpreting, diagnosing and adapting to their environment without the need for human intervention. Cognitive IoT has the cappotential to combine more than one statistics streams, select out patterns, take a look at interactions and internally review their environment and finally be able to extrapolate the intricacies of the IoT and select statistics correlations that could otherwise remain undiscovered. Cognitive computing is, however, a fledgling technology and will need development in order to be employed in this large-scale form.

2.5.3 Power

Although most IoT devices are low-energy devices, most of them rely on batteries and are not associated with energy reassets and in the end run out of energy. The contemporary low type of devices remains capacity in terms of converting run-down batteries. However, with the type of devices predicted to run into billions of devices, a sustainable and long-term energy solution is required.

For manufacturers of IoT devices, making IoT devices with large batteries is usually difficult due to the tradeoff between providing a reliable, steady delivery of energy at the same time as making the devices sleek, attractive and light. Some of the devices require mobility in order to gain whole capacity without being bogged down with the useful resource of the usage of energy needs. Recent hurdles in developing immoderate energy, compact and miniature batteries has seen present-day breakthroughs in developing wireless energy to be anticipated due to the fact the future solution within the path of transformative and steady reassets of energy to allow IoT devices to gain their whole capacity, increase their lifespan independent of electrical outlets. Some of the areas being studied for delivering wireless energy include magnetic resonance, laser, ultrasound/ultrawave and radio frequency. These areas are, however, need research to determine their commercial enterprise viability and the outcomes if any to human health.

2.5.4 Consumer Illiteracy

Most IoT devices have only some updates or patches that update their safety vulnerabilities. These IoT devices encompass production unit set default logins. The device's password may generally be too complex for the ordinary purchaser to understand or the purchaser is not aware of how to change default logins on their devices. This has made it much easier for cyber criminals to make the maximum hack into the devices.

2.5.5 Weak Regulations

There is a massive amount of regulatory and legal problems surrounding the implementation of IoT devices that needs to be addressed. The primary problems regarding IoT devices embody information retention and destruction guidelines and criminal penalties for safety breaches. The guidelines trouble is laid bare while a number of infected IoT devices are used to attack each different network for a time body finally essential to losses for the affected business corporation. In the type of case, are the producers constructing the IoT devices to be held liable for the out of place business corporation due to their inclined devices? Recommendations for concrete criminal penalties are therefore required to deal with such exposures as this loophole has brought through negligence of some manufacturers in effecting protection functions within the IoT devices they produce. Legislation is therefore required to present stricter recommendations that keep devices off the market until they have got stronger safety or require manufacturers to recall or withdraw products that have been shown to have safety flaws. In specific instances, lack of recommendations in governing IoT devices has hampered the technology in taking off. In some countries, like Kenya, it is illegal to fly unmanned aerial vehicles (UAVs) because the regulatory framework has yet to be drafted. Currently, it seems, technology is advancing at a much faster pace than the associated insurance and regulatory environments.

2.5.6 Fear of Reputational Damage

A breach in the safety of IoT devices has associated costs to the device-producing commercial enterprise corporation or the consumers. Such costs include incident investigation, purchaser perception, logo reputation, falling income and problems to do with public safety. Everywhere a breach takes place which incorporates a large information breach at a monetary group, the monetary group might not be disclosing the said breach in order not to injure its reputation and preserve purchaser confidence. By hiding such information, they hamper future designs and development of new IoT products. Device producing corporations may also cover up information on newly determined bugs in their products in order not to have an impact on the earnings of the said products. As only some corporations are willing to acknowledge losses springing up from cyberspace, this has created a situation wherein most cybercrimes are unreported or underreported. This results in the underestimation of the danger of such incidences taking place, which in turn creates a false sense of safety about the lurking cyber threats.

2.6 Future Prospects

Over the previous few years, hundreds of loads of analog controls in smart grids were modified with digital structures. Digital controls provide better control within the area and transmission of power within the electrical grid. Advances in information generation and operational generation have enabled the new devices but have also made them extra vulnerable to cyber threats. The protection of critical infrastructure in competition to modern-day cyber threats is one of the most exigent disturbing conditions in this era of ubiquitous connectivity. New attack techniques and threats emerge constantly while present ones continuously evolve, making the cyber-protection landscape very dynamic and unpredictable.

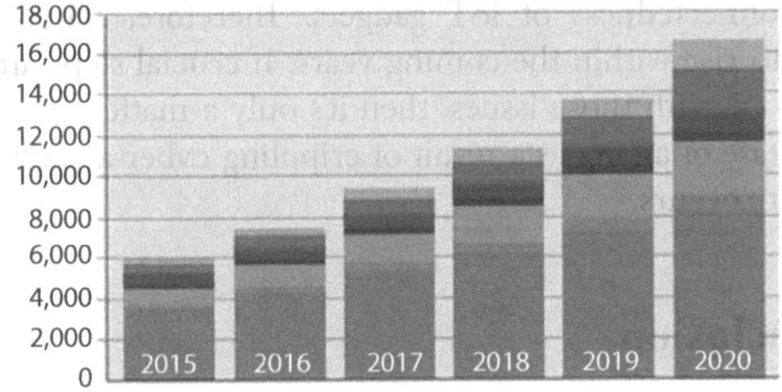

Figure 2.4 IoT gadgets.

Cyberattacks on critical infrastructure cause damage. The vulnerability of nuclear facilities, power grids, dams, and specific critical infrastructure to cyberattack is developing through the methods of the day. Most of their infrastructure is getting old and is underfunded; however, they should protect against advanced cyber threats as their risk will grow exponentially.

While using IoT is presenting prominently in generating future implementation of the SG, it's nevertheless a nascent generation and has a number of disturbing conditions which if not addressed could end up being its Achilles heel. First and foremost is the element of protection. As a critical infrastructure, the SG will now be extra attractive to cyber-assaults, whilst its monitoring and control is carried out over massive net-based protocols and solutions, and may depend upon public communication infrastructure. The reality is that it has loads of devices online and is continually being developed due to the fact the population grows and the power demand increases. As a consequence, an attacker could cause massive economic losses to the software company. With the smart grid being taken into consideration as a precursor and a critical detail of the smart city, its breakdown may possibly cause whole cities to grind to a halt.

Recent studies by means of juniper studies shows that the wide variety of cyber-assaults is forecasted to rise surprisingly because the wide variety of related IoT gadgets will increase and with it will grow the value of breaches to approximately $2.5 trillion globally—nearly 4 instances the predicted value of breaches in 2015 as proven in Figure 2.4 above. Increased penetration of cellular telephones has caused greater publicity to the net through cellular telephones. Increased utilization of the telephones without safety software program collectively with proliferation of reasonably-priced net connections in houses without stable structures will render increasingly IoT structures to be greater susceptibility to cyberattacks. Also the common value of a single cyber safety breach is projected to rise thanks to improved connectedness of IoT gadgets. Therefore, cyber-assaults are anticipated to rise within the coming years. If crucial steps aren't accomplished to cope with those issues, then it's only a matter of time before a nightmare state of affairs as a result of crippling cyber-assaults on crucial infrastructure occurs.

2.7 Conclusion

IoT is the next step towards a globally and pervasive connection to any conversation and computation enabled objects, regardless their get entry to generation, available reassets and location. The smart grid is the maximum crucial deployment of IoT generation, in which smart devices are deployed along the electricity path, all the way from the generation plant to the end-customer. The software program of IoT will beautify the winning energy grids with the resource of the use of providing real-time control and monitoring of the energy grid components. However, cyber safety is seen as one of the important factors impeding the fast and large-scale adoption and deployment of IoT within the smart grid. Ensuring safety for grid devices related to the internet is a great challenge. This is due to the extensive kind of devices related within the network presenting many more opportunities for a cyberattack and functionality, with grave results. With the modern-day IoT based totally absolutely smart grid being implemented, the scope of possible attack ground is growing exponentially.

In this paper, we supplied the software program of IoT as allowing generation for the smart grid. We then supplied a whole survey of the number one safety issues and worrying conditions for the IoT-based totally absolutely SG. We moreover supplied a summary of the maximum crucial worrying conditions in IoT-based totally absolutely Smart Grid and functionality solutions. While the IoT can produce large-scale improvements, like most growing concepts, some technical, legal, and economic elements of the IoT ought to be dealt with carefully in advance, then it turns into a mature, ready-to-use generation for large adoption within the smart grid. As increasingly smart grids and smart meters are installed, the extensive kind of functionality gets entry to elements to grid networks and will grow enormously. It is likely therefore, that cyberattacks will grow over time, and they will end up more sophisticated and capable of causing substantial disruption to town services, the wider monetary machine and society. Despite the worrying conditions, using the IoT is established to beautify the smart grid networks to facilitate better monitoring and control of the energy grids. It is therefore important to popularity on safety issues within the route of the superb tiers of designing, enforcing and integrating of the IoT devices within the smart grid.

References

1. J. Wan, M. Chen, F. Xia, D. Li and K. Zhou, From Machine-to-machine communications towards cyber-physical systems, Computer Science and Information Systems (ComSIS), vol. 10, no. 3, pp. 1105–1128, 2013. doi:10.2298/CSIS120326018W.
   Google Scholar
2. M. Chen, J. Wan and F. Li, Machine-to-machine communications: Architectures, standards and applications, KSII Transactions on Internet and Information Systems, vol. 6, no. 2, pp. 480–497, 2012. doi:10.3837/tiis.2012.02.002.
   Google Scholar
3. A.R. Al-Ali and R. Aburukba, Role of Internet of Things in the smart grid technology, Journal of Computer and Communications, vol. 3, pp. 229–233, 2015. doi:10.4236/jcc.2015.35029.
   Google Scholar
4. J. Gubbi, R. Buyya, S. Marusic and M. Palaniswami, Internet of Things (IoT): A vision, architectural elements, and future directions, Future Generation Computer Systems, vol. 29, no. 7, pp. 1645–1660, 2013. doi:10.1016/j.future.2013.01.010.
   Google Scholar
5. D. Evans, *The Internet of Things: How the Next Evolution of the Internet Is Changing Everything*, Cisco Internet Business Solutions Group (IBSG), 2011.
   Google Scholar
6. D.B. Rawat and C. Bajracharya, Detection of false data injection attacks in smart grid communication systems, IEEE Signal Processing Letters, vol. 22, no. 10, pp. 1652–1656, 2015. doi:10.1109/LSP.2015.2421935.
   Google Scholar
7. M. Centenaro, L. Vangelista, A. Zanella and M. Zorzi, Long-range communications in unlicensed bands: The rising stars in the IoT and smart city scenarios, IEEE Wireless Communications, vol. 23, no. 5, pp. 60–67, 2016. doi:10.1109/MWC.2016.7721743.
   Google Scholar
8. W. Wanga and Z. Lua, Cyber Security in the Smart Grid: Survey and Challenges, in Elsevier, 2012. doi:10.1016/j.comnet.2012.12.017.
   Google Scholar
9. N. Bui, A.P. Castellani, P. Casari and M. Zorzi, The internet of energy: A web-enabled smart grid system, IEEE Network, vol. 26, no. 4, pp. 39–45, 2012. doi:10.1109/MNET.2012.6246751.
   Google Scholar

3

Electrical Load Forecasting Using Bayesian Regularization Algorithm in Matlab and Finding Optimal Solution via Renewable Source

Chinmay Singh, Yashwant Sawle, Navneet Kumar, Utkarsh Jha* and Arunkumar L.

Department of Electrical Engineering, Vellore Institute of Technology, Vellore, Tamil Nadu, India

Abstract

As the smart grid development advances in India, load forecasting of power is a must. Electric power demand forecasting is critical for energy providers and energy consumers in the electrical industry sector, i.e., power generation, transmission, and its distribution to electric markets as it helps organizations maintain a demand-supply power equilibrium. Precise models in electrical load forecasting have been basic to the activity and planning of any company-based organization. Machine Learning algorithms like Artificial Neural Networks (ANNs) are incredibly powerful and are only limited by their implementation and execution. This paper focuses on minimizing the error between predicted and actual load demand for the industry. That means we will use the ANN method to minimize the Mean Absolute Percentage (MAP) Error between predicted load demand and the actual load values to overcome MAP_Error through graphs and show the advantages of using a Bayesian network over other forecasting techniques like using standard analytical functions. We are using a Bayesian algorithm neural network through MATLAB Simulink. Also, we are focusing on implementing this concept into a renewable solution, as it will support a clean environment and will enhance the quality of life.

***Keywords*:** Load forecasting, ANN, Bayesian method, renewable solution, machine learning

**Corresponding author*: jhautkarsh1999@gmail.com

A. Chitra, V. Indragandhi and W. Razia Sultana (eds.) Intelligent and Soft Computing Systems for Green Energy, (41–58) © 2023 Scrivener Publishing LLC

3.1 Introduction

In the present world, electricity has become an essential part of our day and night need. To ensure that electric energy reaches consumers or customers safely and through a monetary method, any electrical industry encounters numerous practical or specialized difficulties. Among these difficulties, load flow study, load analysis, and electrical operation are generally noticeable. Load Forecasting is moreover the best and updated developing area of examination for this important and analyzing field in a neoteric couple of times. Load forecast can be categorized as the amount of correctness of the divergence amongst real and projected valuation of future electrical load demand [1]. It emboldens an electrical utility to be mature in making important decisions by recalling the decisions to buy and generate electricity, exchange the load, and improve their fundamentals. The subject of load forecasting has been present for quite a long time to gauge future interest [2].

Electric load forecasting will help improve the start-up costs of production units and possibly also ready to standby the concern in emerging the necessary number of concentration offices. It can also help examine dangerous activities, inconsistent awareness, encouragement to change supplies, and weakness to disenchantment. Load forecasting basically provides the most approved actual noteworthy datasets to govern conveying-cum-arrangement. It additionally assumes a noteworthy part in energy the board framework. This includes the precise expectation of both the extent and geological areas of electric burden over the various times of the arranging skyline [3-4]. The greatest reserve funds can be accomplished when load forecasting is utilized to control tasks and choices like monetary dispatch/unit duty and fuel distribution/online network examination. The determination of the electricity demand helps to progress the start-up costs for the creation of units and can also be ready to always auxiliary our interest and focus on the latest developments of the detailed number of concentration blocks. It can also help look for dangerous activity, changeable concern, demand to transform business, and weakness into dissatisfaction. It also contains the most important data for controlling transportation and organization. It moreover fulfills a significant function in the energy management system or in any smart grid [1].

Systematically, the load forecasting methods could be classified into two classes, parametric or non-parametric procedures. Linear regression, automatic regressive moving average techniques, i.e., ARMA, universal exponential technique, and stochastic time series methods, are some instances of parametric (measurable) methods. The core disadvantage of this method is its ability to produce unexpected differences in all kinds of climatic or societal fluctuations. In any case, this weakness is overwhelmed by using a non-parametric technique (artificial awareness) given the likelihood of global law enforcement. Amongst these artificial brain-based practices, the fake neural organization has appeared as one of the greatest compelling strategies that have received significantly more attention from analysts. For this reason, we recommend a probabilistic understanding of Neural network learning with Bayesian methods [5]. The Bayesian handling of the display offers critical preferences compared to the conventional NN learning measurement. Among other things, reference can be made to the programmed adaptation of the regularization coefficient or constant using all available information and to the determination of the most important information factors by a certain method known as automatic determination of relevance or, in short, ARD as Automatic Relevance Determination. It also takes into account the consistency of the estimate as the strategy includes an error bar for the performance of the model [6].

Through this paper, we would contend that in order to get a decent model for predicting the charge of electrical energy, it is necessary to accentuate the plane of the NN. At the end of the day, regular NN learning techniques must be improved. The Bayesian way to deal with neural learning is applied to genuine burden information. We apply this method to attain less power usage error through this future load prediction process. And also, we are applying this solution through a renewable solar source (dual-axis solar tracker), which we will explain in detail in the sections below. We ensure that this technology should be applied in the coming future to reduce load over-usage.

3.2 Algorithm

3.2.1 Levenberg-Marquardt Algorithm

ANN has run through a few small interconnected units called neurons. For the best results, we use massive processing of neurons at the same time, which has been artificially interconnected. A continuous repetitive process is carried out by neurons also to connect the input, along with output and unknown layers of the neural network. The relationship among input and output can be measured by neural weight updates obtained through teaching algorithms which are used in ANN. The necessary and produced output fault can be reduced by adjusting the weight and biases. Further details are discussed in [7] and are represented as:

$$A_c = \sum_{c=1}^{k} P_c W_c$$

Where c = 1,2,...,k-1,k; P_c is the cth input, where W_c is the distribution weight for kth term, and Ac is the cth ANN output. ANN has been checked in [8]. In ANN calibration, architectures are selected, and the numbers of neurons are considered as necessary. In addition, ANN training algorithms for updating weights would also be specified. The hustle, correctness, and difficulty of the load forecasting model also depend on the machine learning algorithms used in the model. The feed-forward three-layer multi-perceptron neural network is being applied. Using an activation function for the desired result that would be achieved using veiled neurons. The most common log sigmoid math function is used for teaching. Further details are taken from [9]:

$$F(c) = \frac{1}{1+e^{-c}}$$

In the projected predictive model, the LMA is applied for the hidden layer of weight updates in propose to resolve the non-linear least-squares problem. Numerous algorithms used to form the ANN model of multilayer perception are LMA, the decent gradient, and Bayesian regularization. The planned forecasting approach used the LMA to update the hidden shift weights to achieve the desired results. Initially, the random weights Wc are made in LMA. Then, the squared error of sum E_c is calculated from the initial weights using this below-mentioned formula:

$$E_c = \frac{1}{2} \sum e^2$$

where 1≤rR

1≤q ≤Q

Where $e_{r,q}$ is the learning error that we can calculate using desired output $d_{r,q}$, and although the actual output $A_{r,q}$ is given by:

$$e_{r,q} = d_{r,q} - A_{r,q}$$

After this operation, LMA weights are updated by:

$$W_{c+1} = W_c - (H)^{-1} J_{en}$$

where J is Jacobian matrix and H is the Hessian matrix that could be figured by:

$$H = J^T J + (BI)$$

Where I is an identity matrix and also grouping coefficient, then this value is taken into account when the rationalized error Ec + 1 becomes fewer than the unique Ec. If the updated error Ec + 1 becomes larger than the original En, the process is restarted from the initial generation of random weights [7, 8]. The LMA process can be witnessed in-depth from Figure 3.1 given below.

3.2.2 Bayesian Regularization

Neural networks are proposed for load prediction because of their ability to non-linearly model large multivariate data sets. Researchers have implemented many ways to use these Bayesian methods.

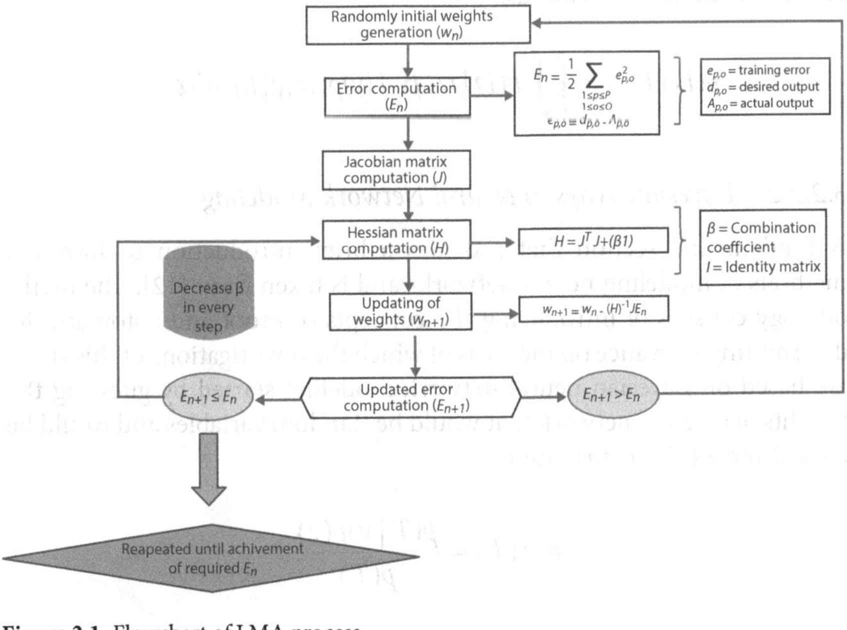

Figure 3.1 Flowchart of LMA process.

Bayesian regularization is easy to contrivance individually for the rest of the Bayesian toolkit. Checking [10, 11] with another technique to take precaution from overfitting in load forecasts.

Cross-validation is an observed method that can be used for many purposes – relating diverse architectures or input sets, evading overfitting, and adjusting training parameters. [12] proves that no guaranteed model chosen by CV is indeed the best. These are easy to comprehend as the countless number of models given is constantly likely to find a model that will outperform both the training and justification set.

3.2.2.1 *Comparison of Bayesian Models*

The Bayesian process of choosing the right number of neurons started through a set H of candidate models. An option of every model Hy is as follows:

$$\frac{p(H \mid Hy)P(Hy)}{p(D)}$$

Let us assume that the hyperparameters are liberated of the use of the Gaussian calculation of their proof, and therefore let us assume that the non-informational priorities are those obtained by the logarithmic proof of a model expressed below as:

$$p(D \mid Hy) = \int\int p(D \mid \alpha, \beta, Hy)p(\alpha, \beta| \, hy)\, d\alpha\, d\beta$$

3.2.2.2 *Bayesian Ways to Neural Network Modeling*

A fundamental section that provides a brief introduction to Bayesian methods of modeling neural networks and is taken from [12]. The methodology consists of introducing the concepts of proof and automatically determining relevance on the basis of which the investigations of this study are based on Bayesian neural network modeling started by guessing the weights of a neural network that would be Random variables and would be considered a prior distribution.

$$p(v \mid T) = \frac{p(T \mid v)p(v)}{p(T)}$$

where v is the weight vector considered and $T = (t)^n$ is the set of target vectors taken.

We have been implementing a new methodology on the basis of the Bayesian approach to load forecasting. This is different from old-style neural network techniques like cross-validation, where the ways are able to compact with the full capability with the model over the use of the proof agenda and model selection are quite effective. It also provided the ability to select the key input variables of the model. Using the Bayesian method, the calculation was able to show error bars in the model output.

3.2.3 Scaled Conjugate Gradient Algorithm

The Conjugate Gradient algorithm has been mainly used to minimize the functions of many variables since there is no need to store a matrix. The convergence rates of the algorithms are only linear unless the iterative procedure is restarted occasionally.

3.2.3.1 Steps of Algorithm

1. Vector v_1 and scalars

$$0 < \rho \leq 10^{-4}, \; 0 < \mu_1 \leq 10^{-6},$$

   Set $B_1 = g_1 = -F'(v_1)$, m=1 with success = true.
2. If success=true, then calculate 2nd order information

$$\rho_k = \rho / |B_m|$$

$$S_m = (F'(V_k + \rho_m B_m) - F'(V_m)) \,/\, \rho_m$$

$$\delta_m = B_m^T S_m$$

3. Scale $\delta_m : \delta_m = \delta_m + (\mu_m - \mu'_m)|B_m|^2$
   And so on as in reference [11].

3.2.4 Gradient Descent

The gradient descent, also known as the sharpest succession, is the greenest training algorithm. It needs information from the gradient vector, and so it is a first-order method.

$$K^{(i+1)} = K^{(i)} - g^{(i)}\eta^{(i)}$$

The parameter η which is used in this is known as the training rate. This value can be established to a fixed value or determined by 1-D optimization beside the learning direction at each step. An ideal value of the learning rate acquired by minimizing lines at each consecutive step is usually desirable. Newton's method is a second-order algorithm because it uses the Hessian matrix. The objective of this process is to find better commands of technics using the second derivative of the loss function.

Let's take $f(K^{(r)}) = f^{(r)}$, $\nabla f(K^{(r)}) = g^{(r)}$ and $(K^{(r)}) = H^{(r)}$.

Take the quadratic guesstimate off at using Taylor's series expansion.

$$f = f^{(0)} + g^{(0)}.(K - K^{(0)}) + 0.5\,.(K - K^{(0)})^2\,.H^{(0)}$$

$H^{(0)}$ is the Hessian of f evaluated at the point $K^{(0)}$, Newton's method iterates as follows:

$$g = g^{(0)} + H^{(0)}.\,(K - K^{(0)}) = 0$$

$$K^{(r+1)} = K^{(r)} - H^{(r)-1}\,.\,g^{(r)}$$

3.2.5 Conjugate Gradient

The conjugate gradient or ascent technique can be viewed as something that falls amid gradient descent and Newton's method. It is inspired by a desire to accelerate the generally slow convergence associated with gradient descent. These operations also avoid the info necessities associated with evaluating, storing, and reversal of the Hessian matrix as essential by the Newton method. In the conjugate gradient teaching algorithm, the search is performed with conjugate directions that normally produce faster convergence than gradient descent directions. These learning instructions are combined in relation to the Hessian matrix.

3.3 Methodology and Modelling

The MATLAB code for the process of forecasting [14] of Technology Tower Load at the VIT University comprises sub-divisions such as 'Data transfer', that is, extracting the data from the excel worksheet in serial date format then replacing non-numeric cells with NaN as well as creating output variables and tables and therefore allocating imported array to column variable names.

The TT data is taken for every 15-minute interval, as shown above in Table 3.1, and system load is observed in Watts then the predictor function that adheres to the prediction of forecasts in MATLAB editing mode is shown:

```
function [X, dates, labels] = TTPredictor1(data)
dates = datenum(data.Date);
if all(floor(dates)==dates)
    dates = dates + (data.Minute)/60;
end
Holide = datenum({'01-Jan-17','26-Jan-17'},'dd-
mmm-yy');
Prev_Day_Same_Hour_Load = [NaN(103,1); data.
SYSLoad(1:end-103)];
Prev_Week_Same_Minute_Load = [NaN(721,1); data.
SYSLoad(1:end-721)];
Prev_24Hrs_AvgLoad = filter(ones(1,103)/103, 1,
data.SYSLoad);
dayOfWeek = weekday(dates);
istheWorkingDay=~ismember(floor(dates),Holide) &
~ismember(dayOfWeek, [6 7]);
X = [data.Minute dayOfWeek IstheWorkingDay Prev_
Week_Same_Minute_Load Prev_Day_Same_Hour_Load
Prev_24Hrs_AvgLoad];
labels = {'Minute', 'Weekday', 'IstheWorkingDay',
'Prev_Week_Same_Minute_Load', 'Prev_Day_Same_Hour_
Load', 'Prev_24Hrs_AvgLoad'};
```

Table 3.1 A part of load data for Technology Tower.

Data Time	Date	Minute	SYSLoad
01-01-2017 0:15	01-01-2017	15	156866.00
01-01-2017 0:30	01-01-2017	30	147948.10
01-01-2017 0:45	01-01-2017	45	158323.60
01-01-2017 1:00	01-01-2017	60	155874.20
01-01-2017 1:15	01-01-2017	75	165594.20
01-01-2017 1:30	01-01-2017	90	162549.70
01-01-2017 1:45	01-01-2017	105	157064.40
01-01-2017 2:00	01-01-2017	120	164749.00
01-01-2017 2:15	01-01-2017	135	155710.30
01-01-2017 2:30	01-01-2017	150	162394.40

Before neural Training:

```
[X, dates, labels] = Predictor1(data);
y = data.SYSLoad;
```

Here we use the neural net fitting tool in order to use Bayesian regularization [15]. The reason for using this algorithm is it may acquire a good amount of time but certainly results in better rationalization for all chaotic, difficult, tiny as well as noisy datasets. Training stops according to adaptive weight minimization (regularization) and backpropagation algorithms[16] by selecting the mapping between X and Y rows, and it gets completed in 272 iterations, and ten epochs were processed.

After neural Training:

```
Y_predicted = sim(net,X')';
plot(Y_predicted,'DisplayName','Y_predicted');
hold on;
plot(y,'DisplayName','y');hold off;
r = y-Y_predicted;
map_error = nanmean(abs(r./y*100))
```

Now the graph is being plotted between the predicted demand Y predict and y, i.e., the actual load; hence we determine the difference between them very accurately and precisely. All the tasks are associated with the Neural Net Fitting tool of MATLAB. The map_error (mean absolute percentage error) is observed between the plots, and that much is overcome through distributed generation techniques through solar PV generation [17–21].

The number of iterations achieved is shown below in Figure 3.2.

Now, here comes an example to illustrate the system-level model of a photovoltaic generator (Figure 3.3) that can produce solar power in order to provide the requirements of decentralized generation and overcoming error between the predicted and actual loads, which was forecasted earlier. The DC to AC converter is used as a device that converts battery power or solar power

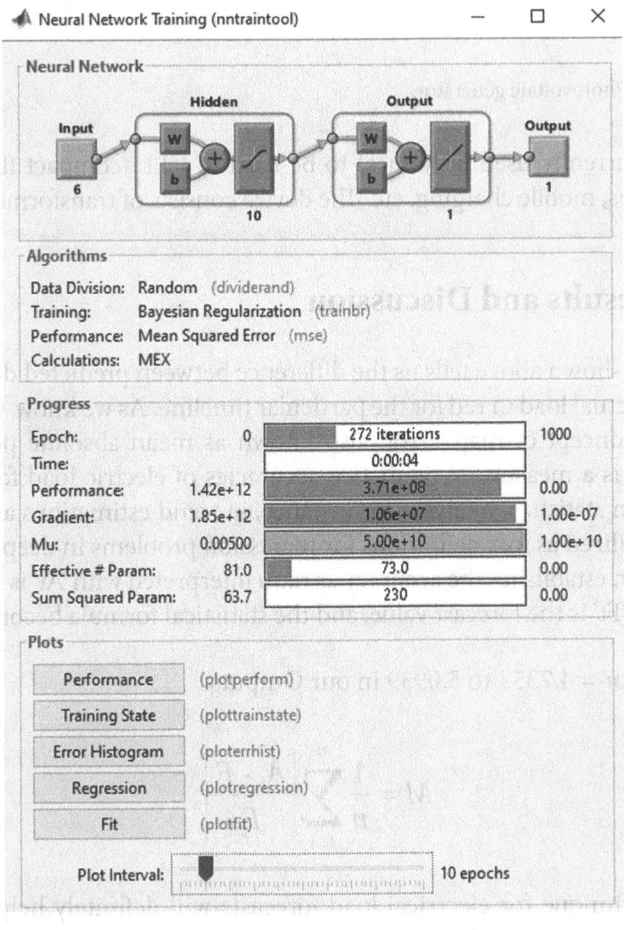

Figure 3.2 Neural network training.

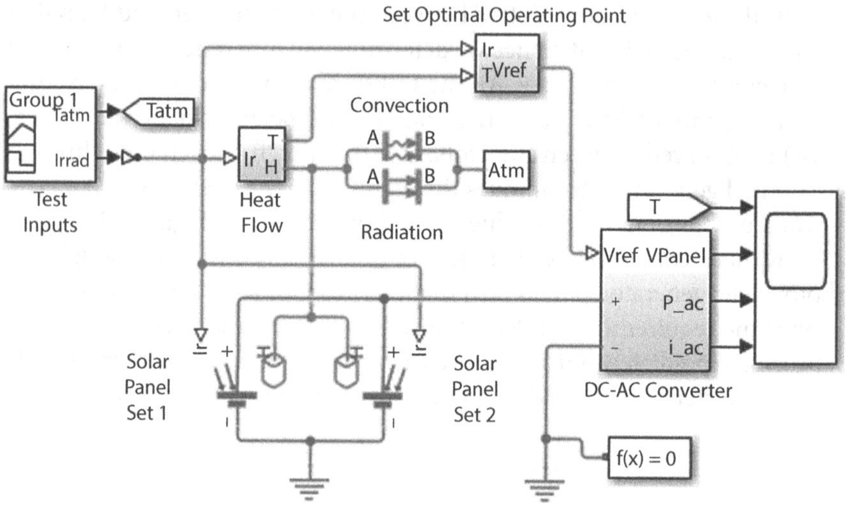

Figure 3.3 Photovoltaic generator.

into ac (currently used in homes) to be used in CFL (compact fluorescent lamp) bulbs, mobile charging, etc. The device consists of transformer, IC, etc.

3.4 Results and Discussion

Figure 3.4 shown above tells us the difference between predicted demand in blue and actual load in red for the particular timeline. As we know, the mathematical concept of map_error, also known as mean absolute percentage deviation, is a measure of predictive accuracies of electric load forecasting methods in statistical analysis, for instance, in trend estimations and analysis, also utilized as loss definitions for regression problems in deep learning. It, however, establishes the accuracy as ratio interpreted with 'At' is the actual value and 'Ft' is the forecast value, and the statistical formula becomes:

MAP_Error = 4.7353 to 5.0939 in our Outputs.

$$M = \frac{1}{n} \sum_{t=1}^{n} \left| \frac{A_t - F_t}{F_t} \right|$$

Our technique for electrical load forecasts will definitely help all such power or energy providers and distributors to calculate the power/energy

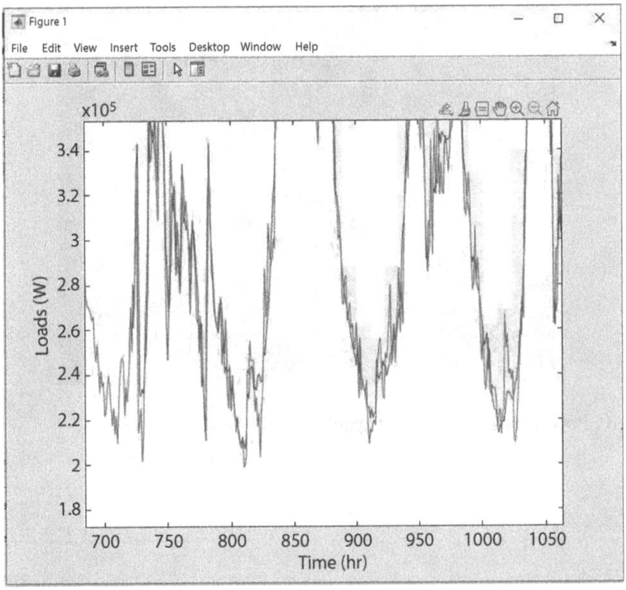

Figure 3.4 Comparison of load demands.

needed to meet the demand and supply steadiness. The accurateness and precision involved in forecasts are certainly of great importance for the operations and managerial actions in the loading of an electrical profitability company and energy investors.

The simulation results for the photovoltaic generator to help in supplying the insufficient production (that is, the error obtained), and the power obtained plots are displayed below in Figure 3.5, Figure 3.6, and Figure 3.7:

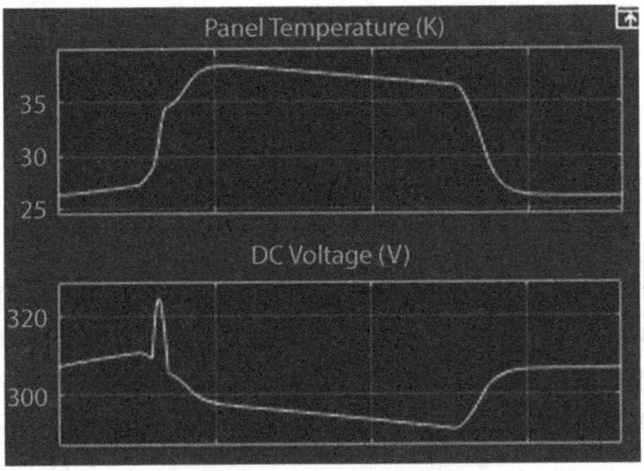

Figure 3.5 Panel temp. and DC voltage graphs.

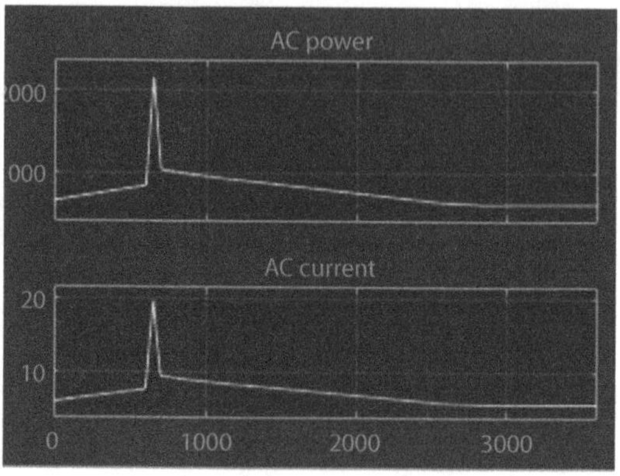

Figure 3.6 AC Power and AC current graphs.

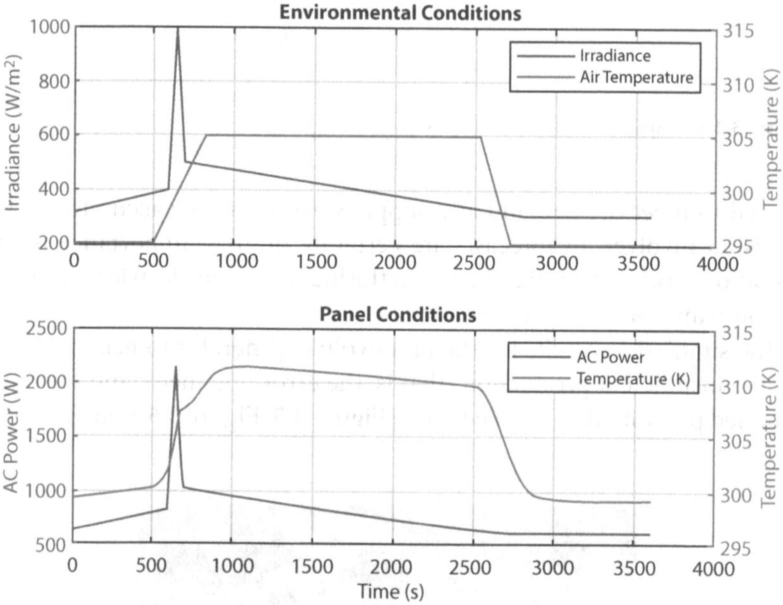

Figure 3.7 Environmental and panel condition graphs.

The plot obtained thus displays output power, DC voltage, and AC power as well as current and temperature in Kelvin of panel used over for 'an hour testing'. The surrounding external temperatures vary over time, thus totally affecting its conversion efficiencies.

3.5 Conclusion

Machine Learning algorithms like Artificial Neural Networks (ANNs) are incredibly powerful and are only limited by their implementation and execution. Load forecasting better helps in modeling power generation and transmission by helping organizations maintain a demand-supply equilibrium. In this paper, we have used its versatility to help solve load forecasting problems. This paper has shown the advantages of using a neural network-based load forecasting over other methods that are currently used like extrapolation, correlation, or the combination of both. Since the neural network-based model can learn from past data, factors affecting load forecasting like the weather impact will not significantly impact the model as the weather is based on seasonality. Also, since the load data is non-linear, neural networks can model it and predict future outcomes better than any other technique. Neural networks are likewise invulnerable to noise in the figures and are very much fault-tolerant. In further research, we can combine different machine learning techniques together like an ensemble neural network, boosting techniques, and recurring neural nets like long- and short-term memory and gated recurring units in order to increment accuracy to a great extent.

References

1. Hippert, H. S., & Taylor, J. W. (2010). An evaluation of Bayesian techniques for controlling model complexity and selecting inputs in a neural network for short-term load forecasting. *Neural Networks, 23*(3), 386–395.
2. Baliyan, A., Gaurav, K., & Mishra, S. K. (2015). A review of short term load forecasting using artificial neural network models. *Procedia Computer Science, 48*, 121–125.
3. Kumari, Divya, *et al.* "Prediction of alcohol abused individuals using artificial neural network." *International Journal of Information Technology* 10.2 (2018): 233–237.
4. Singh, A. K., Khatoon, S., Muazzam, M., & Chaturvedi, D. K. (2012, December). Load forecasting techniques and methodologies: A review. In *2012 2nd International Conference on Power, Control and Embedded Systems* (pp. 1–10). IEEE.
5. Su, P., Tian, X., Wang, Y., Deng, S., Zhao, J., An, Q., & Wang, Y. (2017). Recent trends in load forecasting technology for the operation optimization of distributed energy system. *Energies, 10*(9), 1303.

6. Kiartzis, S., Kehagias, A., Bakirtzis, A., & Petridis, V. (1997). Short term load forecasting using a Bayesian combination method. *International Journal of Electrical Power & Energy Systems, 19*(3), 171–177.
7. Hippert, Henrique Steinherz, Carlos Eduardo Pedreira, and Reinaldo Castro Souza. "Neural networks for short-term load forecasting: A review and evaluation." *IEEE Transactions on Power Systems* 16.1 (2001): 44–55.
8. Du, Yi-Chun, and Alphin Stephanus. "Levenberg-Marquardt neural network algorithm for degree of arteriovenous fistula stenosis classification using a dual optical photoplethysmography sensor." *Sensors* 18.7 (2018): 2322.
9. Yu, H., & Wilamowski, B. M. (2011). Levenberg-Marquardt training. *Industrial Electronics Handbook, 5*(12), 1.
10. Chan, Z. S., Ngan, H. W., Rad, A. B., David, A. K., & Kasabov, N. (2006). Short-term ANN load forecasting from limited data using generalization learning strategies. *Neurocomputing, 70*(1–3), 409–419.
11. Gaikwad, Ninad K., and Sagarkumar M. Agravat. "On The Development of Solar & Wind Energy Forecasting Application Using ARIMA, ANN and WRF in MATLAB."
12. Cataltepe, Z., Abu-Mostafa, Y. S., & Magdon-Ismail, M. (1999). No free lunch for early stopping. *Neural Computation, 11*(4), 995–1009.
13. Singh, S., Hussain, S., & Bazaz, M. A. (2017, December). Short term load forecasting using artificial neural network. In *2017 Fourth International Conference on Image Information Processing (ICIIP)* (pp. 1–5). IEEE.
14. Nguyen, S. T., Nguyen, H. T., Taylor, P. B., & Middleton, J. (2006, September). Improved head direction command classification using an optimised Bayesian neural network. In *2006 International Conference of the IEEE Engineering in Medicine and Biology Society* (pp. 5679–5682). IEEE.
15. Musakulova, Z., Mirkin, E., & Savchenko, E. (2018, October). Synthesis of the backpropagation error algorithm for a multilayer neural network with non-linear synaptic inputs. In *2018 IEEE International Conference on Electrical Engineering and Photonics (EExPolytech)* (pp. 131–135). IEEE.
16. Dhlamini, N., & Chowdhury, S. D. (2018, June). Solar photovoltaic generation and its integration impact on the existing power grid. In *2018 IEEE PES/IAS PowerAfrica* (pp. 710–715). IEEE.
17. Sawle, Y., Gupta, S.C. and Bohrer, A.K. "Socio-techno-economic design of hybrid renewable energy system using optimization techniques." *Renewable Energy* 119 (2018): 459–472.
18. Sawle, Y., Gupta, S.C. and Bohrer, A.K., A Novel Methodology for Scrutiny of Off-Grid Hybrid Renewable System, John Wiley & Sons, *International Journal of Energy Research*. 2018; vol. 42 pp. 570–586.
19. Sawle, Y., Gupta, S.C. and Bohrer, A.K., Review of hybrid renewable energy systems with comparative analysis of off-grid hybrid system, *Renewable and Sustainable Energy Reviews*, vol. 81, June 2017, pp. 2217–2235.
20. Sawle, Y., Gupta, S.C. and Bohrer, A.K., PV-wind hybrid system: A review with case study. *Cogent Engineering Journal*, (2016). 3(1), 1189305.

21. Hobbs, B. F., Helman, U., Jitprapaikulsarn, S., Konda, S., & Maratukulam, D. (1998). Artificial neural networks for short-term energy forecasting: Accuracy and economic value. *Neurocomputing, 23*(1–3), 71–84.
22. Lauret, P., Fock, E., Randrianarivony, R. N., & Manicom-Ramsamy, J. F. (2008). Bayesian neural network approach to short time load forecasting. *Energy Conversion and Management, 49*(5), 1156–1166.
23. Ranaweera, D. K., Karady, G. G., & Farmer, R. G. (1997). Economic impact analysis of load forecasting. *IEEE Transactions on Power Systems, 12*(3), 1388–1392.
24. Saini, L. M. (2008). Peak load forecasting using Bayesian regularization, Resilient and adaptive backpropagation learning based artificial neural networks. *Electric Power Systems Research, 78*(7), 1302–1310.
25. Titterington, D. M. (2004). Bayesian methods for neural networks and related models. *Statistical Science*, 128–139.
26. Gross, G., & Galiana, F. D. (1987). Short-term load forecasting. *Proceedings of the IEEE, 75*(12), 1558–1573.
27. Park, D. C., El-Sharkawi, M. A., Marks, R. J., Atlas, L. E., & Damborg, M. J. (1991). Electric load forecasting using an artificial neural network. *IEEE transactions on Power Systems, 6*(2), 442–449.
28. El-Keib, A. A., Ma, X., & Ma, H. (1995). Advancement of statistical based modeling techniques for short-term load forecasting. *Electric Power Systems Research, 35*(1), 51–58.
29. S. K. Sheikh and M.G. Unde "Short-Term Load Forecasting using ANN Technique," *International Journal of Engineering Science and Emerging Technologies*, Vol. 1, issue 2, pp. 97–107, Feb, 2012.
30. A. Azzam-Ul-Asar and J. R. Mcdonald, "A Specification of Neural Networks in the Load Forecasting Problem," *IEEE Transaction on Control System Technology*, Vol. 2, pp. 135–141, 1994.

4

Theft Detection Sensing by IoT in Smart Grid

N. Siva Mallikarjuna Rao[1]*, M. Ramu[2] and Lekha Varisa[3]

[1]*Dept. of EECE, GITAM (Deemed to be University), Hyderabad, India*
[2]*Dept. of EECE, GITAM (Deemed to be University), Visakhapatnam, India*
[3]*Dept. of ECE, GITAM (Deemed to be University), Visakhapatnam, India*

Abstract

A smart grid is the combination of electrical networks and communication-based facilities which include smart metering, automation facilities and time-based monitoring techniques with the existing electrical network. A smart grid represents a two-way power transfer capability from utility to consumer and from consumer to utility. Smart grid infrastructure is also used in industries for continuous monitoring and safety of the electrical equipments and regulation of electrical network. The smart grid does this through a network of transmission lines, smart meters, distribution automation, substations, transformers, sensors, software and more that are distributed to a grid across the city. Internet of Things (IoT) technology in sensible grid provides an approach to collect and retrieve information from all the equipments connected in the system and speed up the informatization of facility system; it is helpful for effective management of the power grid infrastructure. The smart grid with part of IoT Framework give us facilities like remotely monitoring and managing everything from lighting, faults warnings, early maintenance warning and early detection of things like power influxes as the result of earthquakes and extreme weather changes.

***Keywords*:** Smart Grid (SG), Internet of Things (IoT), Power Internet of Things (PIoT), Parallel redundancy protocol (iPRP), Phasor Data Concentrator (PDC), User Datagram Protocol (UDP), Active Distribution Network (ADN)

**Corresponding author*: mnandana@gitam.edu

A. Chitra, V. Indragandhi and W. Razia Sultana (eds.) Intelligent and Soft Computing Systems for Green Energy, (59–68) © 2023 Scrivener Publishing LLC

4.1 Introduction

The term Internet of Things (IoT) is associated with an intelligent network that provides the facilities to communicate with the devices which are far away using wireless technology. The IoT has become universal to focus on the vision that all the physical objects must communicate with each other. The smart grid (SG) is more advanced than the normal electric grid. The demand of electric power is increasing day by day. Theft of electricity is a very big problem in the electricity sector, especially in developing countries; many times, companies' losses are so high that the government must give them subsidies. In some regions power theft is frequent; this can be reduced if we detect the area of power theft [4]. The VAR acquisition is the only way in a smart grid for identification of the theft location. By victimization the IoT technology in the grid leads the digitalization and sight changes in it. The system developed must establish communication between the consumer and the utility.

4.1.1 Power Theft Identification

We can identify any kind of power theft by implementing IoT into our power distribution network. The arduino in the circuit of a smart energy meter provides information about the amount of power measured to the raspberry pi hub through zigbee network. Smart transformers provide information to raspberry pi hub about the amount of power it has delivered to the loads. Now the raspberry pi hub compares the amount of power delivered by smart transformer with the actual amount of power measured by smart energy meter. If there is a difference then raspberry pi informs the monitoring center about the power theft, then the monitoring center stops the power supply to the power theft area [8].

4.1.2 Basic Structure of Smart Grid

Power Internet of Things (PIoT) is the integration of various perception devices in generation, transmission and distribution of power to observe and control the changes in power systems in different conditions. PIoT is divided into three layers: perception layer, network layer and application layer [6]. In the perception layer the sensors gather information about the changes in position of objects and environment. The information gathered by sensors is collected by backbone node and the information is transmitted to the monitoring center through a wireless network. Some of the sensors which monitor the transmission of power are:

A. Tension sensor
   These sensors gather information about icing of the conductor because snow accumulates on the conductors in cold areas and this causes flashover between the conductors.
B. Micrometeorology sensor
   These sensors gather information about rise and fall of temperature, humidity, rainfall and solar irradiation. According to the information provided by the sensors, the monitoring centers operate the transmission of power to the loads.
C. 3D acceleration sensor
   These sensors gather information about velocity of wind and galloping of conductors so that monitoring centers decide whether to continue the power supply or not.
D. Leaning sensor
   These sensors provide information about the tilt angle made by transmission towers or electric poles. According to the information, monitoring centers take preventive measures.

Network layer helps in transmitting the information gathered by sensors to monitoring centers in the form of packets. The functions of the network layer are:

1. Routing
   This layer makes different paths in between sources and destinations for transmitting different packets of information.
2. Logical addressing
   This layer adds header to the packets which includes logical address of both source and destination.
3. Internetworking
   This layer provides the logical connections between different types of networks by using gateway devices and routers.
4. Fragmentation
   This layer breaks the packets of information into small individual units so they can travel through different networks. This is done because maximum transmission unit size varies router to router.

Application layer provides interface between networks and applications. The functions of the application layer are:

1. This layer is responsible for the packets to be identified and accepted by receiving devices.

2. This layer ensures the required communication interface to transmit packets.
3. This layer also makes sure about the problem recovery, integration of data and privacy.
4. This layer provides certain rules so as to transmit the packets of information in between sources and destinations.
5. This layer helps us to access information from our systems and to make modifications in the information.

4.2 Problem Identification

4.2.1 Power Theft Methods

1. Bypassing meter
   In this power theft method the consumer connects the phase and neutral terminals of the load directly to the phase and neutral of distributed overhead lines at the input of the energy meter. Therefore, the energy meter cannot measure the power consumed by the load.
2. Direct hooking from the line
   In this power theft method, the consumer connects the phase and neutral terminals of the load directly to the phase and neutral of distributed power lines to supply power to the load.
3. Injecting foreign element in the energy meter
   In this method an electronic circuit is installed inside the meters, so it's very easy to slow down or fasten the meter any time using remote. This modification is also not detected by authorities during external inspection because there is not any problem in the meter until the remote is not operated.
4. Different theft techniques for meter tampering
   In this type of tampering method the phase from distributed overhead power line is connected to the output phase terminal instead of input phase terminal of energy meter to supply power to the load. So current does not flow through current coil and hence current cannot be determined. Hence power measured is zero.

Figure 4.1 represents the normal working condition of meter. P in and P out are phase input and phase output, respectively. N in and N out are neutral input and neutral output respectively. CC is current coil and PC is pressure coil.

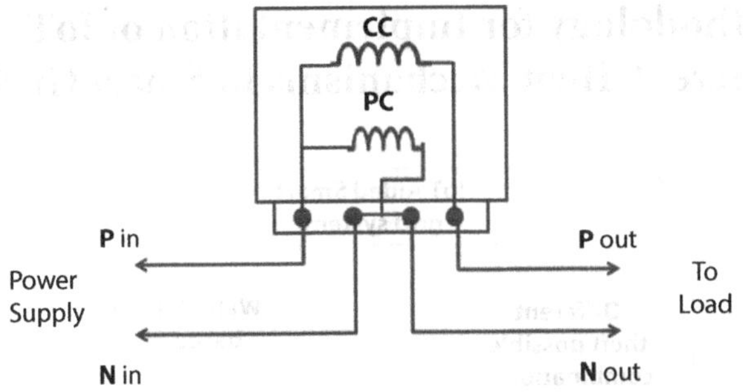

Figure 4.1 Block diagram of energy meter.

In this type of tampering method the phase from distributed overhead power line is connected to the input phase terminal of energy meter and the neutral from distributed overhead power line is grounded. The output phase terminal of energy meter is connected to load and the load is grounded. Therefore, neutral is not connected to pressure coil then voltage cannot be measured. Hence total power measured by energy meter is zero.

In this type of tampering method, the load is connected in between the phase and neutral which are from distributed overhead power line at the input terminals of energy meter. Here the neutral point is grounded. Therefore meter cannot measure voltage as the pressure coil is connected to neutral. As the output terminals of energy meter are open circuited current does not flow through current coil. Therefore, current cannot be measured. Hence power measured by energy meter is zero.

Another type of tampering method is when the phase and neutral from overhead power line of distributed network is connected to input terminals of energy meter directly and some part of load is connected in between the output terminals of energy meter, and the remaining part of load is grounded as shown in Figure 4.2. So voltage cannot be measured across the remaining part of load. So power measured across the remaining part of load is zero [5].

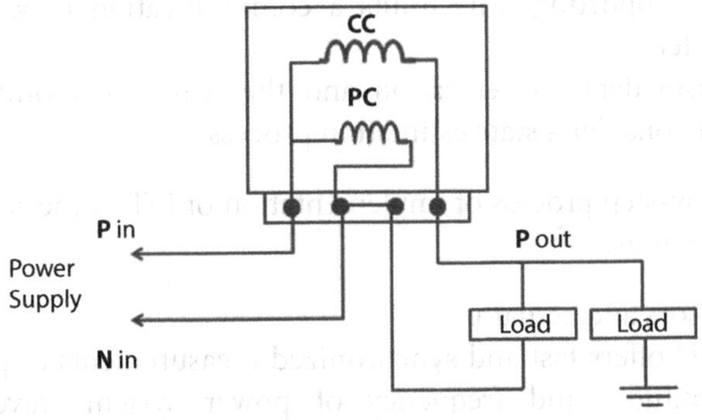

Figure 4.2 Partially tampered energy meter.

4.3 Methodology for Implementation of IoT to Different Theft Mechanisms in Smart Grid

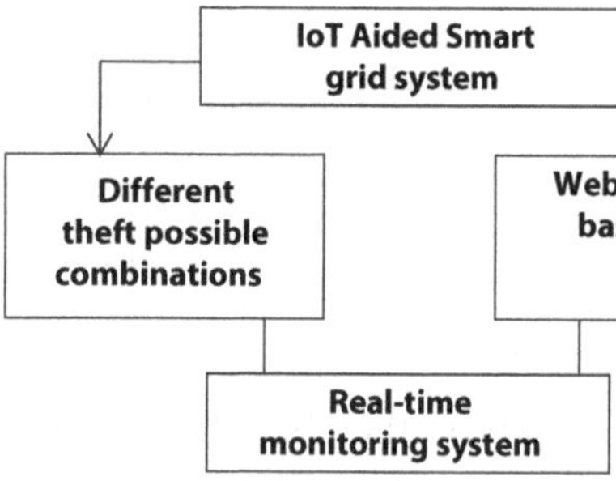

Figure 4.3 Schematic diagram of implementation of IoT to theft mechanisms in smart grid [2].

Here we tried to implement monitoring in real time of medium and high grids of voltage and this is the base of our prototype that we used in this paper. The block schematic of the implementation logic is as shown in Figure 4.3. The estimation of network state of the active distribution system by monitoring system using Phasor Measurement Unit (PMU). Use of PMU here provides the significantly low latency and also high state estimation frames rates during monitoring. Short line combined power, heat generation units, high volatile loads and injections of active power are generated using the photovoltaic panels offered by it. This prototype offers validation of the precision and time latency in real time for a three-phase state estimation process deployed in a real active Distribution network (ADN).

The three main components of such a type of system are:

1. Secondary network substation, on its medium voltage, PMUs are connected with the help of specific current transducers.
2. For supporting time limits a communication network is needed.
3. Phasor-data concentration and the real-time monitoring that consider a state estimation process.

The step-by-step process of implementation of IoT in the smart grid in our prototype is as:

1. Sensors and actuator
   PMU offers fast and synchronized measurements of phase, amplitude, and frequency of power system waveform as it considered as an AMI. Synchrophasor estimation

algorithm which allow improvement of the performance of interpolated-DFT method which use specific compensation method for spectral interference that generated by negative image of the spectrum; due to this reason this PMU is adopted for our prototype. To feed the power system to the analog input module of the PMU the high voltage/current have to be transformed into considerable lower values current and voltage signals with very minimum distortion of the phase as well as in amplitude for this purpose we use ALTEA CVS-24 having 0.1-class complaint.

Dedicated shielded cables used for the connection between the PMU and sensors for every substation and RG213 was used for the connection between the PMU and GPS antenna and as well as for PMU synchronization, Global positioning system (GPS) had been used for every substation.

2. IP gateway for communication

   The IPv6 used for building the communication network which used the twisted pair cables already exist and through which with the help of single pair high-speed digital subscriber line (SHDSL) technology communication takes place. It is needed to store all the data coming from PMUs, hence all the collected Data centered near the SHDSL concentrator is placed inside the PBX box room. As the bit-rate speed of the optical fiber is high at 100MB/s was used for connection between the Phasor Data Concentrator (PDC) and PBX room and not using SHDSL due to its low bit-rate speed of 2Mb/s that is not sufficient for our purpose.

3. Date management and storage unit

   An IP version with a protocol of parallel redundancy is developed for this prototype that is known as iPRP (Parallel redundancy protocol) that checks the duplication of UDP packets at PMUs and also removes these duplicate packets at PDC. Previously iPRP was used as a transport layer solution for Linux operating system used in PDC and PMUs, which provides us very precious advantage of not to apply any new modification in any network devices used for PDC or PMUs. When the PMUs estimate the data for frequency as well as rate of change of frequency, nodal absorbed or injected power and the data of synchro phasor, all these data are collected by a device known as PDC. The PDC used to communicate with PMUs and their datagram was capsulated by it. As the synchro phasor are time aligned and buffer which is circular, aggregated subsequent subset of these measurements is further given to the state estimators. To identify the accurate amount of reception time of the data packets is also too difficult due to diverging report rates among devices and data streaming which induces the measurements to delay.

4. User and visualization interface
   For development of our prototype use of adaptive algorithm is suitable for us that distinguishes the time for each dataset required to wait for the rest of phasors that were based on an event timeout having an identical timestamp. After triggering of events, the current created datasets is transmitted to application of state estimator. Hence we considered that this procedure assures the available estimation to forward at streaming of 50 Fps to reach under acceptable time range of 20 ms of PMUs that increase the chances for acceptance of the process. Here the PDC is hosted by machine and state estimator are synchronized by GPS, which allows us to identify and solve the bottleneck of whole chain that cause higher delay. The PDC feed its output data to state estimator through which state of the system is identified.

The Discrete Kalman Filter is adapted in this prototype for providing more accuracy in estimation as compared with weighted least square method. The no control input required for Discrete Kalman filter gives the solution for identify the state of a discrete time system process which is useful for the three-phase system and majorly depends on the synchrophasor measurement yielded with the help of the PMUs.

In the case of Power outage, the whole system is durable for 8 hours. This archetype appears to be harmonious for real-time assurance with all control, functionalities, that is presumed to be advanced for active distributed networks (ADN).

4.4 Conclusion

After completion of our prototype, we can measure the real-time data of electrical quantities and we also are able measure the power output data from grid versus usage of power at utility level. If any unbalancing of data is recognized at the monitoring level then the software automatically detects the area of theft to be abandon with GPS location or outage to undergo repair. Here we conclude that the implementation of the IoT devices and structure like PMU, PDC, synchro phasor, etc., in the power system for any purposes especially for theft detection, identification, and controlling provides us a very satisfactory result and it's also very compatible with software uses in the electricity industry. The real-time monitoring systems greatly helps in overcoming the different problems related to transmission, distribution at both substation as well as utility levels. This prototype is also used as the monitoring device for power outage at the grid and in lines.

4.5 Future Work

Moreover, it is important to note that this is not overall the latest technology; however, particular ones are more fitting to singular SG utilization than others. In general, wired technologies before-mentioned as DSL, PLC, and optical fiber are pricy for wide-area deployments, notably in provincial areas. Nevertheless, they can maximize both communication potential and security. Wireless technologies, on the other hand, could diminish installation expenses although they have bandwidth and security restrictions.

References

1. *Smart Grids Infrastructure, Technology, and Solutions*, edited by Stuart Borlase.
2. Xi Chen, Jianming Liu, Xiangzhen Li, Limin Sun, Yan Zhen, Integration of IoT with Smart Grid, Proceedings of ICCTA, 2011.
3. Qinghai Ou, Yan Zhen, Xiangzhen Li, Yiying Zhang, Lingkang Zeng, Application of Internet of Things in Smart Grid Power Transmission, *2012 Third FTRA International Conference on Mobile, Ubiquitous, and Intelligent Computing.*
4. Dr. V. Saravanan, J. Deepika, M. Gopika, Design of Smart Electric Meter Based on IoT, *International Journal for Research in Applied Science & Engineering Technology, (IJRASET)* Volume 5 Issue III, March 2017, ISSN: 2321-9653.
5. Mitali, MahadevRaut, Ruchira Rajesh Sable, Shrutika, RajendraToraskar, Internet of Things (IoT) Based Smart Grid, *International Journal of Engineering Trends and Technology* (IJETT), Volume 34 Number 1, April 2016.
6. B. Nageswara Rao Yadav, Chandra Mohan Reddy Sivappagari, IoT Based Smart Grid for Load Assessment and Abnormality Sensing, *International Journal of Innovative Research in Information Security (IJIRIS)* Issue 09, Volume 3 (December 2016).

7. S S Nagendra Kumar, S Koteswara Rao, M Suresh Raju, S Trimurthulu, K Sivaji, T Ram Manohar Reddy, IoT Based Control and Monitoring of Smart Grid and Power Theft Detection by Locating Area, *International Research Journal of Engineering and Technology (IRJET)* Volume 04, Issue 07, July 2017.
8. Priyanka Bhausaheb Deshmukh, Prof V.M. Joshi, IoT based Smart Grid to Remotely Monitor and Control Renewable Energy Sources Using VLSI, *International Journal of Innovative Research in Science, Engineering and Technology*, Vol. 6, Issue 11, November 2017.
9. Maninder Kaur and Dr. Sheetal Kalra, A Review on IoT Based Smart Grid, *International Journal of Energy, Information and Communications* Vol. 7, Issue 3 (2016), pp. 11–22.

5

Energy Metering and Billing Systems Using Arduino

M. Ramu[1]*, Lekha Varisa[2] and N. Siva Mallikarjuna Rao[3]

[1]*Dept. of EECE, GITAM (Deemed to be University), Visakhapatnam, India*
[2]*Dept. of ECE, GITAM (Deemed to be University), Visakhapatnam, India*
[3]*Dept. of EECE, GITAM (Deemed to be University), Hyderabad, India*

Abstract
The discovery of electricity forever changed the future of mankind. It now plays a significant role in our day-to-day lives and it has been a significant contributor to the development of various fields of science and technology. India has seen a significant increase in its capability to generate power. But its measurement technologies and mechanisms remain inadequate, in view of its growing population. When left unchecked, this might cause serious repercussions leading to a deficit of power and natural resources. This power deficit may eventually lead to a blackout in the future. Further, these old methods and techniques have paved the way to power theft, which is one of the major problems concerning the Central Ministry of Power. We have designed a wireless prototype, a smart electricity meter utilizing concepts of IoT. This helps us in calculating the bills more efficiently by reducing human errors; it also minimizes power thefts and allows us to regulate our own power consumption by providing real-time data related to energy and pricing.

***Keywords*:** Energy meter, smart meter, Arduino, Zigbee

5.1 Introduction

There has been a significant shift in global technological advancements pertaining to automated wireless technologies. These advancement are at the heart of progress, as they include and integrate the latest and important

**Corresponding author*: rmutyala@gitam.edu

A. Chitra, V. Indragandhi and W. Razia Sultana (eds.) Intelligent and Soft Computing Systems for Green Energy, (69–78) © 2023 Scrivener Publishing LLC

technologies of IoT and artificial intelligence. These smart and intelligent systems can act autonomously without instructions or human involvement. There are two types of meters, electromechanical and electronic. The former are used in rural areas and have become outdated. The latter, however, are widely used nowadays. These consist of an LCD/LED to display the reading. The LCD is calibrated to indicate the instantaneous values of power and energy. A bill is generated by the ratings produced by these meters, thereby reducing human efforts and errors. This IoT-based smart meter is regulated by using an Arduino mega. The micro-controlling board is extremely efficient, reliable and has an impeccable memory and GPIO. This data can be transferred to the cloud via the internet.

This data can be transferred without any interference of noise and can be recovered accurately and efficiently, free from the problems of distance or human interference. However, there are many other technologies such as Bluetooth and Zigbee, which are used within a certain range because of the restrictions due to distance. This project helps consumers get updated with essential data which helps us to realize and modify our consumption patterns. This also reduces many drawbacks faced by the electricity boards.

The consumption of electricity has seen a tremendous upsurge in recent years due to the demands of a growing population and electrical gadgets inundating the market. Despite increase in production, there is a deficit of power in many developing and underdeveloped regions of the world. This calls for using our resources warily. The aim of this research work is to provide smart and innovative methods to track and analyze our consumption patterns, thereby helping us regulate our consumption and save power by using a Smart energy meter. The Smart energy meter provides consumers with information about their power consumption and makes efficient billing calculations, thereby avoiding excess reliance on human effort and minimizing man-made errors [1–8].

1. Energy meters are used in various industrial domains, thereby proving to be an invaluable tool in saving and utilizing power. However, little effort has been made to track and analyze the household consumption which is a potential source to save power. In this research paper we have focused on architectures which are directed towards implementation of smart network meters which are consumer friendly.
2. Smart energy meters provide various advantages compared to traditional energy meters. Yet the design of a smart energy meter needs to be properly made in order to tailor fit the needs of the organizations and consumers. This research

goes into detail regarding implementation of proper communication network along with other major issues of security and maintenance of the equipment and infrastructure.

3. Though smart energy meters equip us with a tremendous advantage over conventional meters, there is an enormous amount of data being recorded which study consumer behavioural patterns and habits. In this research we have provided some mechanisms which ensure the safety of data with or without the involvement of third parties. Mechanisms to measure the degree of privacy also have been discussed.
4. Energy consumption has become a prime need. People are tending towards a transparent and reliable metering system rather than depending on traditional means. Timely feedback plays a major role in going towards energy conservation and management. So, energy consumption and errors can be eradicated by providing fallible feedbacks and real-time data.
5. There has been tremendous support in Europe to shift from conventional energy meters to smart energy meters. But we have to uphold our commitments to minimize the impacts of energy generation on the environment. In this research we discuss the transition of Europe from traditional energy metering to smart energy systems, the knowledge that can be learnt and implemented in developing and underdeveloped countries which may be undergoing drastic changes in these metering technologies, whilst upholding our commitment to fighting climate change.
6. Smart metering is an essential and advanced software system. This system is mainly employed at the consumer level but can also be implemented at the production side. It is highly used for fraud-detection and theft management. In these rapidly changing times, a smart meter is a breakthrough revolution. With proper research and development, it could be a potential game changer. It has lower cost and is highly efficient.
7. In the near future there will be considerable advances which will revolutionize the smart metering technology by home automation and communications technology. Many countries have implemented advanced metering infrastructure. In this research we have discussed the developments in AMI and its implementation in various developing countries thereby increasing efficiency, providing consumer information and preventing power theft [9, 10].

8. Over the years, energy grid trends have gradually changed. Recent trends are towards clean and green electricity. It has also been about efficiency over cost. Though there has been a significant cost decline, efficiency has always shown an upward graph. Its evolution has also significantly increased the complexity of the system. This has led to a few loopholes. These downfalls are carefully identified and are being analyzed.
9. This research discusses the architecture which enables the transfer of real-time demand and price based on peak load and therefore controls the smart appliances. Dynamic distribution of demand is employed, in order to ensure lower operating costs as well as minimizing damage to the environment. This also uses advanced OLTP systems and data from AMI's and supply data from SCADA. OLAP systems are used to provide deep real-time data about pricing and consumption patterns [10].

5.2 Smart Meters and Billing Systems

Energy meters are basically devices that are used to measure the amount of energy that is consumed by the end user in a residential or commercial site. The main aim of these is to monitor the real-time bills of the energy that is utilized by electric or electronic devices.

5.2.1 Arduino Mega

The Arduino Mega 2560 is used in designing the smart meter. It has a total of 16 inputs that are analog in nature and 54 that are digital. It is perfect for open face interfacing and has all the necessary functions to support the metering system. We could connect it with a UCB or an AC-DC adapter could be used for the microcontroller to start. This process could also be done using a battery.

5.2.2 LCD

The high-resolution LCD module is used along with the Arduino mega 2560 to display the necessary data. It is has 18 bits, 240x320 pixels control of each and every pixel. This make the LCD compatible and highly efficient to use.

5.2.3 Proteus Software

The Proteus Design Suite is a proprietary software tool suite used primarily for electronic design automation. The software is used mainly by electronic design engineers and technicians to create schematics and electronic prints for manufacturing printed circuit boards. It is a software suite containing schematic, simulation as well as PCB designing.

5.3 Working

Arduino 2560 consists of 16 analog and 54 in-out digital pins, 4 serial ports, an ICSP header, a power jack, USB connection, 4 hardware ports, 16 Mhz crystal oscillator, and a reset button. LCD is interfaced with the help of digital pins. By the USB connector, the setup is connected to a computer. The Arduino is connected to Vcc and the ground terminals. The LCD displays the power and energy reading from the Arduino. These pins of LCD and Arduino are paired in the following order: 3, 4, 6, 12, 13, 14 and 22, 23, 24, 25, 26, 27 respectively. The 5th pin of the LCD is connected to the ground. The microcontroller is connected to a serial monitor, which enables transmission and receives signals within the circuit. The 53rd pin of the Arduino is connected to a relay circuit which is connected to an LED. This LED is then grounded. The LED acts as a load indicator. It is turned ON at a certain limit and when it crosses beyond a threshold value it turns OFF. The commands are sent to the Arduino using a computer. The code has been written in Arduino IDE which is dumped to the microcontroller. The Arduino IDE is designed in a way to debug and modify the code or to control the Arduino from the PC keyboard.

Potential transformers and current transformers reduce the magnitude of voltages and current to enable the smooth functioning of the Arduino which is a low-voltage device. The Arduino receives the low voltage and current signals from the potential transformers and current transformers within the circuit. The potential transformers are connected to the ADC0-ADC2 sections, whereas the current transformers are connected to AD06-ADC4 of the Arduino board. We have used 6 sensors to receive the inputs for 3 PT's and CT's.

In the code initially we have declared the values of LCD and its working in order to display the values. Then we have declared working of the serial board and allocation of output pins. We have set loops for different values of voltages and currents, so that the maximum values are assigned to the variables V1, V2, V3, and I1, I2, I3 respectively. We have declared the required formulas in the code which enable us to compute the cumulative values of power, energy, and the final price.

5.4 Applications

- Energy meters are widely used in private areas for the measurement of electric power utilized by the customers, and these energy meters are commonly used in the industrial sector for regulating the electric power of diverse contrivances according to its reading and for measurement of electric power.
- The system can be used in the Domestic and Commercial areas for electric supply. For Gas supply lines as well as Water supply.

5.5 Time of Use

The Time of Use (ToU) net metering consists of a smart electric meter which directs the control and consumption of electricity throughout the day. The cost of production is usually low when its consumption graph is minimum. Production cost of electricity is usually highest at the time when its consumption reaches the peak value. Therefore, production could be done at a time when its utilization is minimal which is mostly at night. This technique of time of use metering is employed efficiently for renewable sources like in solar power schemes, for example, Italy has installed several photovoltaic cells, that the peak prices have shifted to the evening times. This method helps us in regulating the peak prices of electricity.

5.6 Observations

By observing the results of the experiment, we can observe that, in a basic energy meter, only the reading of energy is measured. Whereas, in the case of a smart meter, we can obtain various parameters like voltage, current, power factor, power and energy consumption with real-time pricing. The load limit is also indicated by the LED. This would help the user to get a comprehensive understanding of the energy usage and the generated cost.

5.7 Equations

Power (P) = current (I)*voltage (V)
$P_t = V_1I_1 + V_2I_2 + V_3I_3$
Energy (E) = P_t * time (t)
Cost (c) = Energy (E) * (cost per unit kWh)

5.8 Results

All the results are observed on the virtual terminal. As we can see, at initial instant the values of current and voltages are indicated. This corresponds to no utilization of power and energy and hence, there is no initial charge. Figures A and B represent the initial output and output at time instant t

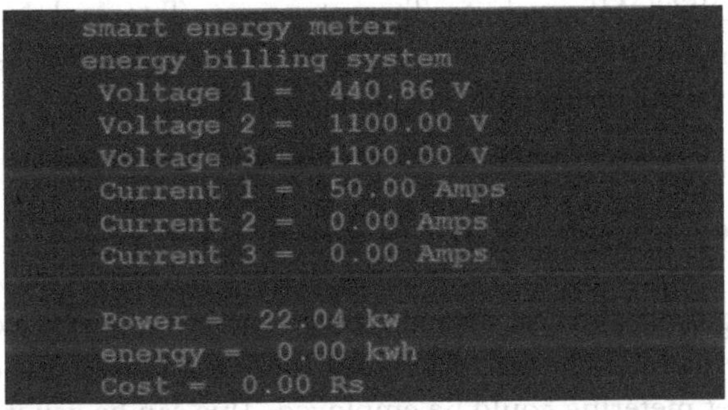

Figure A Initial output at T = 0.

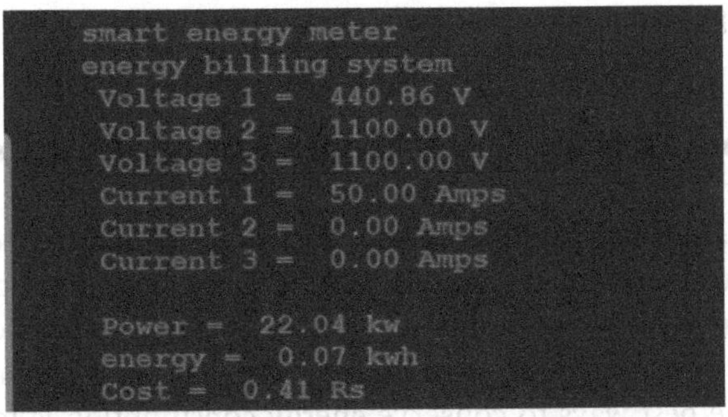

Figure B Output at a later instance time T = t.

respectively. However, in a real case scenario, the metering starts with an initial base-price.

5.9 Adoption in India

The implementation of this system in the country is adequate. The policy, however, varies from state to state. Each state produces its own share of power through its solar arrangement and gets paid for the surplus produced. For instance, the Indian states Karnataka and Andhra Pradesh implemented these policies back in 2014. After careful scrutiny by the respective brands, bi-directional meters are installed and 30% of the distribution transformer capacity has been allowed on a first-come basis and other important criteria. From Sept. 2015, Maharashtra state has also allowed the utilization of solar rooftop grip net metering system. This also has a provision of 40% of transformer capability on solar net metering.

Various DISCOMS such as Torrent power, Tata and MSEDCL are required to maintain net metering. Currently, MSEDCL does not utilize this TOD charging tariffs consumers and net metering.

5.10 Excess Generation of Electricity

Net metering is used in cases when global generation offsets a certain portion of its demand. Therefore, whenever there is an insatiable demand for electricity, net metering could be employed. This can be generated by certain kinds of fuel generators that are renewable in nature. In the case of a TOU, there is a separate determination process for each time-of-use tier.

In other words, it can also be determined as the difference of energy that is delivered to the customer and the energy that is essentially produced by the renewal generation.

5.11 Commercial Use & Home Energy Monitoring

Large companies and industries use electronic meters to record power usage in chunks of thirty minutes. These grids have multiple demand surges within the day; the companies will be inclined to provide compensation to such customers, so as to decrease the demand at such times.

One of the best ways to conserve energy consumption is to utilize the data obtained through smart meters. One of its salient features is that it provides real-time feedback. This real-time data reveals the consumption patters of consumers. This in turn creates awareness among the consumers about how changes in their consumption patterns should be made and implemented. There have been several instances in recent times, where such data has positively impacted the behavior of consumers and brought a significant decrease in the consumption of energy. An experiment in Ontario using consumer readable meter in 500 homes in Hydro One showed an average of 6.5% drop in total electricity in total households. Subsequently, 30,000 customers were provided with power monitors. Many companies, such as Google, have espoused the use of power meters and endorsed their use, thereby enabling us to identify devices that consume maximum power along with those that use excess standby power. Though there are several models available in the market, the underlying principle remains one and the same.

5.12 Conclusion

The proposed Smart metering system integrates both the producer and the users to utilize power efficiently. Consumers can monitor the load at any time. If the consumption is regulated during peak hours this would reduce the need for excess production at that instant, thereby reducing the production cost. The problem of power theft also can be addressed by adhering to proper tariff norms and integrating a proper theft detection system within the circuit. This system can efficiently measure values of current, voltages, and energy utilized, which are then utilized in generating a real-time bill.

References

1. Van der Geer, J., Hanraads, J. A. J., & Lupton, R. A. (2000). The art of writing a scientific article. *Journal of Science Communication, 163*, 51–59.
2. F. Benzi, N. Anglani, E. Bassi, and L. Frosini, Electricity Smart Meters Interfacing the Households, *IEEE Transactions on Industrial Electronics*, vol. 58, no. 10, Oct. 2011, pp. 4487–4494.
3. Z. Qiu, G. Deconinck , "Smart Meter's feedback and the potential for energy savings in household sector: A survey," in *IEEE International Conference on Networking, Sensing and Control (ICNSC)*, April 2011, pp. 281–286.
4. J. M. Bohli, C. Sorge, and O. Ugus, A Privacy Model for Smart Metering, in *IEEE International Conference on Communications Workshops (ICC)*, 2010, pp. 1–5.

5. M.Weiss, F. Mattern, T.Graml, T.staake, and E.Fleisch, Handy Feedback: Connecting Smart Meters with mobile phones, in *8th International Conference on Mobile and Ubiquious Multimedia,* Cambridge, United Kingdom, Nov. 2009.
6. L. O. AlAbdulkarim and Z. Lukszo, Smart Metering for the future energy systems in the Netherlands, in *Fourth International Conference on Critical Infrastructures*, 2009, pp. 1–7.
7. M. Popa, H. Ciocarlie, A. S. Popa, and M. B. Racz, Smart Metering for monitoring domestic utilities, in *14th International Conference on Intelligent Engineering Systems (INES)*, 2010, pp. 55–60.
8. S. Ahmad, Smart Metering and home automation solutions for the next decade, in *International Conference on Emerging Trends in Networks and Computer Communications (ETNCC)*, 2011, pp. 200–204.
9. J. Stragier, L. Hauttekeete, L. De Marez, Introducing Smart grids in residential contexts: Consumers' perception of Smart household appliances, in *IEEE Conference on Innovative Technologies for an Efficient and Reliable Electricity Supply (CITRES)*, Sept. 2010, pp. 135–142.
10. D. Y. R. Nagesh, J. V. V. Krishna and S. S. Tulasiram, A Real-Time Architecture for Smart Energy Management, in *Innovative Smart Grid Technologies (ISGT)*, Jan. 2010.

6

Smart Meter Vulnerability Assessment Under Cyberattack Events – An Attempt to Safeguard

Kunal Kumar[1] and R. Raja Singh[2*]

[1]*Department of Electrical and Electronics Engineering, Vellore Institute of Technology, Vellore, India*

[2]*Department of Energy and Power Electronics, Vellore Institute of Technology, Vellore, India*

Abstract

Global smart meter deployment is gradually increasing throughout the world due to the apparent benefits over traditional meters. The most crucial framework Smart Grid comprises the Advanced Metering Infrastructure (AMI). Smart meters are the most pivotal and revolutionary component of the AMI which are capable of changing the face of the current energy consumption scenario. Smart Meters will not just be a sustainable alternative but will also benefit the various stakeholders economically and socially. While the Smart meter is one of the most revolutionary ideas in the field of energy management we cannot overlook the various threats and security concerns accompanying it. The automation of energy meters and the vast amount of user data that it can access exposes the consumer to breach of privacy and various threats. This is why cybersecurity has been one of the major concerns for companies deploying smart grids and smart metering technologies. In this article, an in-depth look into the many forms of smart meter attacks and vulnerabilities, as well as possible countermeasures, is presented. This paper also provides an opportunistic framework by providing a cryptographically based Attack detection model implemented on Python. Hence this paper will provide a better understanding of the vulnerabilities of a smart meter for researchers and provide a base to propose and improvise protection techniques so that smart meters can be used safely and securely.

**Corresponding author*: rrajasingh@gmail.com

A. Chitra, V. Indragandhi and W. Razia Sultana (eds.) Intelligent and Soft Computing Systems for Green Energy, (79–96) © 2023 Scrivener Publishing LLC

***Keywords*:** Smart meter, cyber security, advanced metering, packet injection, eavesdropping, RSA, encryption, decryption

6.1 Introduction

The key to successful global economies comes from their successful use of advanced technological methods which in turn is dependent on their abilities to harness energy for their usage. However, everything comes at a cost and this massive surge in energy has taken a toll on our environment and at the same time posed new threats in the form of an energy security breach [1]. The measurement of the consumption of energy utilized in a household or an organization is mostly performed via conventional energy meters. The role played by these meters is inevitable as they give consumers an insight into their energy consumption. However, as technical advancements occurred and the need for energy further increased, it just wasn't enough to know how much energy an individual or an organization consumed. Considering the environmental damage, it was also important to utilize energy sustainably as well as in a secured way to ensure minimal energy loss. For a long time, efforts have been made to overcome the hurdles caused due to increased demand for energy. A breakthrough in this domain was made by the introduction of the concept of Smart Grids. They are comprised of network structures that facilitate bi-directional data communication as well as supply and data transfer networks to the end users from the source power generation units. The key advantage of this technology is that it incorporates Information and Communication Technology (ICT) in all of its operational levels right from generation to transmission and distribution to consumption. Thus, the use of ICT has enabled these grids to be an eco-friendly and sustainable, not to mention more efficient, safer, and feasible alternative to traditional methods of energy production and transmission as well as consumption and utilization [2].

Efforts are underway to make the existing network of Smart Grids more receptive and minimize the errors existing in the current models. To understand this, we need to look at the technological composition of the Smart Grid. The most crucial framework of the Smart Grid is the Advanced Metering Infrastructure (AMI). Smart meters are the most pivotal and a revolutionary component of the AMI; they are capable of changing the face of the current energy consumption scenario. The Smart Meters are not just a sustainable alternative but they also benefit the various stakeholders economically and socially.

Conventional energy meters provided one-way measurement and analysis of our energy consumption. These measurements were performed via manual data collection and analysis, thus making it impossible to understand the trends of energy consumption via consumers. This was made viable by the bidirectional communication of a smart meter. This technology allows the amassing of an enormous amount of data that could be employed for advanced data analytics to extract information. This information could help in deriving valuable statistics. Such statistics would not only help in billing the data but also give a detailed analysis of the electricity usage by each device. Thus, the employment of a smart meter could help optimize the electricity generation using different algorithms according to the consumer usage trends.

Moreover, it could also give us an insight into the future demands of the respective consumers and generate electricity accordingly. Thus, this could also serve as a useful tool for predictive maintenance and early fault detection which in turn could strengthen the security of the entire power generation and measuring system. The most crucial aspect of a Smart Grid that could play a key role in its success is smart meter data analytics. These data analytics use the methods of data acquisition, transmission, processing, and interpretation to achieve unerring statistics and automated decision support to handle errors promptly. The raw data collected from these meters could help make giant leaps across the utility enterprises [3]. While Smart Grid is one of the most revolutionary ideas in the field of energy management we cannot overlook the various threats and security concerns accompanying it. The automation of energy meters and the vast amount of user data that it can access can expose the consumer to breach of privacy. Let's say access to one's smart grid data could enable an attacker to understand when the institution is occupied by assessing the operational behavior of the electrical appliances. This could expose the consumer to severe threats. Thus, cybersecurity has been one of the major concerns for companies deploying smart grids and smart metering technologies.

Moreover, there is still uncertainty as to what extent these meters are exposed to cybersecurity threats. With the growing technologies, correlatively also increases the risk of cyberattacks on devices and systems. This often leads to several organizations suffering losses that could amount to millions of dollars [4]. Thus, the vulnerabilities of the smart meter technology and the increase in cybersecurity attacks pose a matter of serious concern to all the stakeholders. Thus, it is important to address these security concerns to develop confidence in various stakeholders with regard to this technology and to ensure successful adoption as well as operation of Smart Metering. This research paper will explore the negative nuances of smart metering technology and review different types of attacks on smart meters and advance metering infrastructure as a whole and some practical solutions to mitigate these threats. This paper will also analyze implementation of a secured algorithm using Python 3.6 to detect an attack during communication of data through cryptography between smart meter and utility. The paper is structured as follows: section (6.2) contains an introduction to advance metering infrastructure architecture, smart meter working and block diagram. In section (6.3) a review of different type attacks on advance metering infrastructure is done and a review of some sustainable countermeasures is also presented. In section (6.4) cryptography-based algorithm is explained, in section (6.5) implementation of algorithm and other discussions are done. Section (6.6) concludes the whole work of the paper and suggests future work that can be done.

6.2 Advanced Metering Infrastructure Architecture

Advanced metering infrastructure is a system comprising contemporary electronic-digital hardware and software. The combination facilitates data computation intermittently as well as enables continuous remote communication. These frameworks have a foundation that can make precise estimations, collect periodic information regularly, and coney it to the stakeholders as required [5]. The architecture of the advance metering infrastructure is presented in Figure 6.1.

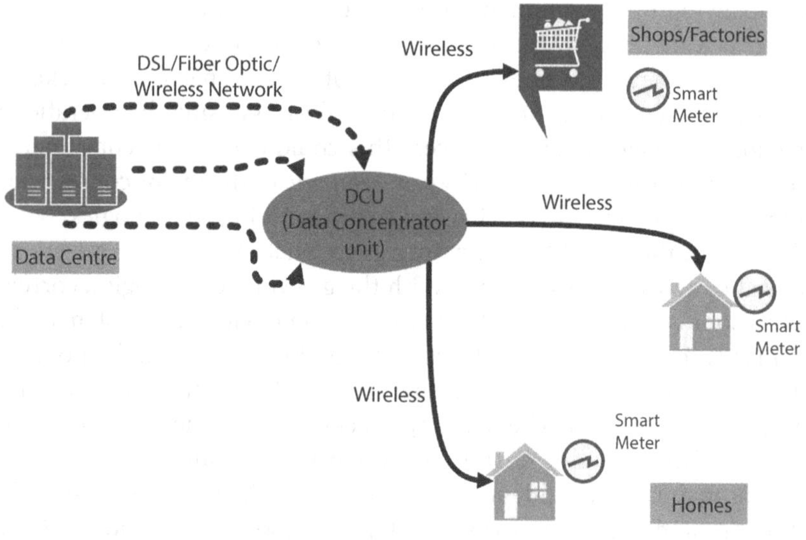

Figure 6.1 Architecture of advance metering infrastructure.

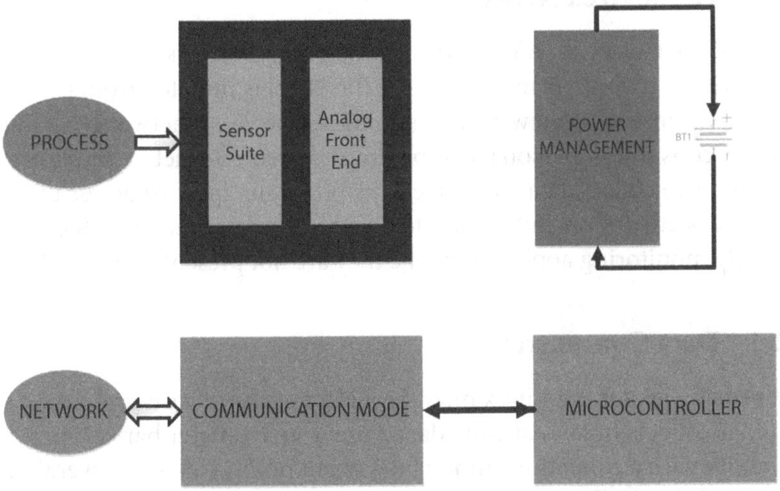

Figure 6.2 Schematic representation of a smart meter design.

6.2.1 Smart Meter Architecture and Design

The gathering of data in the smart meters is done locally and the transmission to the collector is accomplished using the LAN (local area network). However, when the data is needed to be transferred to the utility central collection central point to enable the processing of data and its use in applications of business, it is transmitted via wide-area network (WAN). Due to the bidirectional path of communications, the direct sending of signals or commands to the meters, customer premises or distribution devices is facilitated quite conveniently. Figure 6.2 is a typical schematic representation of a smart meter design that shows the primary and auxiliary components of a smart meter [8]. For system management, many types of sensors, front end electronics, an energy source with related power management circuitry, a communication node, and a microcontroller are employed [6, 7].

6.2.2 AMI Communication Network

The AMI communication network consists mainly of the Smart Meter Data Collector (SMDC), smart meter, and a structure linking the above two. The linking structure in general comprises either Frequency Hopping Spread Spectrum mesh or another cellular network [9].

6.2.3 Home Area Network

The customer facet of AMI is the Home Area Network (HAN). It has a wireless connection, normally ZigBee. The HAN is initiated from the smart meter to a consumer gateway system in the home. It is a network that linked with the consumers' personal computers and their internet. This allows the consumers to collect data and inspect, by providing internet access utilizing advanced statistics on individual devices for energy use, or by themselves remotely monitoring appliances while they are not present at home [10].

6.2.4 Data Concentrator

The Data Collector is a Linux microkernel-based hardware computing system that collects huge real-time data from a smart meter bank. It provides the utility with a collection cum management point. Utilizing several mesh radios, per data collector handles a few thousand smart meters. It then provides, via Ethernet or some other WAN backbone, a direct connection to the utility. The data collector is normally responsible for the smart meter's cryptographic key management as well [9].

6.3 Possible Attacks on AMI

Broadly attacks on AMI are classified as of two types: Manual Attacks and Cyberattacks.

6.3.1 Manual Attacks

a) Meter spoofing: The ID number of smart meters is procured by attackers via physical access.
b) Meter manipulation: Any external influence/element introduced in meters or in the metering circuit, which results in the loss of energy measurement or fallacious energy measurement.
c) Physical Attacks: This affects the SM data collectors and disrupts the interaction between the energy provider and the power user [11].

6.3.2 Cyberattacks

a) Eavesdropping: Eavesdropping allows the intruder to listen to the information that is being transmitted or transferred between the smart meter and the service providers. Such attacks on consumers' privacy are carried out quite easily via a wireless transmission channel or a power line. It is very difficult to detect eavesdropping, since it does not hinder communication [12].

b) Packet Injection Attacks: These attacks are initiated by injecting the network with false packets. Such false packets can be in the form of false commands. These attacks are usually carried out by cutting back the amount of power to certain parts of the smart metering system or by tampering with the billing mechanism in order to produce erroneous bills. These attacks thus result in enormous monetary losses [13].

c) Denial of Service (DoS) Attacks: DoS attacks target the entire smart meter system, the smart grid or a portion of it. A competitor performs DoS attacks by dispatching various commands to the smart meter gateways or across the other end of the utility servers than is necessary. The attack inundates the smart meter system to the point that it is no longer in a position to respond to valid requests. Subsequently, the grid or portions of the grid for critical utilities would shut down [13].

d) Malware Injection Attacks: These attacks are carried out by injecting malware into the network. The attack compromises the transmission between the grid equipment, which results in erroneous billing and reporting process. The load on the grid is destabilized by the disturbance of the demand/consumption condition of the grid.

e) Remote Disconnect/Connect Attacks: The smart grid or its components can be stopped by utilizing the smart grid's remote disconnection/connection feature. This attack thus is able to affect users by disconnecting them from electricity, gas and water supplies.

f) Firmware Manipulation Attacks: Manipulation of a smart meter's firmware or a metering gateway could interrupt the meter's billing and accounting process. This involves manipulating the pre-payment feature or disclosing to the remote readout entity incorrect usage status. The attack is carried out by rigging a smart meter via physical access or through the WAN on the condition that remote firmware updates are supported by the gateway. Such attacks typically affect a single person, but they have the potential to have a large impact [14].

g) Man-in-the-Middle Attacks: In such attacks, an attacker establishes a link with one of the targeted communicating parties. The party's messages are then collected by the attacker and transmitted to the other party, in a way that it sees that the parties are communicating directly. This allows the attacker to easily eavesdrop or alter the data transmitted between the communicating parties. A man-in-the-middle attack may be launched within the WAN in the context of smart metering, which can result in incorrect measurements being transmitted [4].

Below is Table 6.1 showing the different types of attacks discussed above, the attack Surfaces of these attacks and the impact these attacks can have on AMI or grid.

Table 6.1 AMI attack, attack surfaces and impact.

Attack	Attack surface	Impact
Denial of service (Dos)	Via Wide Area Network	This can affect part or all of the grid, including AMI
Firmware manipulation	Giving physical access to the SM. Upgrade remotely via the gateway via Wide Area Network	This affects the smart meter's meteorological and non-meteorological components, which can affect single to many gateways
Man in the middle	Wide Area Network or Inside local meteorological network (LMN or WAN)	False measurement of one of the gateways within the LMN or of all the meters connected to the gateway through WAN
Remote connect/ disconnect	Through Wide Area Network	It can range from one premise to the entire grid
Wi-Fi/Zigbee	Through HAN	Can facilitate denial of service and replay attack
Packet injection	Through Wide Area Network	False billing for both suppliers and customers
Eavesdropping	Remotely through Wide Area Network or power line	Tough to detect and can expose customer's privacy

6.3.3 Threats and Countermeasures of Attacks on Smart Meter

In Table 6.2 below are some of the effective countermeasures of attacks on the smart meter.

6.4 RSA Attack Detection Model

Data that is critical to the AMI's operation is the major target for attackers looking to disrupt the smart meter's and AMI's smooth operation. To avoid cyber dangers, it is critical to recognize any unauthorized manipulation of data that is being transferred. The suggested model shown in Figure 6.3 is a cryptographic-based methodology for detecting data theft in smart meters that can be applied to a variety of data flows in the AMI [19].

Let x be the ideal measurement of the data for period of time t. A cypher message x with the help of an encryption function f is generated by the equation:

$$\dot{x} = f(x) \tag{6.1}$$

Table 6.2 Threats and countermeasures of attacks on smart meter.

Threats	Countermeasures
Energy Theft	Multiple measuring stages to measure the delivered energy and compare the reading predict the tampering based on the behavior of the smart meter [15]
Identify Spoofing	Implementing good Encryption Algorithm using advanced authentication schemes implementing good Communication protocols [13]
Denial of Service	Using intrusion protection systems properly configuration packet filtering rules [16]
Sniffing and Traffic Analysis	Implementing a good Encryption Algorithm [17]
Malware Trading	Implementing periodic security updates [18]
Eavesdropping of Private Information	Encrypted data before transmission using encryption standards like AES

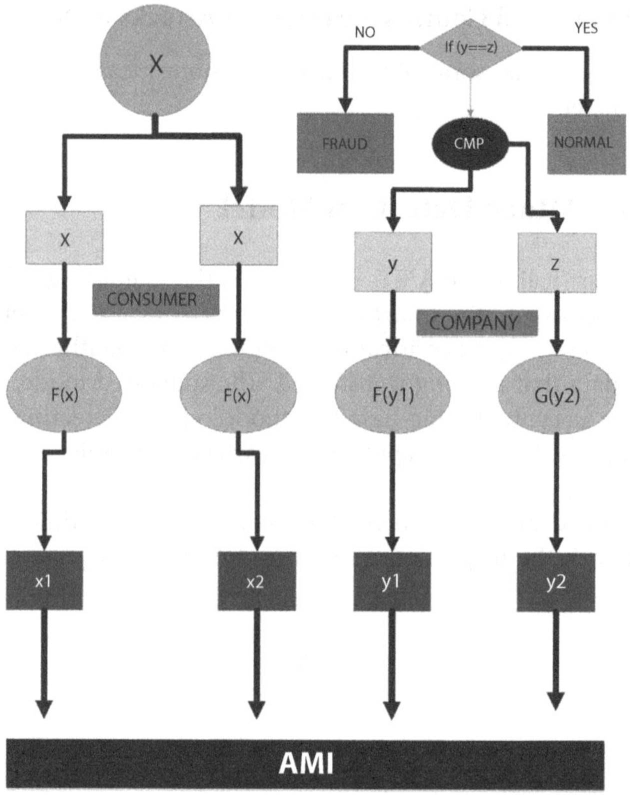

Figure 6.3 RSA attack detection model.

Now the inverse function f' i.e., the decryption function of cryptogram $\dot{x}$ reports the original value

$$x = f'(\dot{x}) \tag{6.2}$$

The proposed Attack Detection Model's core idea is to encode smart meter readings prior to sending them to the company, and then have the utility decode the obtained value to get the initial message. However, the main issue is determining whether the value obtained is originally from the decryption function or has been tampered with intentionally or inadvertently during the smart meter-company communication stages. A reference value is required to solve this problem so that the received measurement value can be checked for accuracy. A second encryption procedure on the smart meter side creates another code $\ddot{x}$ for this purpose with the help of function *g* where:

$$\ddot{x} = g(x) \tag{6.3}$$

As mentioned in Eq. (6.2), the original value of x must be returned by the decryption function g' of the code $\ddot{x}$.

$$x = g'(\ddot{x}) \tag{6.4}$$

In normal instances, the recipient (utility) will get two cryptograms, and once they are decrypted, the two decryption functions f and g should report the identical value x. (no manipulation). If the reported values show discrepancy, it signifies that the data exchanged during transmission has been tampered with. The procedure is straightforward: the measured quantity of data (for example electricity usage) is encrypted two times at the consumer end, i.e., at the smart meter, before being relayed across the AMI networks to interested parties (e.g., billing services). At the first stage, one value is split into two codes. The recipient compares the two cryptograms after they have been decrypted. The two obtained values must be the same in the regular procedure because the two codes are derived from the same value; however, if the received messages are different it is indicative of the fact that an unethical action tampered with the data during the communication between the consumer and the utility.

6.4.1 RSA Keys Creation

Figure 6.4 shows the step-by-step procedure to generate public and private keys.

There are three main stages:

1. Keys creation, public and private keys.
2. Message is encrypted by using public key.
3. Message is decrypted by means of decryption key.

6.5 Hash Code for Data Integrity

Hash values are fixed-length numeric values that uniquely identify data. Hash values can be used to check the accuracy of data. To determine whether data was modified or not, comparison between the hash value of received data and the hash value of data as it was sent should be made. To make this algorithm more reliable Hash values of both the decrypted messages from different pairs of keys are compared. If the compared values are

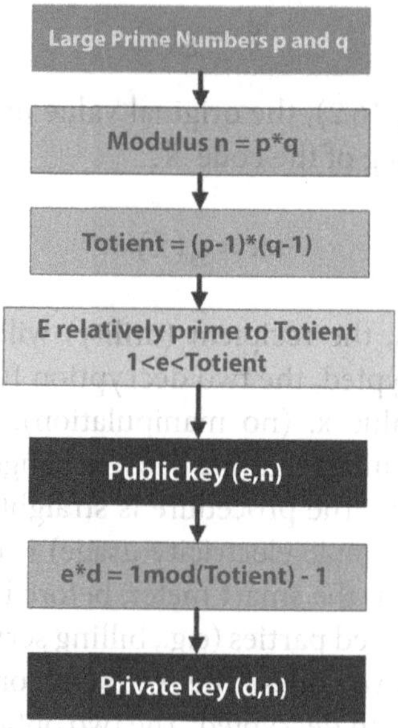

Figure 6.4 RSA public and private keys creation.

not the same that means some unethical action is taken on the data during communication. Hence comparison of hash values ensures data integrity and confidentiality.

6.6 Results and Discussion

6.6.1 Attack Detection System

In this system, two detection codes are developed from the smart meter reading before they are sent to the company for any task, as mentioned in the Attack detection model (example billing task). As a result, the amount of data is encrypted two times using two different encryption keys (e, n) and (e', n'), so the resulting Eq. (6.1) and Eq. (6.3) as mentioned in ADM become

$$\dot{x} = (x) => \dot{x} = x^{e} \text{ mod}(n) \tag{6.4}$$

$$\ddot{x} = g(x) => \ddot{x} = x^{e}\text{mod}(n) \tag{6.5}$$

The two codes $\dot{x}$ and $\ddot{x}$ are received by the utility company. By performing the decryption operation by using the different decryption keys

$(d; n)$ and (d', n') associated to their encryption keys (e, n) and (e', n') respectively, Eq. (6.2) and Eq. (6.4):

$$x = f'(\dot{x}) \Rightarrow x = (\dot{x})^{d} \bmod(n') \tag{6.6}$$

$$x = g'(\ddot{x}) \Rightarrow x = \ddot{x}^{d} \bmod(n') \tag{6.7}$$

As a result, the conclusive outcome of the bi-decryption operations should be the same value x, which corresponds to the initial data (electricity usage data) sent. However, if the values do not correspond, it is symptomatic of the fact that either one or both of the conveyed values $\dot{x}$ and $\ddot{x}$ were altered, which appears to be unethical behavior. The following output depicts the entire attack detection model based on the cryptographic RSA algorithm.

6.6.2 Python Implementation

To implement the above algorithm on Python 3.6, take two different sets of a prime number (11,13,17) and (23,29,37) to generate two different public and private keys. The program will generate two pairs of public and private keys. Give some message (example money) that will be encrypted by both the pair of keys, and then both the messages are decrypted by using their respective private keys. Since no attack has been done both decrypted messages will be the same. Further hash values of both decrypted messages are compared by using the SHA512 inbuilt function to ensure the data integrity.

```
import random
import hashlib
# Find GCD of 2 numbers
def gcd(x, y):
   while y != 0:
      x, y = y, x % y
   return x
# Find multiplicative inverse
def egcd(x, y):
   if x == 0:
      return y, 0, 1
   else:
      g, b, a = egcd(y % x, x)
      return g, a - (y // x) * b, b
def inverse(a, m):
   g, x, y = egcd(a, m)
   if g != 1:
      return -1
   else:
      return x % m
def check_prime(n):
   if n == 2:
      return True
   if n < 2 or n % 2 == 0:
      return False
   for i in range(3, int(n**0.5)+2, 2):
      if n % i == 0:
         return False
   return True
```

```
# function which generates key pairs
def key_pair(p, q ,r):
  if not (check_prime(p) and check_prime(q) and check_prime(r)):
      raise ValueError('All numbers must be prime.')
   elif p == q or p==r or q==r:
        raise ValueError('Numbers cannot be equal')
   n = p * q * r
   ph = (p-1) * (q-1) * (r-1)
   e = random.randrange(1, ph)
   g = gcd(e, ph)
   while g != 1:
      e = random.randrange(1, ph)
      g = gcd(e, ph)
      d = inverse(e, ph)
   return ((e, n), (d, n))
def encrypt(public, text):
#Unpack the key
   k, n = public
   #Convert letters using a^b mod m
   ci = [chr((ord(c) ** k) % n) for c in text]
   return ci
def decrypt(private, text):
#Unpack the key
   k, n = private
    #Generate plaintext using a^b mod m
   p = [chr((ord(c) ** k) % n) for c in text]
   return ''.join(p)
if __name__ == '__main__':
   print("Enter first set of 3 prime numbers, each should be different from one another")
```

```
    key1 = list(map(int,input().split()))
   print("Enter second set of 3 prime numbers , each should be different from one another")
    key2 = list(map(int,input().split()))
   print("\nGenerating your public/private keypairs for first input")
         public1,private1 = key_pair(key1[0], key1[1] ,key1[2])
     print("Your public key is ", public1 ," and your private key is ", private1)
   print("\nGenerating your public/private keypairs for second input")
         public2,private2 = key_pair(key2[0], key2[1] ,key2[2])
     print("Your public key is ", public2 ," and your private key is ", private2)
     message = input("Enter a message to encrypt with your public key: ")
      encrypt1 = encrypt(public1, message)
   print("\nYour encrypted message from first key is : ")
print(''.join(map(lambda x: str(x), encrypt1)))
      encrypt2 = encrypt(public2, message)
   print("\nYour encrypted message from second key is : ")
   print(''.join(map(lambda x: str(x), encrypt2)))
      print("\nDecrypting message with private key of first ", private1 ," . . .")
```

msg1 = decrypt(private1, encrypt1) print("Message from key pair 1",msg1) print("\nDecrypting message with private key of second ", private2 ," . . .") msg2 = decrypt(private2, encrypt2) print("Message from key pair 2",msg2) print("\nComputing hash Values") h1 = hashlib.sha512(msg1.encode())	h2 = hashlib.sha512(msg2.encode()) **#print(msg1,msg2,h1.hexdigest(), h2.hexdigest())** ifh1.hexdigest()==h2.hexdigest(): print("computed hash values from both decrypted messages are same") #print("Hash value of first",h1) #print("Hash value of second",h2) else: print("Values are different")

6.7 Conclusion

Smart meters prove to be an invaluable alternative to the conventional meters still in use, but in the present era of digitization, the potential cybersecurity threats presented by smart meters cannot and should not be ignored. Thus, it is pertinent to classify and peruse the types of potential cybersecurity threats possessed by smart meters and find different and effective ways to mitigate them. It is also necessary to know about the different vulnerabilities which expose the smart meters to such threats. In this paper, an attempt is made to review the different types of attacks a smart meter is subject to and to understand the vulnerabilities of the smart meter. Some sustainable protection techniques are also presented to minimize the risk. This thesis provides an opportunistic framework by providing a new ADM idea. This technique encrypts data from the smart meter twice before delivering them to the utility. In the utility the data is decrypted and examined to ensure that the given values are correct. The RSA method is used to construct the Attack Detection System and implemented on Python 3.6 because of its robustness and power against hacking. Hash values of data are also compared in the algorithm which ensures data integrity and makes the algorithm more reliable. As the world progresses towards automation, cyberattacks become more and more serious. For future work, this attack detection system (ADS) can be used as a base to enhance the cybersecurity of smart meters, and some more extensive mitigation plan could be proposed by studying different vulnerabilities of smart meters to minimize the breach of privacy by different types of attackers. This would ensure that the implementation of Smart Meters can be performed in a hassle-free and, most importantly, safe way.

References

1. Wennersten, R., Sun, Q., Li, H., The future potential for Carbon Capture and Storage in climate change mitigation - An overview from perspectives of technology, economy and risk, *J. Clean. Prod.*, 103, pp. 724–736, 2015, doi: 10.1016/j.jclepro.2014.09.023.
2. Tweneboah-koduah, S., Tsetse, A., *et al.*, Evaluation of Cybersecurity Threats on Smart Metering System. 2018, doi: 10.1007/978-3-319-54978-1.
3. Alahakoon, D., Yu, X., Smart Electricity Meter Data Intelligence for Future Energy Systems: A Survey. *IEEE Trans. Ind. Informatics*,12, 1, pp. 425–436, 2016, doi: 10.1109/TII.2015.2414355.
4. Menon, V.D., Kumar, T.K, *et al.*, Cyber Security for Smart Meters. *IEEE Int. Conf. Intell. Tech. Control. Optim. Signal Process. INCOS 2019*, pp. 1–5, 2019, doi: 10.1109/INCOS45849.2019.8951407.
5. Akpolat, A.N., Dursun,E., Advanced Metering Infrastructure (AMI): Smart Meters and New Technologies. *IATS'17*, Elazığ, Turkey, 2017
6. Ajenikoko, G. A., Olaomi, A. A., Hardware Design of a Smart Meter. *Int. J. Eng. Res. Appl.*, 4, 9, pp. 115–119, 2014.
7. Koponen, P., Rochas, C., *et al.*, Definition of Smart Metering and Applications and Identification of Benefits, https://www.researchgate.net/publication/235709839, 2008.
8. Carriero, C., Bissanti, M., Wireless Technologies for Smart Meters. pp. 1–4, https://www.analog.com/en/technical-articles/wireless-technologies-for-smart-meters.html#:~:text=Several%20technologies%20are%20available%20when,%2Dline%20communication%20(PLC), 2011.
9. Foreman, J. C., Gurugubelli, D., Cyber Attack Surface Analysis of Advanced Metering Infrastructure. http://arxiv.org/abs/1607.04811, 2016.
10. Parks, R. C., Advanced metering infrastructure security considerations. *SANDIA REPORT, Sandia Natl. Lab.*, pp. 0–34, http://cms.doe.gov/sites/prod/files/oeprod/DocumentsandMedia/20-AMI_Security_Considerations.pdf, 2007.
11. Hahn, A., Cyber security of the smart grid: Attack exposure analysis, detection algorithms, and testbed evaluation. Iowa State Univ., pp. 140. http://lib.dr.iastate.edu/cgi/viewcontent.cgi?article=4105&context=etd, 2013.
12. Wei, L., Rondon, L. P., *et al.*, Review of Cyber-Physical Attacks and Counter Defense Mechanisms for Advanced Metering Infrastructure in Smart Grid. *2018 IEEE/PES Transm. Distrib. Conf. Expo.*, pp. 1–9, 2018.

13. Khattak, A. M., Khanji, S. I., Khan, W. A., Smart meter security: Vulnerabilities, threat impacts, and countermeasures, *Adv. Intell. Syst. Comput.*, 935, pp. 554–562, 2019, doi: 10.1007/978-3-030-19063-7_44.
14. Chinnow, J., Bsufka, K., *et al.*, A simulation framework for smart meter security evaluation. *SMFG 2011 - IEEE Int. Conf. Smart Meas. Grids Proc.*, pp. 1–9, 2011, doi: 10.1109/SMFG.2011.6125758.
15. Marah, R., Gabassi, I. E., Larioui, S., Yatimi, H., Security of Smart Grid Management of Smart Meter Protection. *2020 1st Int. Conf. Innov. Res. Appl. Sci. Eng. Technol. IRASET 2020*, 2020, doi: 10.1109/IRASET48871.2020.9092048.
16. Y,i, P., Zhu, T., *et al.*, A Denial of Service Attack in Advanced Metering Infrastructure Network. *IEEE ICC 2014 - Communication and Information Systems Security Symposium*, pp. 1029–1034, 2014.
17. Salpekar, M., Protecting smart grid and advanced metering infrastructure. *Proc. Int. Conf. I-SMAC (IoT Soc. Mobile Anal. Cloud), I-SMAC 2018*, pp. 22–26, 2019, doi: 10.1109/I-SMAC.2018.8653793.
18. Jiang, R., Lu, R., *et al.*, Energy-Theft Detection Issues for Advanced Metering Infrastructure in Smart Grid. *Tsinghua Sci. Technol.*, 19, 2, pp. 105–120, 2014.
19. Naim, K., Khelifa, B., Fateh, B., A cryptographic-based approach for electricity theft detection in smart grid. *Comput. Mater. Contin.*, 63,1, pp. 97–117, 2020, doi: 10.32604/cmc.2020.09391.

7

Power Quality Improvement for Grid-Connected Hybrid Wind-Solar Energy System Using a Three-Phase Three-Wire Grid-Interfacing Compensator

Boopathi R. and Dr. Indragandhi V.*

School of Electrical Engineering, VIT University, Vellore, India

Abstract

This paper describes a grid-interfacing power quality (PQ) compensator and grid-connected hybrid wind-solar renewable energy sources and nonlinear loads. It is connected to the electrical power system and plays a major role to providing a stable and safe energy supply to critical communities, as well as communities in remote locations. Power quality problems within the power distribution system were caused by distributed power generation sources integrated with a large distribution system, which provides many edges. The power quality issue emerges as a result of the integration of renewable energy source's intermittent nature with sophisticated power electronics device technology. The availability of nonlinear and unbalancing loads in the distribution system also appears to provide an impact on the Power quality of energy providers in the power distribution system. The distribution system on the power system is analyzed for power quality impacts such as variation of electrical power, voltage fluctuation, Total Harmonic Distortion (THD), and Unbalance voltage levels. This proposed compensator scheme is interconnected with the battery energy system at a point of common coupling (PCC) to minimize power quality issues. Its battery energy storage system (BESS) is designed mainly to assist the actual power sources when wind, Photovoltaic generating power is fluctuating and nonlinear load generates harmonics condition.

Keywords: Power quality, solar power generating system, wind generating system (WGS), STATCOM, battery energy storage system

**Corresponding author*: indragandhi.v@vit.ac.in

A. Chitra, V. Indragandhi and W. Razia Sultana (eds.) Intelligent and Soft Computing Systems for Green Energy, (97–110) © 2023 Scrivener Publishing LLC

7.1 Introduction

Integration of an environmentally friendly renewable power system generating power (RES) into electrical power systems is becoming a required and important concern in order to meet the enormous increase in electrical energy consumption while also reducing pollution issues generated by the usage of conventional energy sources [1, 2]. In order to generate electricity, various types of RES such as wind, photovoltaic (PV), biomass and fuel cell are used. Wind and PV are the two most common RES integrated into electrical systems due to their numerous benefits, as evidenced by the yearly increase rate of RES. Hybrid renewable energy systems primarily based on wind energy systems and PV systems are unable to provide the desired reactive power for the duration of fault incidents within the system. As a result, the voltage fluctuates at the PCC. The performance of the electrical network, particularly system stability, power factor, and power quality, is affected by such voltage fluctuations. Moreover, if these voltage fluctuations are not effectively controlled, they will reach dangerously high levels, resulting in the disconnection of these RESs. Nowadays, improving PQ is a major challenging issue in the field of distribution networks. Arc furnaces, Variable speed drives, switched mode power supplies (SMPS), uninterruptible power supplies (UPS), and different power electronic device using equipment are all nonlinear characteristics in nature. These loads are the source of different PQ issues, including voltage regulation, stability, and harmonics in source and load voltages and currents, all of which have a negative effect on system efficiency, safety, and stability. As a result, many researches have been carried out using various custom power devices (CPD) to improve the system's PQ [3, 11]. To address these power quality issues, the proposed scheme interfaces a Static Synchronous Compensator (STATCOM) with BESS at the PCC, which maintains voltage profile at PCC due to reactive power variation in wind generator, fluctuating real power from solar power and nonlinear load generate harmonics condition [4, 12, 13]. The proposed grid interconnected hybrid system with STATCOM controller is used to improve power quality at the PCC as shown in Figure 7.1. The STATCOM performance mainly depends on the choice of harmonics extraction algorithm, current control algorithm and DC link voltage control, which is implemented and explained with help of the flowchart shown in Figure 7.2.

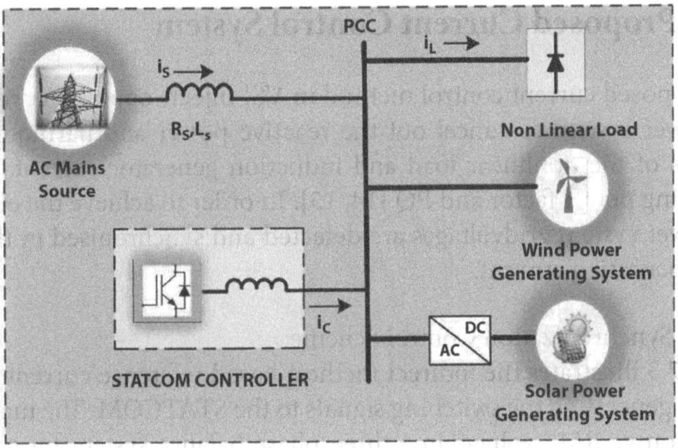

Figure 7.1 Proposed hybrid system with controller.

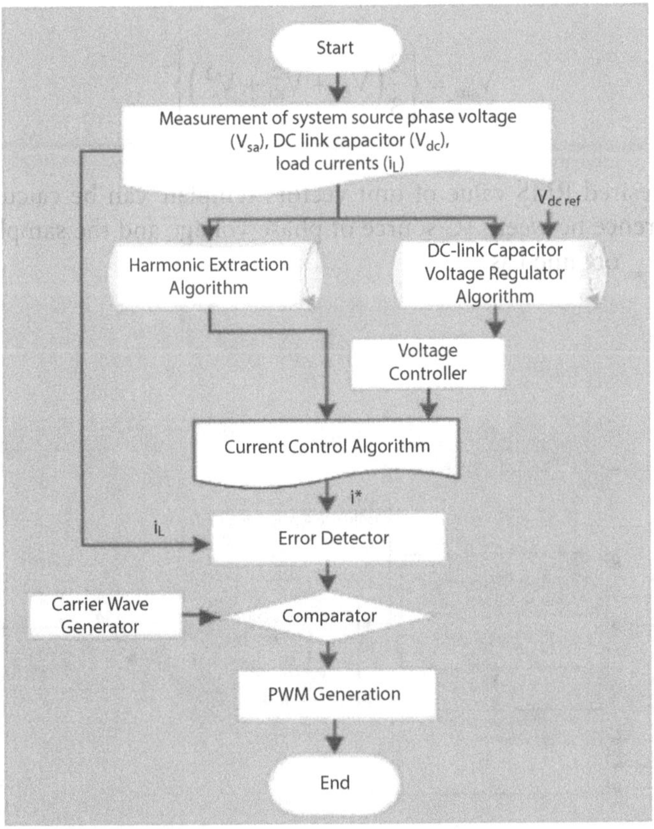

Figure 7.2 Flow chart of control algorithm.

7.2 Proposed Current Control System

The proposed current control method in VSI injects opposite current into the power system to cancel out the reactive power and harmonic components of the nonlinear load and induction generator current, thereby improving power factor and PQ [14, 15]. In order to achieve the objective, the power system gridvoltages are detected and synchronised in the form of compensating current.

a) Grid Synchronization Control Scheme

Figure 7.3 illustrates the indirect method-based reference current generation for generating the switching signals to the STATCOM. The magnitude of the source RMS voltage in a three-phase balance system for grid synchronisation is estimated at the sampling frequency from the three-phase source voltage (Vsa,V_{Sb},V_{Sc}) [3, 5]. Represented sample peak voltage V_{sm} as

$$V_{sm} = \left\{\frac{2}{3}\left(V_{sa}^2 + V_{sb}^2 + V_{sc}^2\right)\right\}^{\frac{1}{2}} \tag{7.1}$$

The desired RMS value of unit vectors template can be calculated by the difference between AC source of phase voltage and the sampled peak voltage V_{sm} obtained as

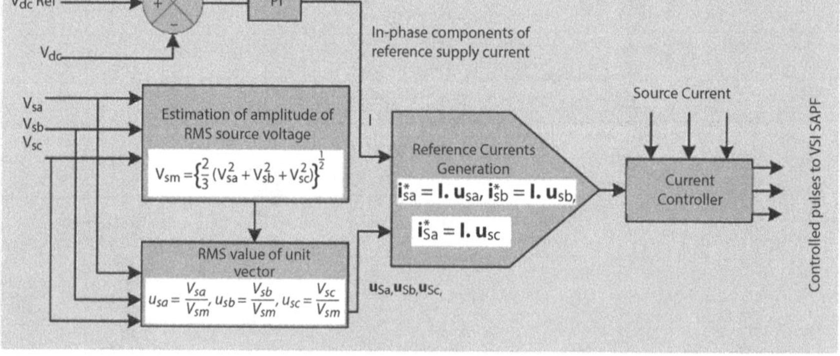

Figure 7.3 Indirect method-based reference current generation.

$$u_{Sa} = \frac{V_{Sa}}{V_{Sm}}, u_{Sb} = \frac{V_{Sb}}{V_{Sm}}, u_{Sc} = \frac{V_{Sc}}{V_{Sm}}, \tag{7.2}$$

Where, U_{sa} is RMS value of unit vector, the calculated reference currents are using an in-phase unit voltage template as a reference value are obtained as

$$i^*_{Sa} = I.u_{Sa}, i^*_{Sb} = I.u_{Sb}, i^*_{Sc} = I.u_{Sc} \tag{7.3}$$

b) Hysteresis Current Controller

The primarily control technique depends on injecting the currents into the grid and to using a conventional hysteresis current controller (HCC). The HCC can maintain current variable in between the hysteresis boundaries and generate the required control pulse by comparing the actual current to reference current for obtaining an error signal. The resulting generates the requisite control PWM switching pulse for the operation of STATCOM shown in Figure 7.4. For phase 'A' the switching function SA is written as If $i_{sa} < (i_{sa}^* - HB)$, upper switch S_A is switch OFF ($S_A = 0$) and if $i_{sa} > (i_{sa}^* - HB)$, Lower switch S_A^* is switch ON ($S_A^* = 1$) [5], [18]. Similarly, the switching characteristic S_B, S_C can be obtained from the hysteresis current-band "b" and "c" phases.

c) STATCOM - Battery Energy Storage System (BESS)

The BESS will normally keep a constant dc-link voltage and is especially suitable for STATCOM because it quickly injects or absorbs requirement of reactive power to stabilize the power system. When there is a power variation in the network, the BESS will be used to stabilize the fluctuation by charging and discharging [6, 10, 17]. BESS can also work properly as

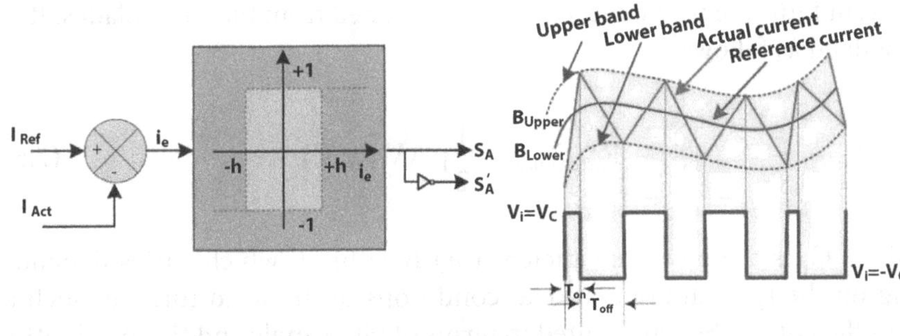

Figure 7.4 Single-band hysteresis current-control scheme.

an active power storage medium, supplying or consuming the appropriate amount of active power to maintain Squirrel cage induction generator (SEIG) frequency constant.

d) Grid-Connected Photovoltaic Systems
According to the photovoltaic effect, the PV cell is an electrical device modeled as a P-N junction that converts sunlight into electricity falling on its surfaces into electrical energy. Because no battery is utilized in this type of system, capital costs are reduced, and we use a grid-interconnected system. If PV generated energy is incorporated into the conventional grid, it can meet load demand from morning to evening, which is the time period when solar photovoltaic systems can be delivered to the power network. The grid-connected inverter converts DC power generated by PV modules into AC power; the Inverter technology is very important for a PV system's grid interconnection to be safe and reliable operation [7, 16].

e) Grid-Connected Wind Energy Generating Systems
When a wind turbine generating system (WTGS) is disconnected from the connected load, the asynchronous generator self-excites. There's a possibility of self-excitation when WTGS has a compensating capacitor, Reactive power compensation is provided by a capacitor linked to the in. However, the system's balancing determines the voltage and frequency. Self-excitation has some drawbacks, including safety and the balancing across real and reactive power [5, 8, 9]. the potential power of a wind energy system is shown as follows. In Eq. 7.4,

$$P_{\text{wind energy}} = \frac{1}{2}\rho A V_{1\,\text{wind}}^{3} \tag{7.4}$$

Wind turbine mechanical power is extracted from the rotor blades. It is obtained as follows

$$P_{\text{mechanical}} = \frac{1}{2}\rho A V_{1\,\text{wind}}^{3} C_1 \tag{7.5}$$

where C_1 is the power coefficient (or) Betz limit, which varies depending on the type and operational conditions of the wind turbine. Such a co-efficient can be represented in terms of pitch angle and tip speed ratio.

7.3 Simulation Analysis and Discussion

The grid-interconnected hybrid system with STATCOM Controller depicted in Figure 7.1 is modelled in MATLAB/simulink using the system configuration listed in Table 7.1. Nonlinear loads inject harmonic current into the source current waveform, and induction generators require reactive power demand at startup, which affects the PQ issue in the grid.

Table 7.1 Simulation parameters of grid-connected hybrid system.

S. no.	System parameters	Specifications
1	Grid Voltage	3-phase, 415V, 50Hz
2	PV system parameters	Maximum PV power=210.166W, Max. power point voltage=46.6, Max. power point Current=4.51, No. of series cells (Ns)=5, No. of parallel cells (Np)=5, PV Array=7
3	Induction Motor/ Generator	3.5KVA, 415V, 50Hz, P=4, Speed=1440rpm, R_s=0.016Ω R_r=0.015Ω, L_s=0.06H, L_r=0.06H, L_m=3.5H
4	Line Series Inductance	0.05mH
5	Inverter Parameters	DC side Voltage =800V, DC link Capacitance =100μF, Switching frequency =2 kHz
6	IGBT Rating	Collector Voltage=1200V, Gate voltage =20V, Forward Current =50A, Power dissipation =310W
7	Battery	Batter type=Nickel-Metal-Hydride, Nominal voltage = 800V, Rated capacity = 500Ah, Initial state-of-charge = 100%, Battery response time = 30 Sec.
8	PI Controller	K_p=0.5, K_i=0.16
9	Snubber resistance Rs	5000 Ω
10	Snubber capacitance Cs	250 μf
11	Load Parameter	Non-linear Load of C = 1μf, R=10Ω,

The STATCOM generate opposite current hormonic waveform such that the grid will purge the harmonics generated by the nonlinear load and reactive power demand of SEIG. The SEIG is connected to grid at 0.1s. It starts to vary voltage and current at the time of starting as shown in Figure 7.5.

The STATCOM Performance Under Distorted System Voltage Condition controller is turned ON at 0.18 s. It starts to reduce the demand of reactive power as well as present in harmonic current. The result of before and after compensation of supplied source voltage and current, nonlinear load generate harmonic current is shown in Figure 7.6. As can be seen in Figures 7.7 and 7.8, the results of injecting current from STATCOM controller are shown.

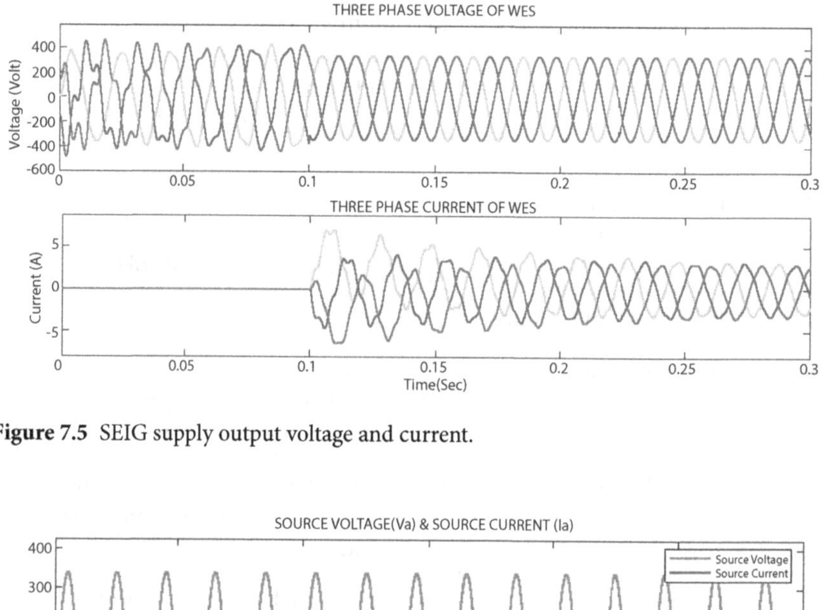

Figure 7.5 SEIG supply output voltage and current.

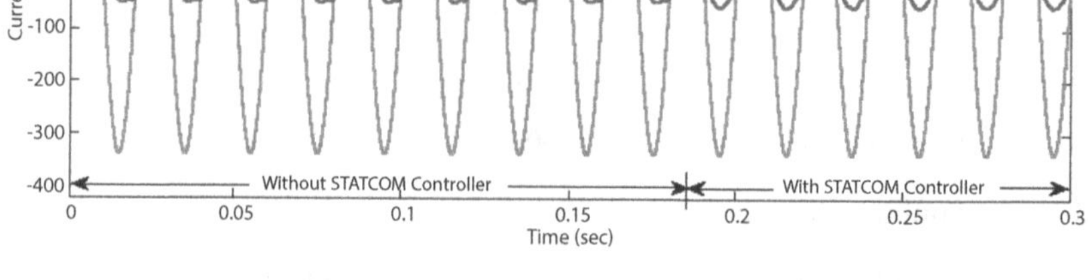

Figure 7.6 Source voltage and source current.

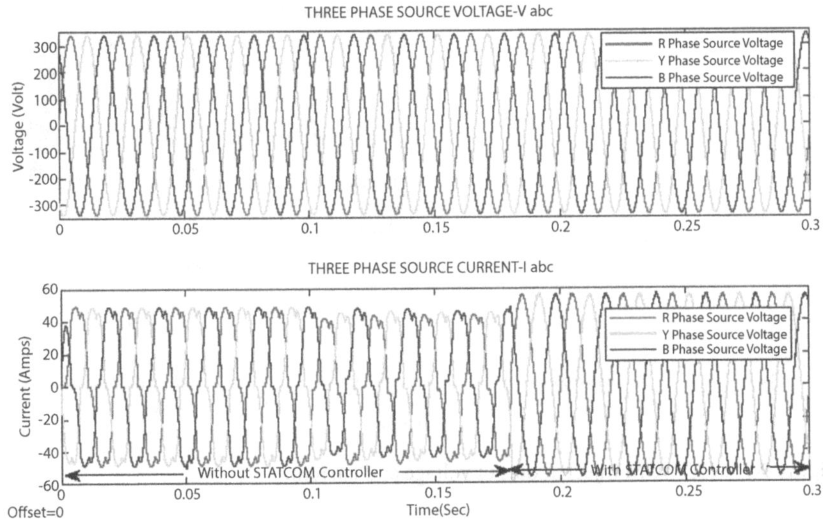

Figure 7.7 Three-phase source voltage and source current.

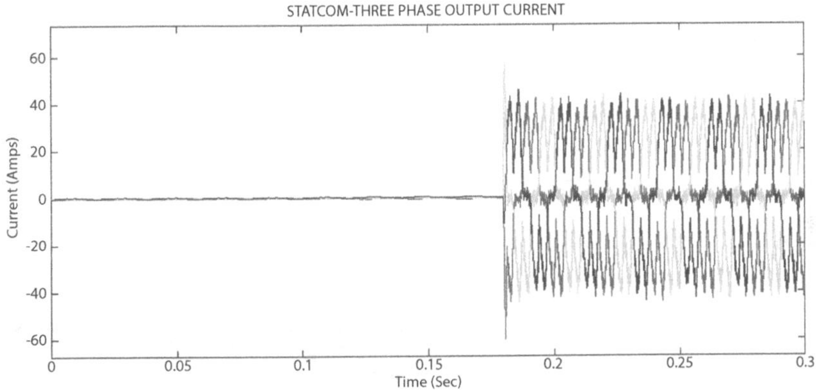

Figure 7.8 STATCOM output current.

It is observed that the output of STATCOM controller When nonlinear loads and induction generators have an impact on the source current on the grid, STATCOM's output is shown in Figure 7.9.

The source continues to deliver average active and reactive power flow to the nonlinear load, the STATCOM delivers reactive power demand, and the induction generates reactive power demand monitored with and without the controller at the PCC in the grid, as shown in Figure 7.10.

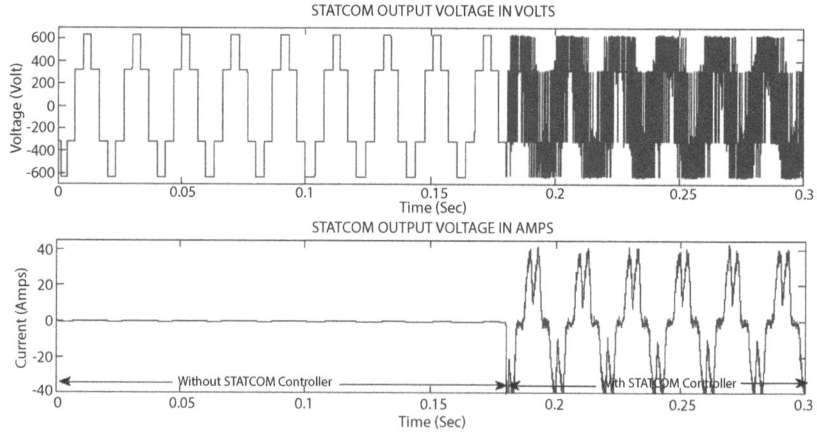

Figure 7.9 STATCOM injected output voltage and current.

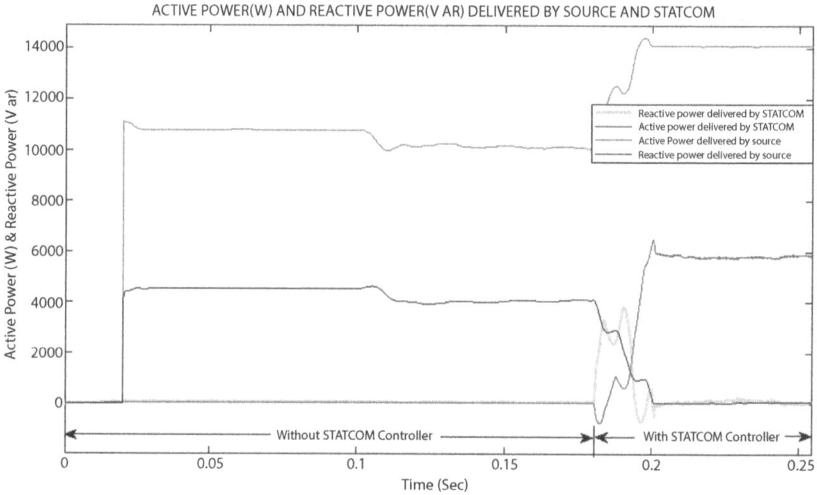

Figure 7.10 Active and reactive power delivered by source and STATCOM.

The source current is analysis without STATCOM controller with help of the Fourier analysis is expressed, the source current waveform at PCC the THD is 22.94 percent as shown in Figure 7.11. When the STATCOM controller is turned on at 0.18s, the source current at PCC with STATCOM is 1.70 percent of THD Value, so that the PQ improves at the PCC as shown in Figure 7.12. It is demonstrated that the THD has been significantly improved and good harmonic mitigation is within the standard IEEE 519 the standard norms.

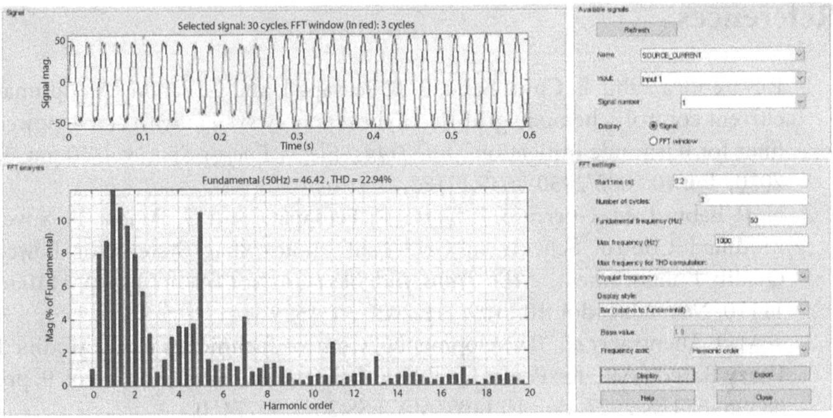

Figure 7.11 FFT analysis of source current before compensation.

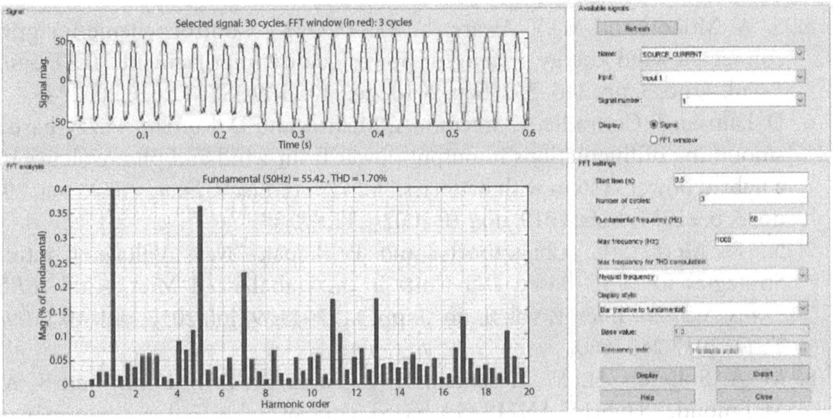

Figure 7.12 FFT analysis of source current after compensation.

7.4 Conclusion

This paper analyses power quality issues generated by hybrid wind-solar energy conversion and nonlinear loads. The current harmonics produced by nonlinear loads were significantly reduced to use an Indirect method-based Reference current generation control strategy in this work. It compensates the reactive power requirement for the induction generator and nonlinear loads at PCC in the power system by keeping the source voltage and current in phase. By eliminating harmonic current, the performance and efficiency are enhanced. The power factor is also maintained at the source.

References

1. P. Narendra Babu, B. Chitti Babu, R. B. Peesapati, and G. Panda, "An optimal current control scheme in grid-tied hybrid energy system with active power filter for harmonic mitigation," *Int. Trans. Electr. Energy Syst.*, vol. 30, no. 3, 2020, doi: 10.1002/2050-7038.12183.
2. N. P. Babu, J. M. Guerrero, P. Siano, R. Peesapati, and G. Panda, "A Novel Modified Control Scheme in Grid-Tied Photovoltaic System for Power Quality Enhancement," *IEEE Trans. Ind. Electron.*, vol. 68, no. 11, pp. 11100–11110, Nov. 2021, doi: 10.1109/TIE.2020.3031529.
3. T. M. T. Thentral *et al.*, "Development of Control Techniques Using Modified Fuzzy Based SAPF for Power Quality Enhancement," *IEEE Access*, vol. 9, pp. 68396–68413, 2021, doi: 10.1109/ACCESS.2021.3077450.
4. X. Liang and C. Andalib-Bin-Karim, "Harmonics and Mitigation Techniques Through Advanced Control in Grid-Connected Renewable Energy Sources: A Review," *IEEE Trans. Ind. Appl.*, vol. 54, no. 4, pp. 3100–3111, Jul. 2018, doi: 10.1109/TIA.2018.2823680.
5. S. W. Mohod and M. V. Aware, "A STATCOM - Control scheme for grid connected wind energy system for power quality improvement," *IEEE Syst. J.*, vol. 4, no. 3, pp. 346–352, Sep. 2010, doi: 10.1109/JSYST.2010.2052943.
6. D. Lamsal, T. Conradie, V. Sreeram, Y. Mishra, and D. Kumar, "Fuzzy-based smoothing of fluctuations in output power from wind and photovoltaics in a hybrid power system with batteries," *Int. Trans. Electr. Energy Syst.*, vol. 29, no. 3, p. e2757, Mar. 2019, doi: 10.1002/ETEP.2757.
7. N. R. Merritt, C. Chakraborty, and P. Bajpai, "New Voltage Control Strategies for VSC-Based DG Units in an Unbalanced Microgrid," *IEEE Trans. Sustain. Energy*, vol. 8, no. 3, pp. 1127–1139, Jul. 2017, doi: 10.1109/TSTE.2017.2657660.
8. M. A. Soliman, H. M. Hasanien, H. Z. Azazi, E. E. El-kholy, and S. A. Mahmoud, "Hybrid ANFIS-GA-based control scheme for performance enhancement of a grid-connected wind generator," *IET Renew. Power Gener.*, vol. 12, no. 7, pp. 832–843, May 2018, doi: 10.1049/IET-RPG.2017.0576.
9. S. W. Mohod and M. V. Aware, "Micro wind power generator with battery energy storage for critical load," *IEEE Syst. J.*, vol. 6, no. 1, pp. 118–125, Mar. 2012, doi: 10.1109/JSYST.2011.2163015.
10. Y. K. Chauhan, S. K. Jain, and B. Singh, "Static Volt Ampere Reactive Compensator for Self-excited Induction Generator Feeding Dynamic Load," http://dx.doi.org.egateway.vit.ac.in/10.1080/15325000802046843, vol. 36, no. 10, pp. 1080–1101, Oct. 2008, doi: 10.1080/15325000802046843.
11. M. I. Mosaad, H. S. M. Ramadan, M. Aljohani, M. F. El-Naggar, and S. S. M. Ghoneim, "Near-Optimal PI Controllers of STATCOM for Efficient Hybrid Renewable Power System," *IEEE Access*, vol. 9, pp. 34119–34130, 2021, doi: 10.1109/ACCESS.2021.3058081.

12. P. Acuna, L. Moran, M. Rivera, J. Dixon, and J. Rodriguez, "Improved active power filter performance for renewable power generation systems," *IEEE Trans. Power Electron.*, vol. 29, no. 2, pp. 687–694, 2014, doi: 10.1109/TPEL.2013.2257854.
13. S. K. Chauhan, M. C. Shah, R. R. Tiwari, and P. N. Tekwani, "Analysis, design and digital implementation of a shunt active power filter with different schemes of reference current generation," *IET Power Electron.*, vol. 7, no. 3, pp. 627–639, Mar. 2014, doi: 10.1049/IET-PEL.2013.0113.
14. S. Biricik, S. Redif, Ö. C. Özerdem, S. K. Khadem, and M. Basu, "Real-time control of shunt active power filter under distorted grid voltage and unbalanced load condition using self-tuning filter," *IET Power Electron.*, vol. 7, no. 7, pp. 1895–1905, Jul. 2014, doi: 10.1049/IET-PEL.2013.0924.
15. E. Sundaram and M. Venugopal, "On design and implementation of three phase three level shunt active power filter for harmonic reduction using synchronous reference frame theory," *Int. J. Electr. Power Energy Syst.*, vol. 81, pp. 40–47, Oct. 2016, doi: 10.1016/J.IJEPES.2016.02.008.
16. A. Deihimi and A. Rahmani, "Application of echo state networks for estimating voltage harmonic waveforms in power systems considering a photovoltaic system," *IET Renew. Power Gener.*, vol. 11, no. 13, pp. 1688–1694, Nov. 2017, doi: 10.1049/IET-RPG.2017.0046.
17. Y. Hoon, M. A. M. Radzi, M. K. Hassan, and N. F. Mailah, "Operation of Three-Level Inverter-Based Shunt Active Power Filter under Nonideal Grid Voltage Conditions with Dual Fundamental Component Extraction," *IEEE Trans. Power Electron.*, vol. 33, no. 9, pp. 7558–7570, Sep. 2018, doi: 10.1109/TPEL.2017.2766268.
18. H. Komurcugil, "Double-band hysteresis current-controlled single-phase shunt active filter for switching frequency mitigation," *Int. J. Electr. Power Energy Syst.*, vol. 69, pp. 131–140, Jul. 2015, doi: 10.1016/J.IJEPES.2015.01.010.

8

Energy Trading in Virtual Power Plant Enabled Communities Using Double Auction Technique and Blockchain Technology

Radhika Yadav, Balla Manoj Kumar, Saurav Baid and Padma Priya R.*

School of Computer Science and Engineering, Vellore Institute of Technology, Vellore, India

Abstract

Recently it has been observed that the occurrences of power outages are increasing. Two plausible reasons for this are the boom in power consumption due to the proliferation of more interconnected devices and the inflexibility of traditional power grids to adapt to changing demands. Smart grids have been an alternative for traditional grids. However, the inherent intermittent characteristics of smart grids are a challenge. Virtual Power Plant (VPP) is a demand-side management approach that manages Distributed Energy Resources (DERs) by virtual aggregations. However, in such transactions the data being handled may be compromised, which leads to several vulnerability issues. Hence in this paper we have focused on proposing an energy trading approach which is both efficient and secured with concerns to embrace social welfare requirements as well. In our model we have considered three objectives: a) to maximize the profits for buyers and sellers in the market through a Double Auction based algorithm, b) to ensure the most optimal route to be used for end-to-end transmission of energy supply among the chosen sellers and buyers, and c) to provide protection for the transactions being made during auction through asymmetric encryption (ElGamal) in combination with blockchain-based architectural concepts. Our experimental results depict that the suggested scheme is successful in maximizing social welfare while minimising energy transmission time.

Keywords: Energy trading, blackout, smart grid, double auction, blockchain, virtual power plant, end-to-end line routing

**Corresponding author*: padmapriya.r@vit.ac.in

A. Chitra, V. Indragandhi and W. Razia Sultana (eds.) Intelligent and Soft Computing Systems for Green Energy, (111–126) © 2023 Scrivener Publishing LLC

8.1 Introduction

Microgrids are networks which allow two-way supply of electricity and information between their customers [1]. Smart grids, specifically microgrids, provide efficiency and are more environmentally friendly than their counterparts [2]. The disadvantages of repeated use of fossil fuels have shifted attention towards renewable forms of energy to be utilized via these microgrids. But a major problem is the irregularity of power generation through these renewable sources. This is where distributed technologies come in as feasible solutions [3]. Nowadays, the individual microgrids in a grid system store their surplus energy to be utilized in need. Further through peer-to-peer energy trading the microgrids can even act as traders and sell or buy energy from others in a community [4]. Energy trading means the buying and selling of various energy items like electrical power, coal, crude oil and natural gas. However, power shortages caused by human error or natural calamities have severe consequences on these grids. The year 2019 was one of the worst for power supply systems, leaving almost 70% of Venezuela powerless as well as 100 million people in Indonesia and surrounding countries. In 2020, the number was 30 million. Still, a majority of solutions in this domain are preventive, as discussed in section 8.2.

Microgrid communities also face certain security risks. The entire auction process can be compromised or individual bids can be manipulated. An efficient solution is implementing blockchain methodologies. Blockchain is a decentralized distributed system in which the blocks containing the ledgers are growing lists interlinked through cryptographic algorithms. If someone attacks a node and manipulates the data then as the next block is generated it cross checks with all the other nodes in the database and if there is any discrepancy then the attacked block is replaced with the original block. The blocks in the blockchain contain a ledger which is the backbone of the entire system [5]. After the virtual power plant securely stores the data, it can be mined and decrypted for conducting the trading process. The bids stored and encrypted on the VPP in our mechanism include a unique trader ID, no. of energy units to be bought/sold, price for each unit and a timestamp when the bid was sent to the VPP.

The main contributions of this paper can be summarized as:

- A Double Auction (DA) mechanism to decide winning traders in an energy trading market through trading between microgrids with insufficient supply (buyers) and surplus supply (sellers).

- Assigning each winning buyer to the closest seller that can satisfy its demands in the least time, i.e., take location awareness into consideration.
- Blockchain-based architectural concept to provide protection for the transactions being made among the bid traders during auction and also for the transactional data generated and stored in VPP through asymmetric encryption (ElGamal).

The remainder of this paper is organized as follows: in Section 8.2, we elaborate on the related work. In Section 8.3, we introduce the architecture and methodology of the scheme. In Section 8.4, we present and explain the results. Finally, we give conclusions in Section 8.5.

8.2 Related Work

In this section, we discuss the existing literature related to our work. In [6], Dileep G. presents an in-depth analysis of the various enabling technologies and future applications of a smart grid which is the base of our work. In [1], the authors discuss the implementation of such technologies in a developing country. The economic aspects of such an adaptation are well explained in [7]. Authors in [8] have worked upon modelling load demand response based on real meteorological values using PSO approach to estimate the economic cost to be incurred in establishment of a medium-sized microgrid satisfying 4MWh industrial plants for Tamil Nadu, India. Now most grids have renewable energy sources but their huge disadvantage is the irregularity in energy production. In [9] the authors propose an energy function virtualization-based optimized method to build smart grids that are fault tolerant and reliable in an emergency situation. [10] and [11] propose preventive measures to estimate the occurrence of a power shortage or blackout in advance by using a Grey Wolf-pattern search algorithm and a Markov chain model, respectively, and then find the paths in the grid that are most likely to fail and protect them.

But it is difficult to predict such circumstances. Here energy trading is a useful solution. It is commonly used to solve the previously discussed problem of irregularity of renewable sources.

In [12] Donghe Li *et al.* use a modified double auction energy trading mechanism to ensure a fair trade among Electric Vehicles. The mechanism also maintains privacy of the traders. In [13] a privacy-preserving protocol while conducting double auctions among microgrids is suggested where the bids are encrypted using the Paillier cryptosystem and every trader is given a pseudo-identity. But the entire process still has some computational overheads. The importance for secure storage and transfer of data is further highlighted in [14] where the author presents an architecture for data aggregation on a smart grid system using fog and blockchain technologies noting that it is a more secure and effective way of collecting data. In [15] the author has introduced a mechanism for secure data transfer in the grid using a multi-tiered blockchain architecture. There is a lack of study of energy transmission and distribution between microgrids in a community through a safe and secure multi-unit location aware double auction mechanism in a power shortage. The paper aims to work towards eliminating such issues.

8.3 Proposed Methodology

Figure 8.1 depicts the system architecture for the proposed mechanism. The demand area consists of trader (buyer and seller) microgrids which represent their respective communities. These microgrids send their encrypted bids which include a unique trader ID, no. of energy units to be bought/ sold, price for each unit and a timestamp, to a VPP or the master wallet, which acts as the central resource pool/ controller. The bids are encrypted using an ElGamal algorithm and no traders can access information about others. These bids are then sent to a double auction mechanism and based on an equilibrium price, winning traders are identified. Further, depending on the equilibrium price as well as the distance between the traders, the most optimum seller microgrid is selected for each buyer microgrid. The least distance is calculated based on a distance matrix. This helps in maximizing trader profits and minimizing time of flow of power, hence overall maximizing social welfare while also ensuring the transaction data remains secure.

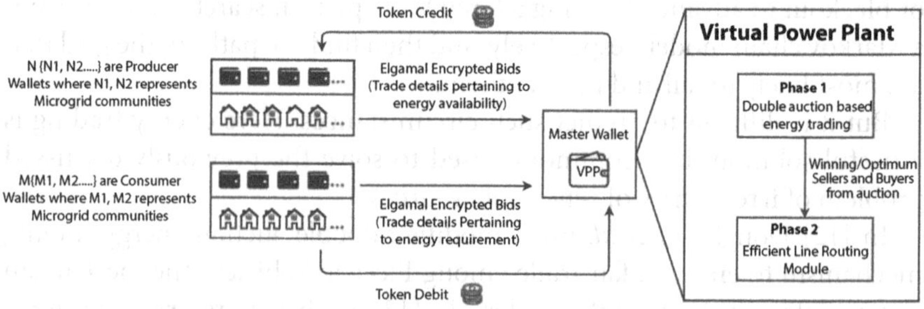

Figure 8.1 Conceptual architecture diagram.

8.3.1 System Model

The system model is designed to ensure maximum social welfare (MSW) by maximising incentives as well as ensuring privacy of each bidder. The privacy of each bid is ensured using the ElGamal encryption and MSW is ensured using the VPP.

The VPP has two objectives namely, 1) Optimum reimbursements for both the traders, and 2) shortest line routing system. The workflow of the proposed methodology of VPP auction-based decision-making methodology is portrayed in Figure 8.2. The most optimum seller is matched with the buyer to maximise social welfare. In this proposed mechanism, some

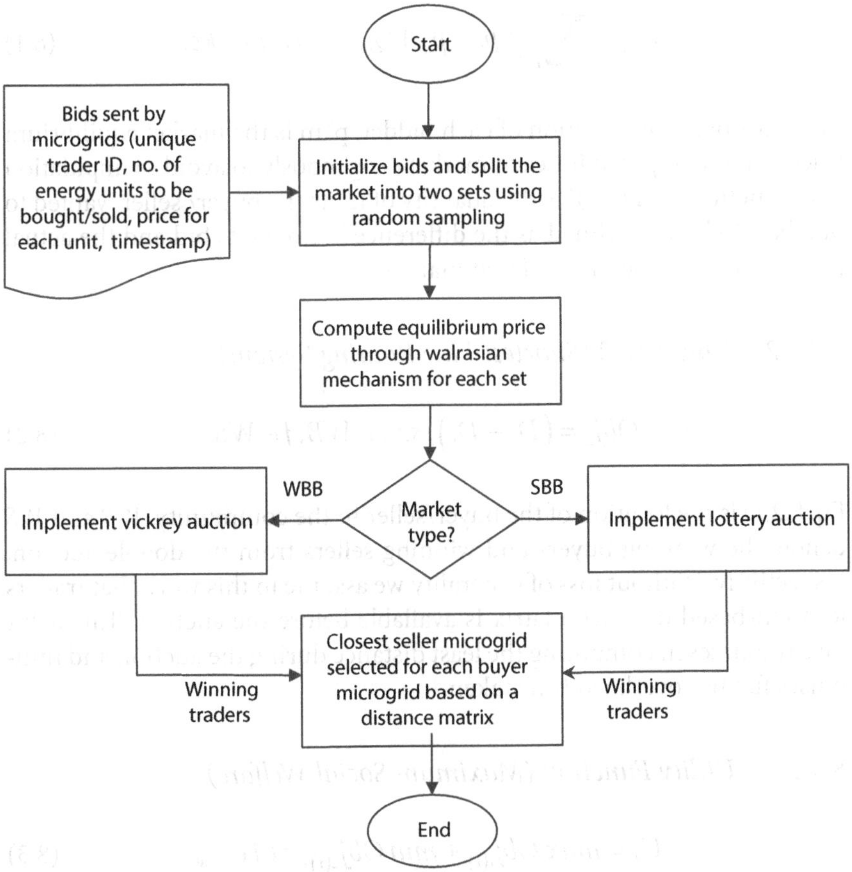

Figure 8.2 Represents the workflow of the proposed work. Here WBB and SBB represent the two different market types.

assumptions have been made for convenience: (a) the power loss during transmission is negligible, (b) auction market is open whenever there are both sellers and buyers, (c) each trader in the market has a Diminishing Marginal Utility (DMU), and (d) each bid is a unique bid. These winning bids are passed onto phase 2 to compute the efficient seller for each buyer using a shortest line routing mechanism.

8.3.2 Problem Formulation

In this section, firstly we define the objectives of the paper.

8.3.2.1 Objective 1 (Optimum Reimbursements for both the Traders)

$$Obj_1 = \sum_{i=1}^{T} \left(p_i - p_m^*\right).Q_i, \; s.t \; m \in market \tag{8.1}$$

In Eq. 8.1 pi is the valuation of each bidder, p*m is the market equilibrium price that is computed for each market exogenously to avoid manipulation of the auction results, Qi is the quantity of units the buyer/ seller wanted to purchase/sell. It is defined as the difference between its bid and the actual payment or remuneration of that market.

8.3.2.2 Objective 2 (Shortest Line Routing System)

$$Obj_2 = \left(D_j - D_i\right), \; s.t \; i \in WB, j \in WS \tag{8.2}$$

Eq. 8.2 *D* is the location of the buyer/seller in the community. *WB* and *WS* denote the winning buyers and winning sellers from the double auction, respectively. Without loss of generality we assume in this work that traders location-based distance matrix is available before the auction. This helps save resources in computing the least distance during the auction and minimises further overhead time taken.

8.3.2.3 Utility Function (Maximum Social Welfare)

$$U_i = max \; Obj_{1(i)} + min \; Obj_{2(i)}, \; s.t \; i \in T_w \tag{8.3}$$

The MSW of each trader in the market is denoted through its utility function U_i. Eq. 8.3 Obj_1 and Obj_2 denote the two objectives and T_w represents the winning trader. Here profit for the traders is maximised and time/distance of flow of power is minimized hence maximising U_i or MSW.

Along with these two objectives, the proposed double auction scheme must also ensure the following:

- Dominant Strategic Incentive Compatibility (DSIC): indicates traders are able to achieve the maximum social welfare if and only if bids are truthfully reported.
- Individual Rationality (IR): indicates that traders will earn non-negative profit.
- Budget Balance (BB): indicates that the total payment of all the buyers should be equal to or larger than the total reimbursement.
- Prior free: Even if there is a lack of statistical information on the valuations, the auction should prevail in the worst-case scenario (conflicting bidding).

8.3.3 Our Approach

Firstly, we have designed a multi-unit double auction system that helps in determining the winning traders. The traders' bid values are encrypted through ElGamal algorithm and stored on the VPP beforehand. It is later mined and decrypted and bid values are used to conduct the auction so as to identify winning traders. These winning buyers and sellers participate in a line routing scheme to find the most optimum seller for each buyer.

8.3.3.1 Double Auction

Unlike a one-sided market where there is a 1 : N relation, in double auction, there is a N : N, i.e., there exist multiple buyers and sellers. Therefore, there is a need to find an equilibrium point at which the market agrees and no trader makes a non-negative utility to ensure IR.

Equilibrium price – To keep the integrity of the market, an equilibrium price is proposed at which the majority of the market ensures individual rationality. Denoted as p^*, to find the equilibrium price we use the walrasian mechanism. We classified all the buyers who have bid above the equilibrium price and all the sellers below the equilibrium price as active buyers and active sellers, respectively. This is evident as buyers who have quoted a price less than the equilibrium price cannot match the price to participate in the auction and sellers who have bid a higher price will not receive profits in this market.

Market type - In a realistic competitive scenario, markets are classified into two types, i.e., a monopolistic market and a free market. A monopolistic market is one wherein the market/market controller makes profits in trading fees. This is a type of a Weakly Budget-Balanced (WBB) market. In other words, the total Gain-From-Trade (GFT) of the market is always higher since only the most profitable selections occur. A free market is a type of market wherein the market makers do not make profits during the auctions. This is a strongly budget balanced (SBB) market. The total GFT is always equal to traders GFT. Based on the market type, an auction is proposed.

Winning Determination - Before we introduce the winning determination scheme, we randomly halve the market and for each market we compute an equilibrium price. Random halving and sampling the market, as mentioned above, further ensures a truthful market as the equilibrium prices for each side of the market are computed exogenously. There are three possible scenarios that can occur while trading. In an ideal scenario, the surplus energy could be equal to the required or demanded energy and the market can freely trade at the equilibrium price. Each seller is uniquely matched with a buyer. However, in a realistic world, either the demand or the supply is greater than the other. As those two scenarios are similar, for our convenience, we shall describe the solution to one of the scenarios. In case of demand being greater than the supply, the short side is identified as supply and long side as demand. Since demand is greater than supply, we have more than enough buyers for each seller. The challenge is to match the best buyer for each seller. Different market types have different auction schemes that propose the best solution in their case.

Lottery Auction: Here, are chosen at random and each buyer purchases energy at the equilibrium price defined till the seller's supply is satisfied.

Vickrey Auction: Here, the highest GFT is buyers are chosen, therefore buyers are sorted in descending order and each buyer purchases at the equilibrium price defined till the seller's supply is satisfied. Vickrey auctions are followed with a trading fee that each winning buyer has to pay. This is taken as a reimbursement for the buyers who could not participate in the auction.

In any case, both these auction schemes maximise the incentive of each active trader participating in the auction. It also ensures various economic objectives that we defined earlier. Lottery and Vickrey are both individually rational for the long and short side. For the short side, each trader maximises their gains and chooses the highest GFT. In the long side of the lottery auction, due to the DMU assumption, each trader only sells till its optimal for them. In Vickrey long side auction, the highest valuation traders are chosen for the most optimal gains. Since the scheme choses traders with highest GFT, the trading fee that each winning trader pays is still non-negative and hence maintains IR characteristics. The Vickrey auction is proven to be a DSIC. In second price schemes, truth telling is the dominant strategy, and therefore Multi-unit Vickrey Auctions also ensure truthful bidding.

8.3.3.2 Shortest Line Route Detection

Generally in energy trading, only the winning traders are identified, but here we introduce a route for each buyer to the closest optimum seller. We categorize the winning traders and use a distance matrix to calculate the optimum seller. The minimum distance between the seller and buyer inversely maximises social welfare.

8.3.3.3 Blockchain

Prosumers (producer and consumer) are considered as participants being involved in the energy trading process. There is a chance for vulnerable sniffing during the bids when being exchanged among such participants since such channels of communication are mostly insecure. In our proposed model each Prosumer (producer and consumer) is provided with a unique e-wallet that consists of tokens. Wallets in our work constitute a unique ID, private and public key used to encrypt and decrypt the bids placed by each trader in the market more securely. For every defined region of the Microgrid, a regulatory local controller is present, which is actually involved in the trading business which owns a wallet and represents the corresponding microgrid community as a single entity. All the traders (regulatory local controllers) of the energy market own their individual wallets; however, only the Virtual Power Plant (VPP) owns a special wallet called the master wallet. The master wallet is responsible for providing the initial private key for encryption where bids are encrypted using ElGamal cryptography and further it's been mined into the ledger. In the further section below, details about ElGamal cryptography are described in detail.

The encrypted data is added to the ledger of the blockchain being maintained by VPP. Inside the ledger, time-stamp (when the bid received time from the trader), previous hash and the encrypted bids where in these bids contain details like Unique trader id, energy units being bought or sold and price for each unit) are considered as the attributes for the generation of the final hash value using SHA 256. This is further used as a previous hash for the succeeding block which will be generated. When a bid is submitted to the VPP either for buying or selling energy units the encrypted information is decrypted successfully and phase 1 (double auction algorithm) of our workflow process proceeds. Subsequently after the DA approach identifies the winning trader, phase 2 (efficient line route detection algorithm) of the workflow process proceeds.

At this juncture tokens are deducted from the consumer wallet and are subsequently credited to the producer wallet. These tokens are fixed by the regulatory controllers in proportion to the energy units consumed. Finally, the respective quantity of tokens are duly exchanged between the producer and consumer through VPP and energy flow is directed between the two winning traders in the market. The details on how the tokens are generated is actually beyond the scope of our paper. But our work can be extended to add incentive-based token generations and motivational approaches as future work.

8.3.3.4 ElGamal Cryptography

The ElGamal cryptosystem [16] is an encryption scheme that assists the sender and the receiver while exchanging sensitive information over any insecure channel. ElGamal encryption is a non-deterministic asymmetric key encryption that is based on the Diffie Hellman key exchange algorithm enhanced with additional security layers. The security mechanism (equations) of this algorithm benefit from using private and public keys which are interlinked with the masters and prosumers wallet for encryption. Like all the other cryptography, this approach also has two phases: Encryption and Decryption. Unless both of the keys are known, the prosumer-VPP or the consumer-VPP communication and access to the bid information will be denied.

Encryption Process:
In the encryption scheme, the sender Public Key(E2) is derived from a triplet (E1,E2,P) where E1,E2 and P, where the E1 is a randomly chosen number and P is a prime number (usually a very large number) as specified in the Eq.4. The private key D (chosen between 1 and P-1) is generated by the traders involved in the transactions locally and subsequently used in decryption of the ciphertext also.

$$E2 = (E1) \bmod P \tag{8.4}$$

On the sender side, the sender receives the public key (E2) from the receiver and encrypts the bid details(M) with the receiver's public key and generates a Ciphertext. The sender uses an ephemeral key(R) for the encryption (random integer between 1 and p-1) of the plain text as in Eq. 8.5. Once the sender uses an ephemeral key for encryption of a bid, the key is discarded and a new key is generated. The sender uses encryption algorithm C2 with the public key as in Eq. 8.6.

$$C1 = (E1^\wedge R)\text{mod } P \tag{8.5}$$

$$C2 = (M^*E2^\wedge R)\text{mod } P \tag{8.6}$$

$$\text{Cypher text} = (C1,C2) \tag{8.7}$$

Decryption Process:
This Cypher text (C1, C2) obtained using Eq. 8.7 is sent to the receiver and the receiver uses his private key(D) for deriving the plain text Y with Eq. 8.8.

$$Y = [C2^*(C2^\wedge D)^\wedge\text{-}1]\text{mod } P \tag{8.8}$$

8.4 Performance Evaluation

In this section, we discuss the experimental results of the proposed auction scheme.

8.4.1 Evaluation Methodology

This algorithm is simulated using a static grid setup with distances between them stored. It uses the NYSE Stock Exchange data [17] as bids in the auction market to add a realistic bidding scenario. The dataset used represents similar behavior as the bidders in any market [18]. All VPP simulations are carried on PYTHON, and conducted on a computer with 2.4 GHz Intel Core i7 CPU and 8Gb RAM. The DA algorithm is simulated over 144 auctions for each of the 1,000 traders to reach an empirical distribution of results. To understand the time involved in processing blockchain-based VPP operations, we used various different processors, namely i3 (3.4GHz - 4.2GHz); i5 (1.2GHz-3.6 GHz); i7 (2.9GHz-4.2 GHz) and found out that the mining rate is dependent on the processing capacity of the device.

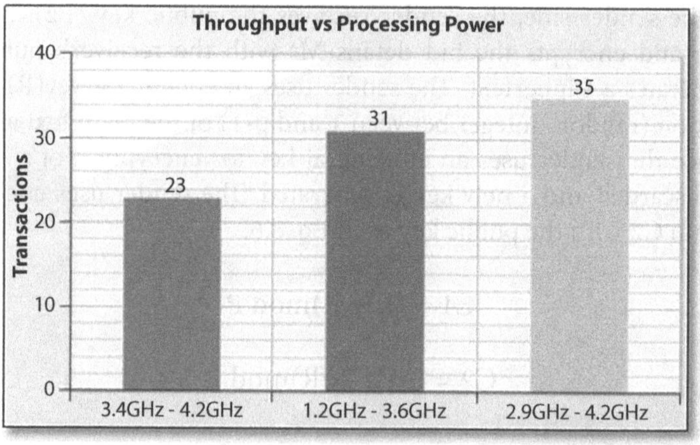

Figure 8.3 Throughput vs. processing power.

We observed as in Figure 8.3 that the number of transactions increased per minute when the processing capacity was increased.

8.4.2 Evaluation Results

In every individual experiment there are n traders, where n = 10, 20, 30 ... 1000. Totally 1000 · 144 auctions were simulated and then averaged the 144 auctions for each value of n. To evaluate the scheme, a social welfare ratio is defined as the cumulative result of the GFT for all traders in the auction based on the market type and the minimum time taken for each buyer to associate with a seller. In Figure 8.4, blue wedges represent the

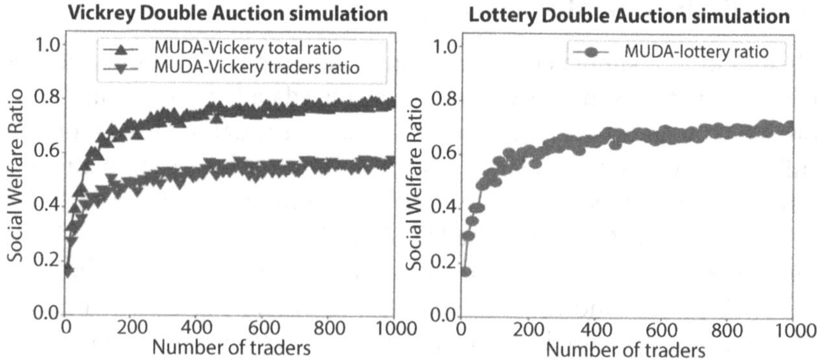

Figure 8.4 Represents Vickrey and Lottery Double Auction graphs, which depict that as the number of traders increases, the probability of traders' demand being met also increases.

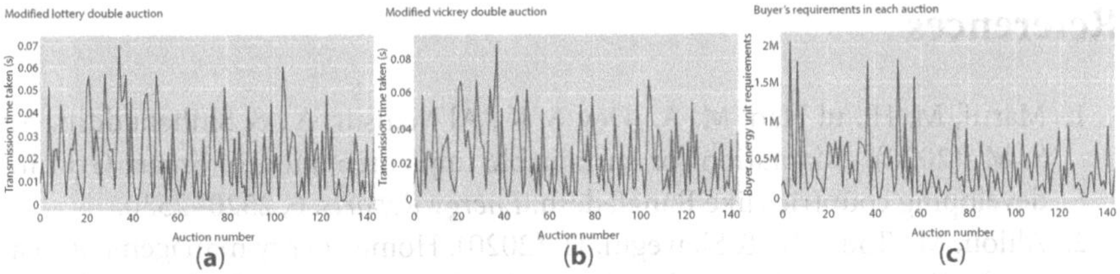

Figure 8.5 The total time taken for the proposed system to run is depicted along the x axis in Figures 8.5(a) and 8.5(b) for lottery and Vickrey double auctions respectively and the y axis represents the auction number. Figure 8.5(c) plots all the buyer energy unit requirements in each auction number from 1-144. This value remains the same for both types of markets. After taking the mean of all 144 values it was found that the program calculates results for lottery double auction in an average of 0.0186s and Vickrey double auction in 0.0216s.

total welfare ratio and the green wedges represent the welfare ratio of the traders in the Vickrey market. The red discs represent the welfare ratio in the lottery market. The difference between the total ratio and traders' ratio is the market fee that each winning trader pays. Both graphs depict that as the number of traders increases, the probability of traders' demand being met increases as well. Additionally, in the 144 simulations, the lottery market takes lesser time on average to complete the auctions compared to the vickrey market. Their run times are represented in Figure 8.5 alongside the buyer's requirements in each auction.

8.5 Conclusion

In this paper, we addressed the energy imbalance during a blackout through an energy trading scheme with complementing objectives. The scheme is executed using a double auction mechanism. We further introduce a line routing mechanism to optimise energy trading and maximise social welfare in the market. Most importantly, for safety and privacy purposes, an encryption methodology for the auctions' information using blockchain is also suggested in the architecture. The experimental results satisfy the objectives of the paper and give a new direction to the recent development of decentralised energy transmission. The paper also opens doors to various paths such as the evaluation of this mechanism with a dynamic line routing scheme and real-world application of ElGamal encryption-based blockchain schemes.

References

1. Maruf, M. H., ul Haq, M. A., Dey, S. K., Al Mansur, A., & Shihabuddin, A. S. M. (2020). Adaptation for sustainable implementation of Smart Grid in developing countries like Bangladesh. *Energy Reports*, 6, 2520–2530.
2. Alilou, M., Tousi, B., & Shayeghi, H. (2020). Home energy management in a residential smart micro grid under stochastic penetration of solar panels and electric vehicles. *Solar Energy*, 212, 6–18.
3. Albataineh, H., Nijim, M., & Bollampall, D. (2020, August). The design of a novel smart home control system using smart grid based on edge and cloud computing. In *2020 IEEE 8th International Conference on Smart Energy Grid Engineering (SEGE)* (pp. 88–91). IEEE.
4. Zhang, C., Wu, J., Zhou, Y., Cheng, M., & Long, C. (2018). Peer-to-Peer energy trading in a Microgrid. *Applied Energy*, 220, 1–12.
5. Fernández-Carames, T. M., & Fraga-Lamas, P. (2020). Towards post-quantum blockchain: A review on blockchain cryptography resistant to quantum computing attacks. *IEEE Access*, 8, 21091–21116.
6. Dileep, G. (2020). A survey on smart grid technologies and applications. *Renewable Energy*, 146, 2589–2625.
7. Wang, R., Hsu, S. C., Zheng, S., Chen, J. H., & Li, X. I. (2020). Renewable energy microgrids: Economic evaluation and decision making for government policies to contribute to affordable and clean energy. *Applied Energy*, 274, 115287.
8. Padma Priya R, Rekha, D. (2020). Sustainability modelling and green energy optimisation in microgrid powered distributed FogMicroDataCenters in rural area. *Wireless Networks*, 1–14.
9. Wang, K., Wu, J., Zheng, X., Jolfaei, A., Li, J., & Yu, D. (2020). Leveraging energy function virtualization with game theory for fault-tolerant smart grid. *IEEE Transactions on Industrial Informatics*, 17(1), 678–687.
10. Mahdad, B., & Srairi, K. (2015). Blackout risk prevention in a smart grid based flexible optimal strategy using Grey Wolf-pattern search algorithms. *Energy Conversion and Management*, 98, 411–429.
11. Zhai, C., Nguyen, H. D., & Xiao, G. (2020). A robust optimization approach for protecting power systems against cascading blackouts. *Electric Power Systems Research*, 189, 106794.
12. Li, D., Yang, Q., Yu, W., An, D., Yang, X., & Zhao, W. (2017, December). A strategy-proof privacy-preserving double auction mechanism for electrical vehicles demand response in microgrids. In *2017 IEEE 36th International Performance Computing and Communications Conference (IPCCC)* (pp. 1–8). IEEE.
13. Sarenche, R., Salmasizadeh, M., Ameri, M. H., & Aref, M. R. (2021). A secure and privacy-preserving protocol for holding double auctions in smart grid. *Information Sciences*, 557, 108–129.

14. Chaudhary, S., Johari, R., Bhatia, R., Gupta, K., & Bhatnagar, A. (2019, April). CRAIoT: concept, review and application(s) of IoT. In *2019 4th international conference on internet of things: Smart innovation and usages (IoT-SIU)* (pp. 1–4). IEEE.
15. Sikeridis, D., Bidram, A., Devetsikiotis, M., & Reno, M. J. (2020, January). A blockchain-based mechanism for secure data exchange in smart grid protection systems. In *2020 IEEE 17th Annual Consumer Communications & Networking Conference (CCNC)* (pp. 1–6). IEEE.
16. ElGamal, T. (1985). A public key cryptosystem and a signature scheme based on discrete logarithms. *IEEE Transactions on Information Theory*, 31(4), 469–472.
17. Hasbrouck, J. (1992). Using the TORQ database. *New York Stock Exchange*, 11(24), 2010.
18. Lee, C. M., & Radhakrishna, B. (2000). Inferring investor behavior: Evidence from TORQ data. *Journal of Financial Markets*, 3(2), 83–111.

9

Sales Demand Forecasting for Retail Marketing Using XGBoost Algorithm

M. Kavitha[1]*, R. Srinivasan[1], R. Kavitha[1] and M. Suganthy[2]

[1]*Vel Tech Rangarjan Dr. Sagunthala R&D Institute of Science and Technology, Chennai, India*

[2]*Vel Tech Multi Tech Dr. Rangarajan Dr. Sakunthala Engineering College, Chennai, India*

Abstract

Demand forecasting helps in estimating the requirements of products in the future, based on past and present data. The future depends on trends and other aspects of the market. Nowadays, organizations are not able to predict future product demand, which leads to unnecessary storage and decay of products. So by analyzing the past and current market data, future demands are well predicted; thus those demanded products can be manufactured in the near future. This paper is mainly focused on predicting the demand of any particular goods and services that are available at a particular time of the year to satisfy consumer needs. Demand forecasting can also be helpful to both the consumer and the retailer; the consumer satisfies his needs and therefore, the retailer gets profits. This paper mainly presents the forecasting of sales of data by using Time Series analysis and XGboost algorithm, where we can see the trend and recession of sales in different times of the year. The accuracy reached is "0.14" by reducing overfitting and underfitting. Here, Root Mean Square Percentage Error Value is used; the accuracy is calculated in the range of 0.1 to 0.9. Value closer to 0.1 is considered as the best-suited model, well trained and so the prediction is accurate.

***Keywords*:** XGBoost, forecasting, machine learning, time series analysis, goods, consumer

**Corresponding author*: mkavi277@gmail.com

A. Chitra, V. Indragandhi and W. Razia Sultana (eds.) Intelligent and Soft Computing Systems for Green Energy, (127–140) © 2023 Scrivener Publishing LLC

9.1 Introduction

Demand Forecasting in retail [2, 4] is the demonstration of using information and bits of knowledge to anticipate the amount of a particular item or service clients will need to buy during a characterized time-frame. This strategy for predictive analytics assists retailers with seeing how much stock to have close by at a given time. While clarify ing a request measure is significant, the appropriate response ranges across a few zones of a retail business. One Retail Frameworks Exploration report found that almost 75% of "winning" retailers rate request determining innovations as "significant" to their business and their success. Excess inventory can also lead to increased storage, labor, insurance costs and quality reduction, etc. Beyond simply having enough product to meet demand, forecasting can also be used to inform staffing decisions. This can reduce unnecessary investment in inventory. By making sure what goods and services are necessary to keep in stock can balance supply chain mechanism and can preclude being out of stock.

Time Series analysis is a statistical technique that deals with time series data. Time series data is the data which shows the series of particular time periods. Exponential Smoothing in Time series analysis is a method that predicts the trend of anything in the future based on past and present trend of values. It involves averaging of data such that the nonsystematic components of each individual case or observation cancel each other out. It is used for short-term prediction. Alpha, Gamma, Phi, Delta are the parameters which estimate the effect of the time series data. Alpha is used on nonseasonality data. Gamma is used when data is a series which shows a trend. Delta is used on seasonality data. A model is applied according to the pattern of our data. Initially, reducing the amount of capital that has been tied up in unneeded inventory is accomplished and therefore less stock will be in hand, then holding costs will be also be reduced. This pre-planning will be beneficial by avoiding out-of-stock, since the prediction gives the amount of goods and services of different products that need to be available at a particular period of time in a year. Using demand forecasting for retail market can help in maintaining a lean and agile business where investing money in in-demand services make the business more profitable. When forecasted demand techniques are used, then the supplier can easily check in prior to see if the business is on target to hit the predicted sales or not. If it looks like the sales are underestimated or not reaching the required point, then the supplier could cross-promote a related product via advertising.

In the past 10 years, a lot of research work has been published demonstrating the capabilities and shortcomings of different demand forecasting techniques [5, 8]. There are conditions where the pattern in the data becomes very complex and demand depends upon a number of factors. The above-mentioned review shows that forecasting has become one of the most influencial techniques practiced by many global companies. Based on the different review papers, there are many machine learning algorithms [11] and statistical methods available to do forecasting, like Time Series analysis, Predictive analysis, Regression analysis, Random Forest, etc.

9.2 Related Work

Business Intelligence integrated with Machine Learning [7] is proposed to predict the demand forecasting which produced accuracy of 92.83%. 'Time Series forecasting models' says 'Automatically defining parameters' because giving any random parameter values can lead to biased output; therefore if the model after fitting can itself tune the parameter values then accuracy will be good. Cost reducing [13] and customer service can be improved by cost benefit analysis by proper planning and scheduling because customer satisfaction is the main concern of any business to run successfully for a long time. Gaussian Naïve Bayes [6] produces the best accuracy in estimating a demand for a specific product. 'Regression Analysis' explains about 'Item Classification and Prediction'. Initially identifying the various predictions of different classified items and services make the market balance equally where this can be done using forecasting. Extreme Learning Machine [9] was experimented on real-time sales data for an e-commerce company in China which gave good accuracy with high speed. The Gray extreme learning machine [1] was proposed to examine the sales data in retail industry which performs well since it is based on back propagation neural network. Forecasting techniques [12] are adapted in many sectors and have been practiced for many years, such as in weather prediction. Time series and Regression Analysis [10] explains the forecasting that can be used for various predictive analysis techniques and methods where it is merely dependent on the type of data the model needs to deal with. Bayesian New Equipment [9] explains the 'Demand (Conventional Technique) Approach'. Bayesian experimental design provides a general probability where all the other conventional methods are derived. 'Regression Analysis' with reliability z-regression model is the most popular and widely used technique where the model gets input of various features [14] or attributes and it predicts the desired output. Time series forecasting Estimation using algebraic methods in time series' indicates that the model that should be used is any regression model where the data is linearly dependent where output variable is continuous [15].

9.3 Methodology

Demand forecasting for the retail market is a method or estimation of the demand and requirements of any goods and services in coming days. This method is effectively practiced by many organizations like Walmart, Reliance and many other marketing companies. There are many models built based on various machine learning algorithms, where the models vary based on input dataset. The dataset used to build, train and test the model is 'Rossman Stores' and 'Rossman Sales,' which are two separate datasets. Model fitting, training and testing is done by Machine Learning algorithm called 'XGBoost,' which is a very sophisticated algorithm used to build and predict the sales of data. In order to increase performance and speed, XGBoost algorithm is developed which is the implementation of gradient boost decision trees.

9.3.1 XGBoost Algorithm

Step 1: Initially distinct leaf tree is created

Step 2: Average target variable is computed for prediction and residuals are calculated using loss function.

Step 3: Computer Similarity Score = $\text{Gardient}^2/H + \lambda$, H is the number of residuals, Gardient^2 is the squared sum of residuals, λ is regularization hyperparameter.

Step 4: Suitable node is selected based on the similarity score, the higher the score the more homogeneity.

Step 5: Information gain is calculated using similarity score, which is the difference between the new similarity and old similarity.

New Residuals = Old Residuals + ρ Σ Predicted Residuals,

Information Gain = Left Similarity + Right Similarity – Similarity for Root

Step 6: Creation of Tree.

Step 7: Predicting Residual Values

Step 8: Using the above formulas, new set of residual values are calculated, where ρ is the learning rate.

Step 9: The process is repeated for all the trees by going back to step 1.

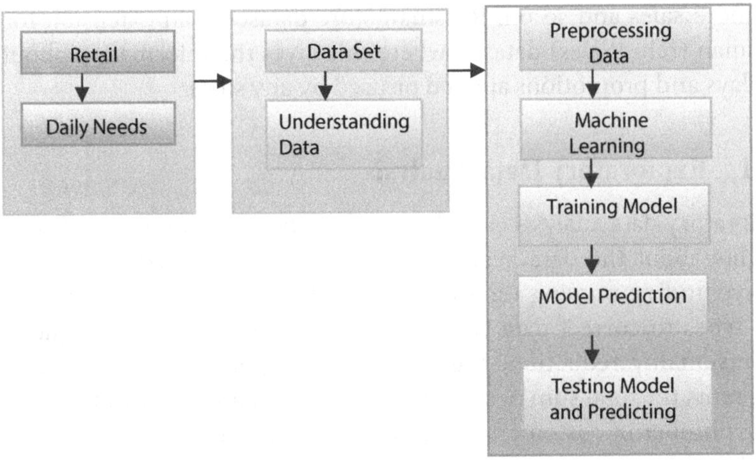

Figure 9.1 Architecture.

9.3.2 Architecture

Figure 9.1 shows the general architecture which shows all the steps involved like preprocessing the data that includes many operations like finding missing values, dropping them from dataset, preparing new data from cleaning old data and applying Machine Learning algorithm for dataset and predicting the output using the trained model.

9.4 Experimental Results

The main modules that are involved in any machine learning are collection of data, importing necessary packages, conducting various exploratory data analysis, applying Machine Learning techniques and finally getting output. The two data sets which are being used for training the model are Rossman data and Rossman Store data. Rossman train dataset has columns like Store Id, Day, Date, Total sales on that day, Total number of customers visited on that day, whether a particular shop is opened or closed. Rossman store dataset is an extension for the Rossman train (Sales) dataset, where this gives the information about the holidays and promotions applied or used by any store. These are the two data sets which are being used for training the model. Rossman train dataset has columns like Store Id, Day, Date, Total sales on that day, Total number of customers visited on that day, whether a particular shop is opened or closed, etc. These can help in understanding how many stores are present in the dataset, which store has more sales and so on. Rossman store dataset is an extension for the Rossman train (Sales) dataset, where this gives the information about the holidays and promotions applied or used by any store.

9.4.1 Exploratory Data Analysis

Exploratory data analysis is exploring the dataset and understanding a few insights about the data available. Operations performed in this section involve finding missing values, removing them, converting all the wrong data types to correct data types, plotting the data and asking questions, dimensionality reduction, feature engineering and so on. Sales column represents the total sum of sales on each particular day; customers column is the number of customers visited on that day, and Open column tells us whether that store is opened (1) or closed (0).

9.4.1.1 Empirical Cumulative Distribution Function (ECDF)

ECDF is a distribution function which takes multiple data samples from the training dataset with 1/n distribution. From Figure 9.2, it is understood that the first plot is ECDF of amount of 'sales', the second plot is ECDF of number of 'customers' and the third plot is ECDF of number of 'Sales per Customer'. The main takeaway from the above graph is that almost 20% of data has 'zero' sales per customer (3rd plot), and almost 80% of time daily amount of sales was less than 1000.

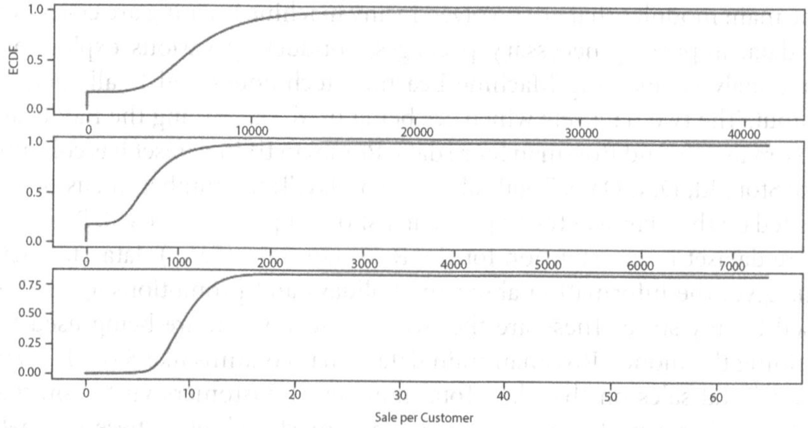

Figure 9.2 Empirical Cumulative Distribution Function.

9.4.1.2 *Exploring the Dataset and Making Visualizations between Months and Sales*

Figure 9.3 shows the plot between month vs. sales with constraint of 'Promo', i.e., how are sales amongst each store (a, b, c, d) in each month with and without any promotion. Interestingly, when plotted to know which store has more sales, store type 'B' has more sales. Almost every store had increased number of sales when they applied promo, i.e., (promo=1) and except store 'B', the other three store types had very poor and similar amount of sales. Another interesting insight found is, every store has increased number of sales at year end (October, November and December).

Figure 9.4 shows the plot between each month and amount of sales per customer; the store type 'B' has least amongst other stores. This means that, store type 'B' has the most number of customers visited and the most turnover is also done by store 'B' only (as shown in above 2, 3 graphs), but why are the sales per customer low in store type 'B'? It is because the amount of money spent by most of the individual customers in this store 'B' is very low, i.e., maybe the customers who visit this store buy small things which cost much less. Therefore, when added up with total number of sales, store 'B' has high sales due to a high number of customers. Similarly, store type 'D' has the highest number of sales per customer; this is because, although store type 'D' does not have the highest turnover (Revenue), the amount of money spent by each individual customer is high. This is maybe because the goods each customer bought were of high cost and so valuable.

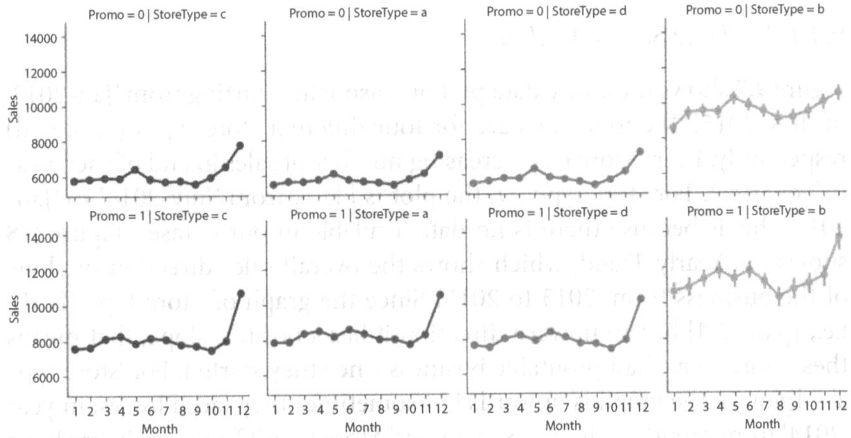

Figure 9.3 Relation between each month vs. number of sales in each month.

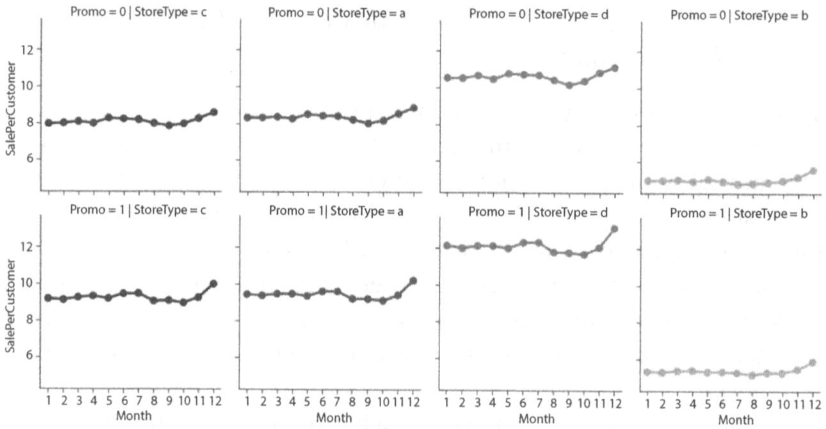

Figure 9.4 Relation between each month vs. 'Sales per customer'.

9.4.1.3 Correlation between each Feature or Attribute

Figure 9.5 shows the relation between each column with each othercolumn in the dataset whether they are positively related, neutral or negatively related. Correlation between the different columns present in the dataset can help us in feature engineering the columns and knowing what features or attributes most contributed to shape the output or result. Here, the relation between customers and sales has a dark blue color, which means they both are positively correlated to each other. Figure 9.6 shows the contribution of each feature or attribute in our dataset to the output that is predicted.

9.4.1.4 Time Series Analysis

Figure 9.7 shows the entire data plot of seasonality starting from 'Jan-2013' to 'July-2015'. The four plots are for four different store types (a, b, c, d) respectively. Every store has increasing number of sales in end of each year (4th quarter). For store type 'D', the plot is blank from 'July-2014' to 'Jan-2015', this is because there is no data available in our dataset. Figure 9.8 shows the Yearly Trend, which shows the overall sales direction or slope of the business from '2013 to 2015'. Since the graph of 'store type (b, d)' i.e., [plots 2,4] has an upward direction it has a positive slope; that means these stores have had profitable business since they started. For 'store type a' – [plot 1], the graph is upwards but sometimes they faced losses, in year '2014' from month of 'July to September'. Store type 'C' in overall, has been at a loss, because since they started in '2013', the graph went downwards

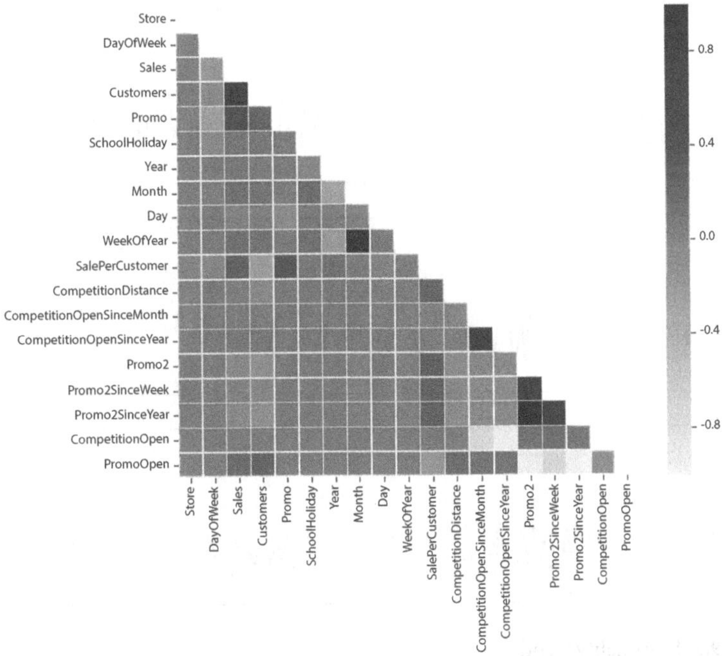

Figure 9.5 Correlation between each feature.

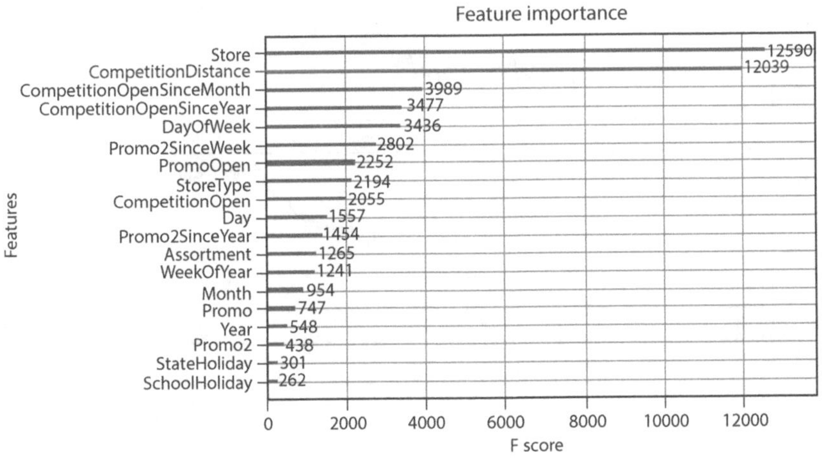

Figure 9.6 Contribution of each feature.

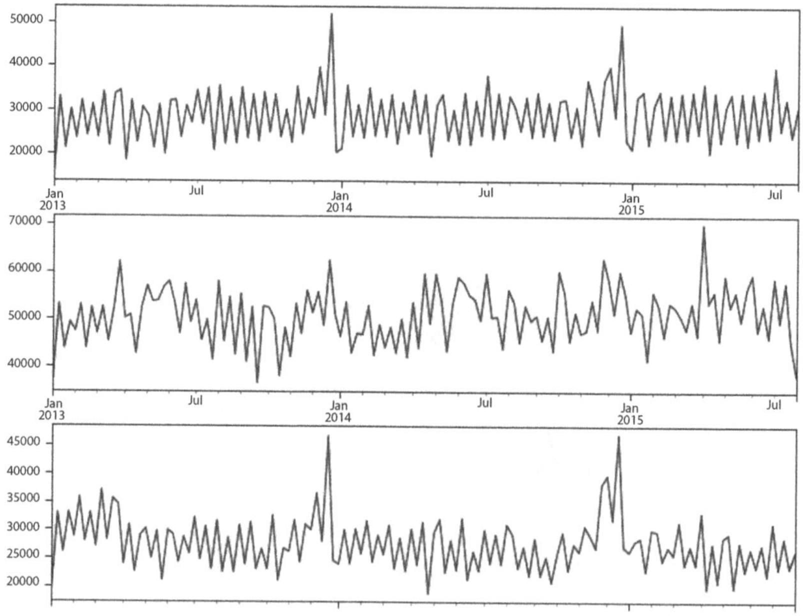

Figure 9.7 Seasonality trend.

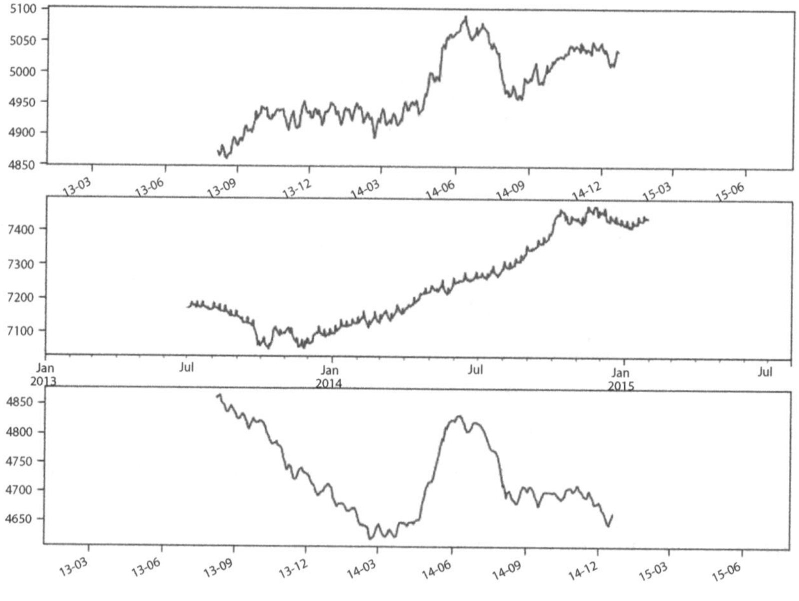

Figure 9.8 Yearly trend.

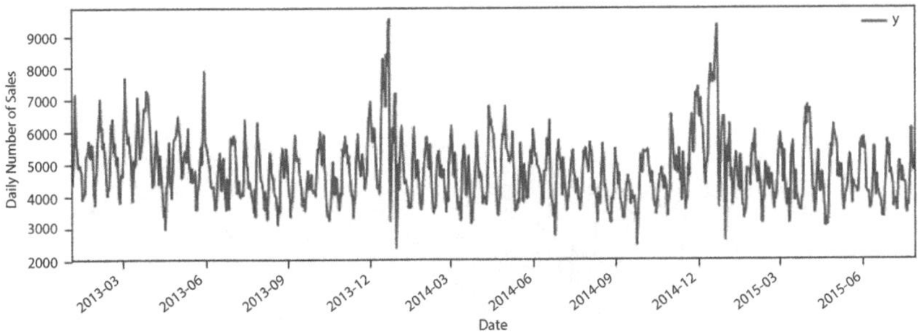

Figure 9.9 Overall sales.

only, even though they got up between in '2014'. Figure 9.9 shows the overall sales direction from 2013.

9.4.2 Model Prediction

Prophet model is used for Time series datasets which contains a huge number of entries of data and if the dataset is too large. Input data is Sales data set and Store data set where they both have columns and are related based on unique store id. Figure 9.10 shows that the sales of testing dataset is being increased because, since the sales in the 4th quarter of every year is being increased, therefore since the test data is started in the month of September, which is in the 4th quarter, the sales will increase.

As the overfitting is high, XGBoost model is used. To reduce the overfitting and underfitting for avoiding biased output for our data, we use grid search for getting the parameter values to the XGBoost model and then

	Id	**Sales**
Date		
2015-09-17	1	4428.240723
2015-09-17	2	4577.308105
2015-09-17	3	4895.477061
2015-09-17	4	5571.973145
2015-09-17	5	4961.958496

Figure 9.10 Sales prediction.

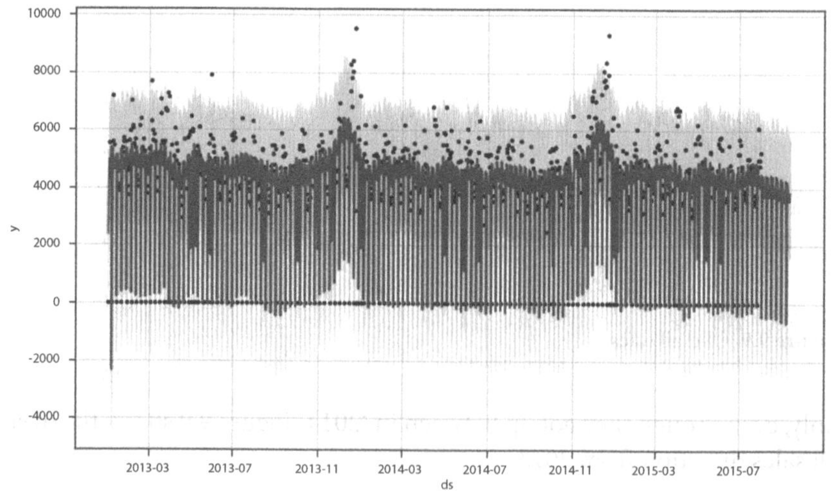

Figure 9.11 Prediction.

give these integrated values to functional testing to get the error value. Assigning the integrated values provided by grid search gives us the best possible model with reduced bias; therefore this gives us the least possible error value. Figure 9.11 shows the output of prediction. The black dots are the observed values by the model and the dark blue line is the trend which is fitted by the model accurately without having any overfitting and underfitting and the shaded blue region is the outlying values which can be varied. XGBoost machine learning algorithm is used for testing the model which is built using Prophet model. This machine learning algorithm is a sophisticated algorithm which will support both classification and regression problems, where you need to give the right parameters for getting good accuracy. The accuracy of approximately (0.1) which is nearly "92% of accuracy" achieved.

9.5 Conclusion

Demand forecasting enhances the operational productivity and reduces losses and wastage. High forecast accuracy helps in formulating established market strategy, stock turnover increase, decrease in supply chain cost, and an increase in customer satisfaction. Here, in the proposed system, Rossman Store sales are increased at the start of every week and then slowly decrease by the end of every week. Also, when we fitted the training data to "Prophet" model, sales decrease, as the graph has the "NEGATIVE

SLOPE" tending downwards. It is also found that, every year at the time of "CHRISTMAS" in the month of "November – December" the sales are increasing and are very high. Overall, the sales of the "Rossman Store" have decreased from "2013" to "2015" and then since the testing data started from '2015-September', the sales initially went up in the months of November and December. Finally, the accuracy we got is "0.14" after reducing overfitting and underfitting. Using "RMSPE" (Root Mean Square Percentage Error Value), the accuracy is calculated in the range "0.1 to 0.9". Value closer to "0.1" means the model is best suited and well trained and so the prediction is accurate.

References

1. Chen F. L. and Ou T. Y., "Sales forecasting system based on grayextreme learning machine with taguchi method in retail industry," *Expert Syst. Appl.*, vol. 38, no. 3 (2011), pp. 1336–1345.
2. Cheriyan S, Ibrahim S, Mohanan S, Treesa S. Intelligent Sales Prediction Using Machine Learning Techniques. In *2018 International Conference on Computing, Electronics & Communications Engineering (iCCECE)* (2018), pp. 53–58.
3. Katkar V., S. P. Gangopadhyay, S. Rathod, and A. Shetty, "Sales forecasting using data warehouse and Naive Bayesian classifier," in *Proc. Int. Conf. Pervas. Comput. (ICPC)*, (2015), pp. 1–6.
4. Krishna A, Akhilesh V, Aich A, Hegde C. Sales-forecasting of retail stores using machine learning techniques. In *2018 3rd International Conference on Computational Systems and Information Technology for Sustainable Solutions (CSITSS)* 2018, pp. 160–166.
5. Lakshmanan, Balakrishnan, Palaniappan Senthil Nayagam Vivek Raja, and Viswanathan Kalathiappan. "Sales Demand Forecasting Using LSTM Network." *Artificial Intelligence and Evolutionary Computations in Engineering Systems*. Springer, Singapore (2020), pp. 125–132.
6. Md. Ariful Islam Arif1, Saiful Islam Sany2, Faiza Islam Nahin3 and AKM, Shahariar Azad Rabby, Comparison Study: Product Demand Forecasting with Machine Learning for Shop, *Proceedings of the SMART-2019, IEEE Conference, 8th International Conference on System Modeling & Advancement in Research Trends* (2019), pp. 171–176.
7. Muhammad Adnan Khan, Shaziasaqib, Tahiralyas, Aneesurrehman1, Yousafsaeed, Asimzeb, Mahdizareei, and Ehab Mahmoud Mohamed, Effective Demand Forecasting Model Using Business Intelligence Empowered with Machine Learning, *IEEE Access*, (2020), pp. 116013–116023.
8. Pavlyshenko, Bohdan M. "Machine-learning models for sales time series forecasting." *Data* 4.1 (2019): p. 15.

9. Samaneh Beheshti-Kashia, Hamid Reza Karimi, Klaus-Dieter Thoben, Michael Lütjen & Michael Teucke, A survey on retail sales forecasting and prediction in fashion markets, *Systems Science & Control Engineering: An Open Access Journal* (2015), pp. 154–161.
10. Tian Y., Liu Y., Xu D., Yao T., Zhang M., and Ma S., "Incorporating seasonal time series analysis with search behavior information in sales forecasting," in *Proc. 21st Int. Conf. Companion World Wide Web* (2012), pp. 615–616.
11. Ma, Shaohui, Robert Fildes, and Tao Huang. "Demand forecasting with high dimensional data: The case of SKU retail sales forecasting with intra-and inter-category promotional information." *European Journal of Operational Research* 249.1 (2016): pp. 245–257.
12. Abolghasemi, Mahdi, *et al.* "Demand forecasting in the presence of systematic events: Cases in capturing sales promotions." *International Journal of Production Economics* 230 (2020): p.107892.
13. Ferreira, Kris Johnson, Bin Hong Alex Lee, and David Simchi-Levi. "Analytics for an online retailer: Demand forecasting and price optimization." *Manufacturing & Service Operations Management* 18.1 (2016): pp. 69–88.
14. Barbosa, N. de P., E. da S. Christo, and K. A. Costa. "Demand forecasting for production planning in a food company." *ARPN Journal of Engineering and Applied Sciences* 10.16 (2015): pp. 7137–7141.
15. Ren, Shuyun, Hau-Ling Chan, and Tana Siqin. "Demand forecasting in retail operations for fashionable products: methods, practices, and real case study." *Annals of Operations Research* 291.1 (2020): pp. 761–777.

10

Region-Based Convolutional Neural Networks for Selective Search

R. Kavitha[1]*, Srinivasan R[1], P. Subha[2] and M. Kavitha[1]

[1]Vel Tech Rangarajan Dr. Sagunthala R & D Institute of Science and Technology, Chennai, Tamil Nadu, India
[2]Sri Sai Ram Institute of Technology, Chennai, India

Abstract
In recent years, image stitching and selective search using neural networks has had an increasingly significant role in various fields, including moving pictures, astronomy and healthcare. Image recognition through selective search consists of complex algorithms, and several cumbersome calculations produce "scans" which then merge together to form a real-life representation of the required area. This paper introduces a low-cost modeling method with user-friendly application that involves the concept of Image Stitching. It also discusses graphics rendering software with simulation of user movement in the scenario created on the computer. This study investigates a type of Harris picture stitching technique that is based on the OpenCV setup environment, in light of the immense scene and high-resolution image stitching challenges. To begin, the feature points are extracted using Harris corner detection. The feature points are then rough-matched using Normalized Cross Correlation, then the algorithm RANSAC is employed to eliminate incorrect matching. Second, to implement the image registration process, a cylindrical projection transformation model is used. Finally, to fuse photos, this study employs an enhanced weighting average fusion technique, which reduces image fusion's computational complexity while also eliminating seams in stitched images.

Keywords: Open3D, OpenCV, RANSAC, Panorama

**Corresponding author*: rkavitha@veltech.edu.in

A. Chitra, V. Indragandhi and W. Razia Sultana (eds.) Intelligent and Soft Computing Systems for Green Energy, (141–150) © 2023 Scrivener Publishing LLC

10.1 Introduction

Modeling is a term used in computer graphics to describe the process of creating a mathematical representation of any surface object using specialised software. It has various uses in different industries like film and architecture and has a wide range of applications. The first real-life similar models were created in the 1960s. Back then, only those professionals in the field of computer engineering and automation who worked with mathematical models and data analysis were involved in 2D panoramic image creation and also with the limitation of resources and technology. The presence of image modeling using stitching in various areas increased greatly with over time. The first step involves the capturing of multiple photos of the desired area using a stereo-vision camera installed in UAVs. These UAVs are programmed to capture all photos identically and in a sequential manner. The end result of it is the set of images which encompasses the entire field of view. The whole process introduces the involvement of the machine learning algorithm which is modified in a way to keep the entire work precise and ahead of schedule. This approach decreases image overlapping during image stitching and improves image registration and mosaic speed. It accomplishes the goal of paying attention to both precision and efficiency [3]. Open3D is an open-source library that helps developers quickly create software that works with 3D data. In both C++ and Python, the Open3D frontend exposes a collection of well-chosen data structures and algorithms. The backend has been extensively optimised and is parallelizable. The open-source community is encouraged to participate. Panorama image stitching is the process of integrating two or more photos of the same scene to create a single high-resolution panoramic image. Image stitching is still a difficult challenge for single and panoramic photos, according to the literature. Many algorithms have been developed in recent years to solve the challenge of panorama image stitching. Image registration, image warping, colour correction, image labelling, and image mixing are the four primary components of the panorama image stitching process. This document provides an overview of the panorama image stitching technique. This paper will go through the basic components of panorama image stitching. On the basis of these ideas, a framework for a comprehensive panorama image stitching system will be shown. Finally, we'll go over some of the current issues with panorama image stitching. RANSAC is a resampling technique that creates candidate solutions by estimating the underlying model parameters using the smallest number of observations (data points) possible. Unlike traditional sampling techniques, which use as much data as feasible to find an initial answer and then remove outliers, RANSAC uses the smallest set possible and then expands it with consistent data points, as Fischler and Bolles point out [13].

10.2 Literature Review

In the paper [1], the image stitching approach is examined. The three primary processes of image stitching are explained here: registration, calibration, and image blending. Image stitching techniques such as direct and feature-based are also discussed. Additionally, an image stitching model is described, which includes image capture, feature identification and matching, image matching, global alignment, blending and composition, and global alignment. Image registration is a feature that matches photos in a group and also searches for image alignments using the direct image alignment method, which minimises the sum of absolute differences between overlapping pixels. Image calibration is a technique for reducing the disparities between an ideal lens model and the lens model utilised on the camera. Through calibration, various optical flaws like distortion and exposure variations were also addressed. Image blending combines the changes found during the calibration stage with the remapping of the pictures to a projected output. Image registration is a feature that matches photos in a group and also searches for image alignments using the direct image alignment method, which minimises the sum of absolute differences between overlapping pixels [14].

In the paper [2], in a variety of subjects and educational contexts, the emergence of additive manufacturing and 3D printing technology is introducing industrial skills shortfalls and opportunities for innovative teaching techniques. As a result, research into these behaviours is appearing across a wide range of education fields, although often without reference to other disciplines' research. To address this issue, this article compiles these disparate sets of research into a comprehensive analysis of where and how 3D printing is being used in the education system. Although evidence of 3D printing-based teaching techniques can be discovered in each of these six areas, adoption is still in its early stages, and recommendations for future research and education policy are given.

Paper [3] offers the visual representation of a 3D printing evaluation, analysis, and classification. 124 papers with a high degree of relevance published between 2014 and 2018 were selected using the CAPES Sucupira platform. These publications were categorised into nine categories: research kinds, affiliations, approaches, study origins, geographic scope, unit of analysis, scope, benefits, and drawbacks. The findings revealed that the number of articles about 3D printing is increasing year after year, indicating its importance and popularity. The majority of scientific research is undertaken and led by individuals affiliated with institutions in Europe, Asia, and the United States.

In the paper [4], powder bed fusion is the most widely utilised 3D printing process in the medical profession. Powder bed fusion is widely utilised in medical equipment because it works with a wide range of materials, including titanium and nylon. It is the most common plastic substance and is a synthetic thermoplastic linear polyamide. Because of its flexibility, durability, minimal friction, and corrosion resistance, it is a well-known 3D printing filament. 3D printing was referred to as "rapid prototyping" in the publication [5]. In 1984, 3D Systems Corporation's Chuck Hull invented the first functioning 3D printer. Dr. Deckard at the University of Texas at Austin created Selective Laser Sintering (SLS) technology later in the 1980s. The technology was developed further in the 1990s with the discovery of a method for solidifying photopolymer, a viscous liquid material, using ultraviolet light. 3D printers were incredibly expensive in the late twentieth century, and they could only produce a limited number of objects. The bulk of the printers were used for research and presentation by scientists and electronics enthusiasts. The printing technology was a blend of modelling both science and building technology, using some of the most cutting-edge technological breakthroughs of the time, even though it was still in its early stages of development.

In the paper [6], rapid prototyping was the term for 3D printing. In 1984, 3D Systems Corporation's Chuck Hull invented the first functioning 3D printer. Dr. Deckard developed Selective Laser Sintering (SLS) technology at the University of Texas at Austin during a project funded by the Defense Advanced Research Projects Agency (DARPA) later in the 1980s. The technology was developed further in the 1990s with the discovery of a method for solidifying photopolymer, a viscous liquid material, using ultraviolet light. According to the paper [7], each rapid prototyping (RP) process has its own set of benefits and drawbacks.

In the paper [8], Chandwani Sharma and Kumar talk about 3D Printing Technology. They discuss the uses and the type of materials that can be used for this technology. The findings were very positive and very informative in regard to the technology and its uses in the advancement of science. According to the article, a lot of different methods can be used in 3D printing. It is also known as additive manufacturing process. This technology can be applied in various fields and industries. Importantly, using this technology we can produce products that can meet human need.

10.3 Existing Method

The present system consists of the X-ray medical image stitching and another one is 3D indoor mapping. Traditional screen-film X-ray images have a limited field of view, making it impossible to show the entire bone structure of large hands on a single frame. Image stitching can be used to combine digitised images from X-ray films to create images that show the entire hand structure. A new medical image stitching method identifies and merges pairs of hands X-ray medical images using minimum average correlation energy filters. The improved algorithm does not detect the SIFT feature points for the entire reference image area but it also utilizes the statistical registration data of every single adjacent picture to process the affine matrix. Consequently, the improved strategy lessens the SIFT feature detection area of the reference picture, and furthermore builds the quantity of matching feature points and improves the proficiency of stitching time.

A mobile mapping system for constructing a geometrically exact 3D model of an unfamiliar indoor space is presented in this research. From small office halls to long, twisting underground mine tunnels, the same general design concept can be used in a number of contexts. Surfaces and features can be properly mapped using images taken by a unique configuration of multiple types of optical imaging sensors and a dead reckoning positioning system. This option ensures that the data obtained contains all of the necessary information to create a 3D model of the area. Our virtual environment facility's system, data collecting and processing procedure, test results, modeling, and display are all described [16].

10.4 Proposed Methodology

The system given here is useful in image stitching by using selective search convolutional neural network. This process will result in 3D mapping of the preferable area. The images will be taken by UAVs in the form of PCD file and then sent to the station where we collect the whole data to process it further. Once all the pictures of the desired area have been taken then the pre-processing of the images will be done and the process will continue further. This project is based on an algorithm which will do every work from image pre-processing to image stitching i.e., we can just transform this algorithm to any system fulfilling the minimum requirements to ensure that the algorithm will run smoothly and effectively and this will do the work completely. So, there is nothing like uses of the high-end machines which are expensive and also, we can use the normal camera to collect the data which can be used to proceed further. This states that our project is totally feasible economically. A brief description of the process is given below.

Feature extraction

To discover and classify images with differentiating features such as image size, rotation, and contrast, the feature selection approach uses feature-point algorithms such as the Scale Invariant Feature Transform. Relationship between the feature and the pixels around it is determined to provide a descriptor that describes the feature data of the feature.

Feature matching

Find the feature with the reference of descriptor. The overlapping image that is most similar to the feature of the current image and feature matching next to the pixel are verified. RANSAC is used to discover the best-fitting model with the most available data within the interior value and to remove outliers from outside area. In the proposed model, a matrix that maps the point in one image with the point in other corresponding images are employed to store the relationship between matching pairs of images.

Model Estimation & Image Warping

Multiple stitching images are mapped from the current image to a coordinate point of the system such as spherical, cylindrical, planar, and annular so that multiple images lead to a single picture, and then projected and stored as a 2D plane image.

Image blending

The seam finder is used to locate the seam between the photos with minimal mistakes, and blending creates a natural relationship between neighbouring pixel values. Several photos are warped and combined into one, but wide view images have resulted in an alternation of the original image to find the differences between the images, as if they were hidden together with a post-it note.

This project uses a dynamic architecture; however, basic functionality of the components remains the same. Image collector may be a single drone or a fleet of drones or vehicle loaded with 2D or 3D Camera. At first the image is made suitable for processing. This process includes removal of any noise or reducing the size of the image to save space. File store system manage the stored images to make it readily available whenever it is needed by any component or even other external processes. Blended and finalized are stored in disk for future reference by user.

We have the image object with attributes like size, file type. Date/time taken and other attributes. Then there are three arrays for x-axis, y-axis and z-axis

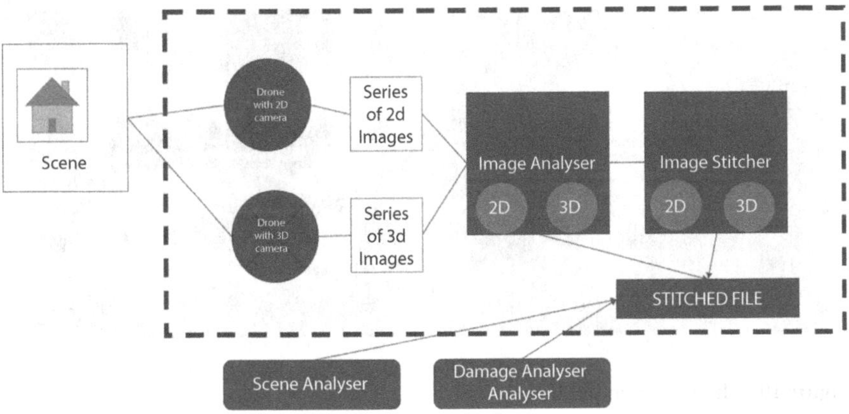

Figure 10.1 Architecture.

describing the x, y, and z axis of a given point, respectively. If the file contains a value with certain x-y-z, that means we have a data point at that certain coordinate and if the value is not present that simply means we do not have the specific point in our image. We can have an unprecedented number of use cases in various fields but we will limit us with only two but the most important use cases. We will have a scene (say a landslide affected area). The image can be captured using drones preloaded with maybe a 2D but preferably a 3D camera shown in Figure 10.1. These drones will send a series of images to our system. After determining the type of file, we will send our image to the analyzer along with file type in the parameter. The image analyzer will check the image to see if it is suitable for stitching or not. Once the overlapping is found we will forward it to the image stitcher core. This core program will be joining the two images with proper orientation and overlapping percentage.

10.5 Implementation and Results

Our input and output may be of several types. Input may range from 2D images with extension .jpeg, .png or it may be a 3D image with extension .pcd. The output of the program will in the same format as the input is. For example: an input with 2d image will give output a 2d image and a 3D image will be the output for a pair of 3D images. Input can be read from a fixed temporary location, an external program keeps changing these images and repeatedly runs the stitcher program. This ensures the real-time joining of the image. The input is always a pair of images, whether 2D or 3D shown in Figure 10.2. The input is always a pair of images, whether

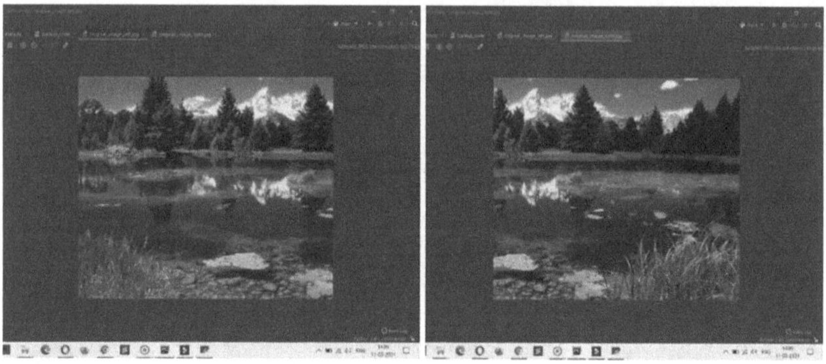

Figure 10.2 Input and output image.

2D or 3D. The output is a single file with panoramic view of the scene. These outputs are securely saved to the other location as copy. This output image is usually one input for the next trigger of the program, especially in the case of real-time stitching shown in Figure 10.3.

The existing system can be used to stitch the medical images like X-rays and in 3D mapping of the indoor environment. This can be done by using the specific devices and a particular set of modules which are very time-consuming as well as space while the proposed system is feasible in every

Figure 10.3 Input and output image.

Figure 10.4 Final image.

manner and can be used in any device fulfilling the minimum requirements. This system is very useful in making the 3D map of the desired environment and to minimize the number of casualties shown in Figure 10.4.

10.6 Conclusion

In the domains of computer vision and computer graphics, image mosaicing is a hot topic of research. It offers a lot of different algorithms for detecting and describing characteristics. The feature detector to use is determined by the problem. In this project, we focused on the real-life-based problems which include natural disasters like landslides and floods. Our study majorly focuses on these two disasters and the destruction and impact caused by them. The following data will used to map the 3D model of the area affected by these natural calamities and then used to rescue the victims and in managing the loss. The data was collected and tested using the image stitching process to generate the panoramic view of the desired area. In the future, we'd like to compare our algorithms against existing feature-based image stitching algorithms, as well as stitching images together to generate dynamic panoramas and stitching images with a lot of parallax.

References

1. Shikha Arya "A Review on Image Stitching and its Different Methods", *International Journal of Advanced Research in Computer Science and Software Engineering*, Volume 5, Issue 5, ISSN: 2277 128X, May 2015.
2. Yaqing Ding, Daniel Barath, Zuzana Kukelova "Minimal Solutions for Panoramic Stitching Given Gravity Prior". Computer Vision and Pattern Recognition.
3. Simon Ford and Tim Marshall "3D Printing in education, Version 1.0 - 17 September 2016.
4. Guilherme Ruggeri Pereira, Fernando Gasi, Sérgio Ricardo Lourenço,"Review, Analysis, and Classification of 3D Printing Literature: Types of Research and Technology Benefits", Vol. 6, issue 6, June 2019.
5. Huang, S.H., Liu, P., Makassar, A. and, and Hou, L., Additive manufacturing and its societal impact: a literature review, *International Journal of Advanced Manufacturing Technology*, Vol. 67, Nos 5/8, 2013, pp. 1191-1203.
6. Chandrashekhar Kalnad, "A Review on 3D Meeting", Vol 5, Issue 7, July 2016.
7. Mr. Akash M. Patil, Miss. Neha U. Deshpande & Prof. V. A. Patil, "Review Paper on 3Dimensional Printing Technology, ISSN: 2277-9655", Vol. 10, Issue 5, 2019.
8. Lakshya Chandwani, Himanshu Sharma, Pankaj Kumar, "3D Printing Technology, Vol. 7, Issue 2, Feb. 2020.
9. Yingen Xiong, "Fast image stitching and editing for panorama painting on mobile phones," *July Proceedings of IEEE Computer Society Conference on Computer Vision and Pattern Recognition Workshops*, 2016.
10. Bang, Seongdeok Kim, Hongjo Kim, Hyoungkwan, "UAV-based automatic generation of high-resolution panorama at a construction site with a focus on preprocessing for image stitching." *Automation in Construction*, Vol. 12, Issue 3, 2017.
11. Ming, Li, Liu Huan, Zhu Xinyan. (2012). Approach to Fast Mosaic UAV Images for Disaster Emergency, *Journal of Catastrophology* 27.3, pp. 139-144.
12. Ham, Y., Han, K.K., Lin, J.J. *et al.* (2016).Visual monitoring of civil infrastructure systems via camera-equipped Unmanned Aerial Vehicles (UAVs), a review of related works. *Vis. in Eng.* 4, 1,p doi:10.1186/s40327-015-0029-z
13. Apoorva Raghunandan, Mohana, Pakala Raghav, H. V. Ravish Aradhya. (2018). Object Detection Algorithms for Video Surveillance Applications, International Conference on Communication and Signal Processing.
14. Zhong Qu, Si-Peng Lin, Fang-Rong Ju, Ling Liu, Lining Sun. (2015). The Improved Algorithm of Fast Panorama Stitching for Image Sequence and Reducing the Distortion Errors. Hindawi Publishing Corporation, *Mathematical Problems in Engineering*, vol. 2015, p. 12.
15. R. Sujeetha, Vaibhav Mishra. (2019). Object Detection and Tracking using Tensor Flow, *International Journal of Recent Technology and Engineering (IJRTE)*, vol. 8.

11

Design and Development of Mobility System for Double Amputees

Dr. Saravanan T S[1]*, Dr. Sagayaraj R[2], Dr. Sivaraman P R[1], Sivamani D[1], Jaiganesh R[3] and Ragupathy P[1]

[1]*Department of Electrical and Electronics Engineering, Rajalakshmi Engineering, College, Chennai, India*
[2]*Department of Electrical and Electronics Engineering, Muthayammal Engineering College, Chennai, India*
[3]*Department of Electrical and Electronics Engineering, K. Ramakrishnan College of Technology, Trichy, India*

Abstract

Mobility refers to a human being's ability to move his or her body in an environment and to manipulate objects. Collectively, these activities enable the individual to pursue life activities of their choosing. An individual's ability to perform any mobility task can be compromised by impaired body functions or structures. The proposed system mainly focuses on the "transfemoral" type of above-knee amputation (AKA) in which both knees are amputated above the knee joint. Prosthetic legs are used by shaping the limbs. But not everyone is capable of using prosthetic legs as they have their own concerns. The proposed system aims to develop a mobility system for such persons with system clearance nearly six inches from ground level, at standstill, so that the amputees can use their hands to get onto the system and perform a transition without the help of the caretaker. Horizontal and vertical transitions are included in the system. The horizontal transition is realized using wheels controlled by an electric motor at constant speed. At standstill condition the amputees can elevate themselves using vertical transition. This aids them in performing necessary activities at a considerable height above ground level. Vertical movement is implemented using scissor lift technique actuated by a permanent magnet DC (PMDC) electric motor. Vertical transition height is restricted to two feet in order to ensure safety.

**Corresponding author*: saravanan.ts@rajalakshmi.edu.in

A. Chitra, V. Indragandhi and W. Razia Sultana (eds.) Intelligent and Soft Computing Systems for Green Energy, (151–162) © 2023 Scrivener Publishing LLC

***Keywords*:** Mobility, amputees, transition, PMDC, wheel hub motor, scissor lift

11.1 Introduction

Double Amputees is a type of a disability caused due to trauma, congenital limb deficiency and cancer. Double Amputees are of two types, Lower Limb Amputation (LLA) and Upper Limb Amputation (ULA). In LAA "transfemoral" is a type in which both legs are amputated above the knee joint [1]. For a transfemoral-type person, mobility is achieved by using the hands. Movement of such a person inside a home requires a caretaker to lift him from one place to another. This project aims to develop a mobility system for such persons with height less than one foot at standstill; thereby the amputees can use their hands to move themselves to the system without the help of the caretaker. Horizontal and vertical movements are included in the system. The horizontal transition is realized using wheels controlled by an electric motor at constant speed [2]. At standstill condition the amputees can lift themselves using vertical movement. This aids them in navigating the things above ground level. Vertical movement is implemented using scissor table technique controlled by an electrical motor. Vertical movement height is restricted to two feet in order to ensure safety.

11.2 Block Diagram

Mobility System for double amputees' block diagram is shown in Figure 11.1. It consists of converter, battery, motor 1 and 2, scissor lift and wheel drive. Electric power is supplied to an electric load using the power supply. The supply here is given to the battery. The battery used is lead acid battery [3] which incorporates both the horizontal and vertical transition with the help of PMDC motor. Motor 1 is employed for horizontal transition of the system whereas motor 2 is employed for vertical transition of the system. Motor 1 is coupled to the wheel through shaft linkage, which is present at the bottom of the system [4]. Motor 2 is used for scissor lift mechanism. Scissor lift is performed to exhibit the vertical transition of the system. A regulated power supply is a power supply circuit, which converts the unregulated DC supply into a constant DC. Rectifier circuit is used to convert AC supply into DC supply. Three terminal voltage regulators IC such as LM7812 are used to produce a constant DC voltage of 12 V. Thus in the proposed system the available ac supply 230V/ 50 Hz is converted into 13.2 V DC supply to charge the battery. A two 12 V, 7.5 Ah lead acid

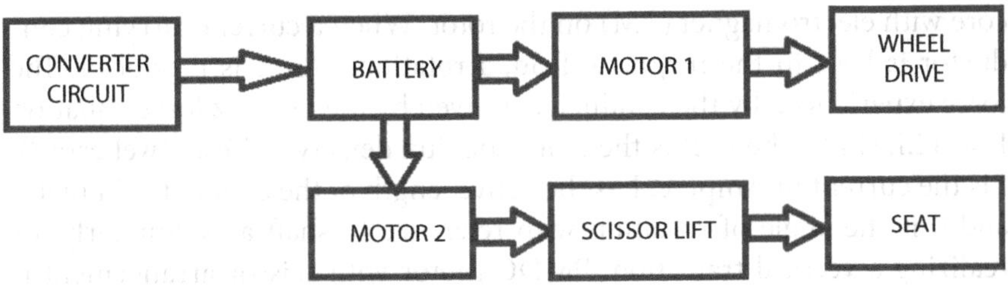

Figure 11.1 Block diagram.

battery is used to power the motor 1 for horizontal transition and motor 2 for vertical transition.

Motor 1 is employed to realize the horizontal transition. The motor is coupled to the wheels of the system to drive the system. This motor is capable of driving the system at low speeds so that the system is capable to provide a mobility system for the double amputees. The supply voltage of the motor 1 is 24V. The motor used is a brushed DC motor of 250W power, running at 1500 rpm at no load. Wheel Drive System Suitable arrangement of wheels is done so that the system is capable to perform the mobility with ease. The selection of wheels is such that it makes low clearance from the ground and provides perfect balance for the overall system [5–7]. The diameter of the wheel is 15 cm, and made of solid polyurethane.

Motor 2 is employed to realize the vertical transition. The motor used is a Permanent Magnet DC Motor, which drives the scissor lift mechanism. This transition is helpful in reaching certain heights, within the household, to perform the daily needful tasks. A 24V, 250 W brushless PMDC motor running at 1500 rpm is used. Scissor Lift Scissor lift mechanism enables a reliable vertical lift for the double amputees so that the major objective of performing activities independently is achieved. It gives a reasonable ground clearance under standstill condition of the system. A linear actuator type scissor lift made of mild steel having the dimension 45 cm X 45 cm X 8 cm. Seat A perfect square seating arrangement is advised so that a perfect balance of the system is achieved. Special care is taken for the comfort of the double amputees when they perform mobility through the system.

11.3 Working Methodology

The main aim of the system is to realize the vertical and horizontal transitions by the help of Electric Motors (especially DC motors). The Vertical transition is realized by a PMDC motor [11–13]. A PMDC motor consists of PM (Permanent Magnet) on the stator and the conventional armature core with electromagnet (EM) on the rotor. When a current carrying conductor is kept in the magnetic field, a rotational force is produced. The force experienced by the conductor is given by the Lorentz force equation $F = B.I.L \sin \theta$ where, B is the magnetic flux density in Tesla (weber/m^2), I is the current in Ampere, L is the active length of the conductor in meter and θ is the angle of ro tation with reference to shaft axis. Similarly for realizing a vertical transition, PMDC motor with scissor arrangement is used. Rotation of the motor in clockwise and anticlockwise direction aids the vertical transition of the system in an upward and downward direction.

A Scissor Lift Mechanism (SLM) is actuated by a PMDC motor. A SLM uses linked, folding supports in a criss-cross 'X' pattern. A pinion arrangement which is coupled to the shaft of the PMDC motor, the movement of the shaft drives the lift mechanism in an upward motion. By reversing the armature terminal, using special switching mechanism aids the movement in the downward motion. A linear actuator driven by the PMDC motor is to be used in the proposed system.

11.4 Design Calculation

The modeling of the proposed system is done with the help of CATIA V5 3D modeling software. This software helps in realizing the structure of the proposed system. With the help of CATIA, the exact dimensions across the whole system are realized [18]. The proposed system comprises the fixed base whose dimensions are 1100 x 600mm. The base is chosen such that it completely houses the scissor lift setup effectively. The scissor lift base (movable base) dimensions are 648mm x 530mm. The setup is capable of elevating up to 530mm. The crossbars, in realizing the elevation 593 x 75 x 12mm. The cross bars are set for both the sides to be capable of holding the load [19–21].

The mechanism used to realize the scissor lift is by the help of a threaded rod 1000mm in length and 15mm in diameter. The threaded rod is coupled to the shaft of the PMDC motor, such that when the threaded rod is rotated clockwise the scissor lift is realized by the movement of the connecting rods of the cross bars from the bottom such that the up movement is realized and vice versa [8]. Figure 11.2 shows the isometric view of the system and corresponding dimensions [22].

Figure 11.3 and Figure 11.4 give a clear picture of the system and the dimensions of the movable base and the fixed base, i.e., its length, breadth and width. It also gives the length of the cross bars, its thickness, so that a reliable system is obtained capable of meeting the load requirements. The length and diameter of the threading rod is also estimated so that proper scissor lift is realized [15–17].

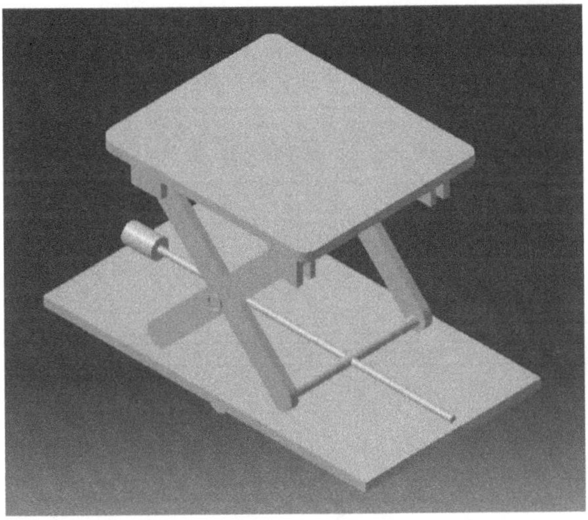

Figure 11.2 Solid isometric view.

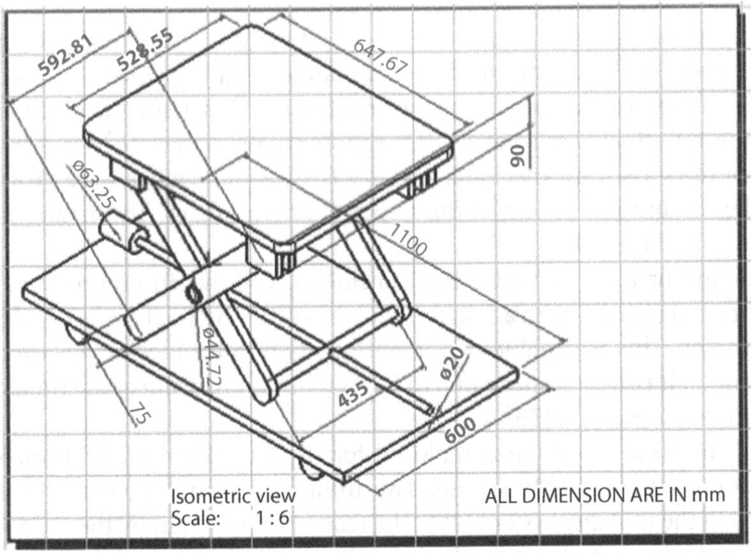

Figure 11.3 Isometric sketch.

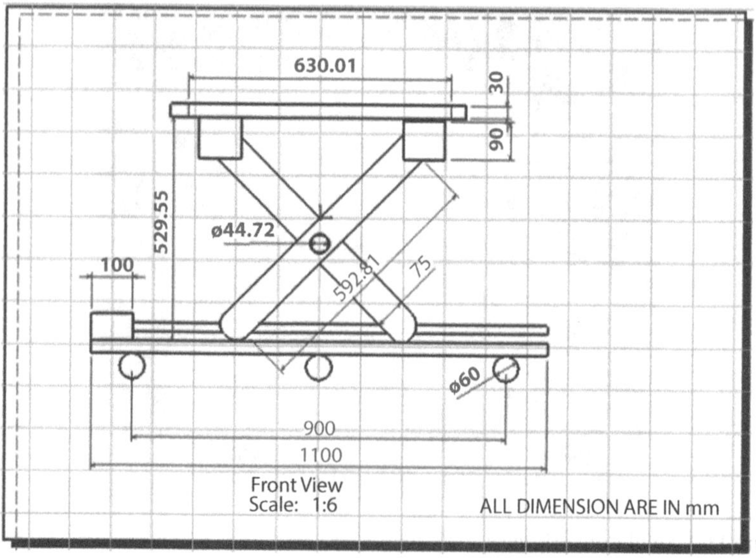

Figure 11.4 Front view Sketch.

11.5 Hardware Implementation

To implement a vertical transition in a system, the thread rod is coupled to the rotor of the dc motor in such a manner that the clockwise rotation of the motor realizes the upward transition of the scissor lift mechanism and the anti-clockwise rotation of the motor realizes the downward transition of the scissor lift mechanism [23, 24]. The guide wheels are provided for the smooth realization of the scissor lift mechanism. The arrangement is shown in Figure 11.5.

The motor used to realize the transitions is shown in Figure 11.6. The motor rating is 24V, with a full load current of 10.4A. The torque developed is 5.23Nm. This is capable of generating enough torque for meeting the objectives. It runs at a speed of 6 to 8 km/hr, thus ensuring a safe transition within the household [25].

Polyurethane wheels are employed for the realization of the horizontal transition of the mobility system. Two front wheels and one rear wheel are placed in the mobility system [26]. The rear wheel is driven by the DC motor and front wheels perform the left–right transition through sliding arrangement coupled to the wheels. A wiring diagram of the mobility system is shown in Figure 11.7. The battery is connected to the lower terminals of the DPDT switches 1 and 2. The motor loads are connected to the center terminals of the DPDT switch. The upper terminals of the switch establish the change in the

Figure 11.5 Scissor lift realization using thread rod.

Figure 11.6 DC motor used in the transition.

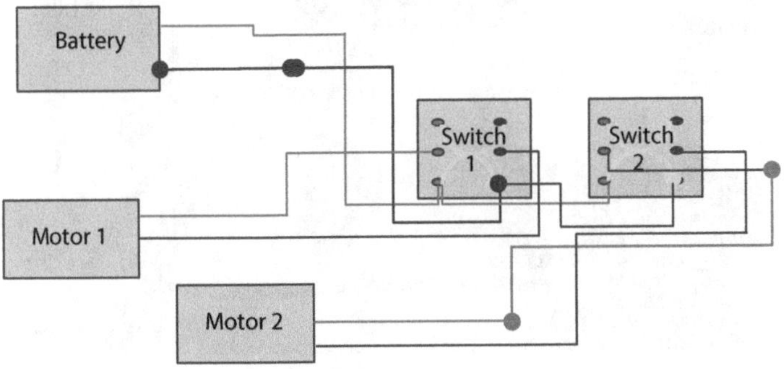

Figure 11.7 Wiring diagram of the mobility system.

polarities of the battery as the positive and the negatives are cross connected to the upper and lower terminals of the switch. Thus the control of the switch is responsible for the transition of the mobility system [27, 28].

A complete mobility system is shown in Figure 11.8. Transition of the system in the vertical movements is shown in Figures 11.9 and 11.10. Figure 11.9 shows the up-lift arrangement, by proper switching, transition system moves upwards through the threaded rod actuated by the motor rotating

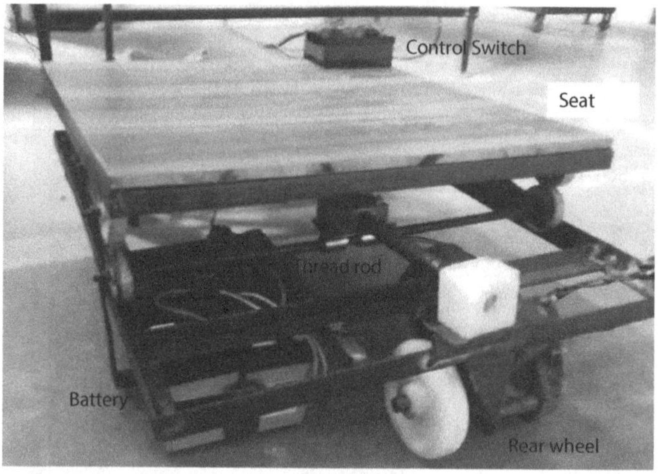

Figure 11.8 Mobility system at standstill condition.

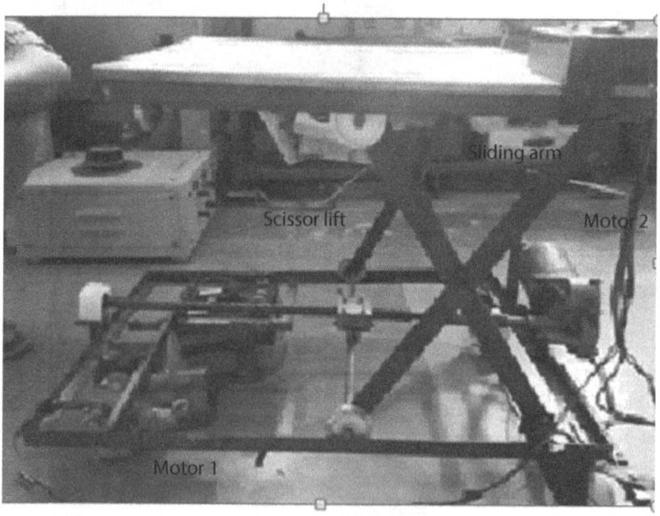

Figure 11.9 Mobility system at up-lift condition.

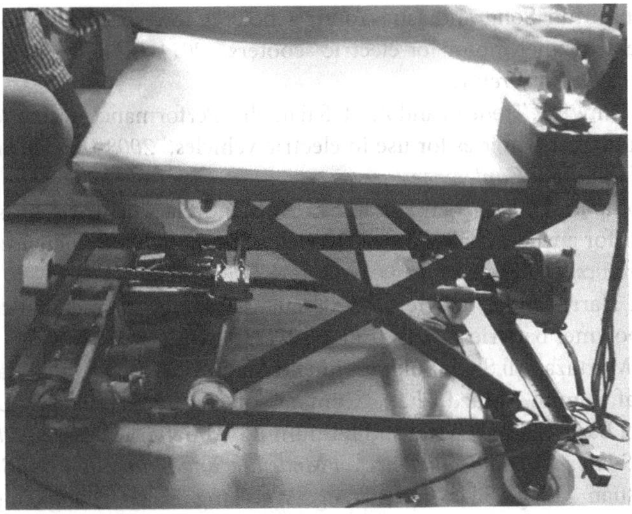

Figure 11.10 Mobility system at down-lift condition.

in the clockwise direction. Figure 11.10, shows the down-lift arrangement achieved by rotating the motor in an anticlockwise direction [29–31].

11.6 Conclusion

This system meets the basic objectives of the mobility solutions. The forward transition, reverse transition, upward and downward transition is performed by a motor with manual switch control. But in future this can be replaced by using the remote controller, the wireless concept. This not only enhances the performances of the system, but also comes with an add-on of reaching to the system. This system can be controlled remotely, without the need of the person to be on the system to execute the task. A battery management system can be implemented for efficient control of charging and discharging of the battery. To reduce the weight of the system lead-acid battery can be replaced by a Li-ion battery. Height of the system can be reduced by using wheel hub motor.

References

1. Yoshiyuki Takahaashi, "Development of a Personal Mobility Vehicle for Short-Range Transportation Support," *2016 7th International Conference on Intelligent Systems, Modelling and Simulation (ISMS).*

2. Byeong-Mun Song and Jang-Young Choi, "A low-speed high-torque permanent magnet motor for electric scooters," *2011 IEEE Vehicle Power and Propulsion Conference.*
3. B. G. Kim, F. P. Tredeau and Z. M. Salameh, "Performance evaluation of lithium polymer batteries for use in electric vehicles," *2008 IEEE Vehicle Power and Propulsion Conference.*
4. P. Gottipati, O. Dobzhanskyi and E. A. Mendrela, "In-wheel brushless DC motor for a wheel chair drive," *2010 Joint International Conference on Power Electronics, Drives and Energy Systems & Power India.*
5. Jason Harris and Dimitrie C. Popescu, "Discharge characteristics of lithium-polymer batteries," *2014 IEEE SOUTHEASTCON.*
6. Syed Murtaza, Ali Shah Bukhari, Junaid Maqsood, Mirza Qutab Baig, Suhail Ashraf, Tamim Ahmed Khan, "Comparison of Characteristics -- Lead Acid, Nickel Based, Lead Crystal and Lithium Based Batteries," *2017 17th UKSim-AMSS International Conference on Modelling and Simulation (UKSim).*
7. Yi-Hsuan Hung, Shao-Wei Chang, Wei-Ting Hsiao, Siang-Ting Huang, Lu-Jung Wang and Cheng-Ta Chung, "Mechanical designs, energy management, and performance assessment of a novel hybrid electric scooter," *2016 International Conference on Advanced Materials for Science and Engineering (ICAMSE).*
8. Md Toufiqul Islam, Cheng Yin, Shengqi Jian and Luc Rolland, "Dynamic analysis of Scissor Lift mechanism through bond graph modeling," *2014 IEEE/ASME International Conference on Advanced Intelligent Mechatronics.*
9. A. NazarAli, D. Sivamani, R. Jaiganesh and M. Pradeep, "Solar powered air conditioner using BLDC motor," *IOP Conference Series: Material and Science Engineering,* Vol 623, Oct 2019.
10. V. Venkatesh, A. NazarAli, R. Jaiganesh and V. Indragandhi, "Extraction and conversion of exhaust heat from automobile engine in to electrical energy," *IOP Conference Series: Material and Science Engineering,* Vol. 623, Oct. 2019.
11. D. Shyam, K. Premkumar, T. Thamizhselvan, A. Nazar Ali and M. Vishnu Priya, "Symmetrically Modified Laddered H-Bridge Multilevel Inverter with Reduced Configurational Parameters," *International Journal of Engineering and Advanced Technology,* Vol. 9, issue 1, Oct. 2019.
12. P. R. Sivaraman and C. Kamalakannan, "Reduction of switching losses for a DC-DC converter using RCD clamp circuit," *Middle East Journal of Scientific Research,* Volume 6, No. 10, 2015, pp. 660-663.
13. A. Nazar Ali, R. Jayabharath, and M. D. Udayakumar, "An ANFIS Based Advanced MPPT Control of a Wind-Solar Hybrid Power Generation System," *International Review on Modelling and Simulations (IREMOS)* 7.4, 2014, pp. 638-643.
14. A. Nazar ali, R. Jayabharath and R. Shanthi Priyadharshini, "A Single phase high efficient transformer less inverter for PV Grid connected power system using ISPWM technique," *International Journal of Applied Engineering Research (IJAER),* Vol. 10, No. 9, 2015, pp. 7489–7496.

15. P. R. Sivaraman, R. Jaiganesh, P. Ragupathy, R. Ramkumar, "HI-TECH Electoral Machine for Election Commission of India," *Biosc. Biotech. Res. Comm. Special Issue*, Vol. 13, No. 3, 2020, pp. 13–17.
16. A. Nazar Ali and R. Jeyabharath, "Ride through Strategy for a Three-Level Dual Z-Source Inverter Using TRIAC," *Circuits and Systems*, 7.11, 2016, pp. 3911.
17. Shyam D, Premkumar K, Thamizhselvan T, Nazar Ali A, Vishnu Priya M "Symmetrically Modified Laddered H-Bridge Multilevel Inverter with Reduced Configurational Parameters," *International Journal of Engineering and Advanced Technology*, Vol. 9, issue 1, Oct 2019.
18. P. R. Sivaraman and S. Chinthana, "Design and Implementation of Three-stage Interleaved Boost Converter for Single-Phase Induction Motor," *International Journal of Emerging Technology in Computer Science & Electronics*, Vol. 7, No.1, 2014, pp. 261–267.
19. A. Nazar Ali and R. Jayabharath. "Performance Enhancement of Hybrid Wind/Photo Voltaic System Using Z Source Inverter with Cuk-sepic Fused Converter." *Research Journal of Applied Sciences, Engineering and Technology*, 7.19, pp. 3964–3970, 2014.
20. A. Nazar Ali, N. Gowri, and P. G. Scholor. "Power factor correction based bridgeless single switch SEPIC converter fed BLDC motor." *Advances in Natural and Applied Sciences* 10.3 pp. 190–198, 2016.
21. A.T. SankaraSubramanian, P. Sabarish and A.Nazar Ali, "A Power factor correction based canonical switching cell converter for VSI fed BLDC motor by using voltage follower technique," *IEEE xplore Digital Library*, pp. 1–8, 2017.
22. R. Jai Ganesh, A. Nazar Ali, S. Kodeeswaran, B. Karthikeyan, L. Nagarajan, "Iot based Water Management System for Highly Populated Residential Buildings," *International Journal Of Disaster Recovery and Business Continuity*, Vol. 11, issue 01, pp. 4018–22, 2020.
23. A. Nazar Ali, R. Jai Ganesh, D. Sivamani, D. Shyam, "Solar powered highly efficient Seven-level inverter with switched Capacitors," *IOP Conference Series: Material and Science Engineering*, Vol 906. Aug 2020, 17578981,1757899X.
24. Vijayalakshmi, S., Sivaraman, P.R., Karthick, R., Nazar Ali, A. "Implementation of a new Bi-Directional Switch multilevel Inverter for the reduction of harmonics," *IOP Conference Series: Materials Science and Engineering*, 2020, 937(1), 012026.
25. Sankara Subramanian, A.T., Priyadharsini, S., Palaniyappan, S., Nazar Ali, A. "A novel IOT based domestic automation system for load monitoring and efficient control," *IOP Conference Series: Materials Science and Engineering*, 2020, 937(1), 012028.
26. Anton Amala Praveen, A., Muthu Kumaran, M. Nazar Ali, A., "Minimization of Power Factor Penalty Charges for Non-Linear Domestic Loads with IOT Technology," *IOP Conference Series: Materials Science and Engineering*, 2020, 937(1), 012011.

27. D. Sivamani, R. Ramkumar, A. Nazarali, D. Shyam, "Design and implementation of highly efficient UPS charging system with single stage converter," *Materials Today :Proceedings*, 10 October 2020.
28. S.R. Paveethra, C. Kalavalli, S. Vijayalakshmi, Dr. A. Nazar Ali, D.Shyam (2020), Evaluation of Voltage Stability of Transmission Line with Contingency Analysis, *International Journal of Scientific & Technology Research*, Vol. 9, issue 02, pp. 4018–22.

12

A Review: Precision Vehicle Control Using Internet of Things

R. Srinivasan*, Kavitha R, Kavitha M and Sridhar K

Department of Computer Science and Engineering, Vel Tech Rangarajan Dr Sagunthala R&D Institute of Science and Technology, Chennai, India

Abstract

A large amount of study work has been carried out on traffic management systems, but intelligent traffic monitoring is still an active study topic due to the up-and-coming technology such as the Internet of Things (IoT). The integration of IoT with precision traffic management system and control techniques is for better and future achievements of urban growth. Software and hardware devices work in tandem to control the vehicle externally. To control the vehicle a command is issued from the authorized person through the universal or region-specific interface, and the given command is verified and broadcasted to the devices located all over the field. Once the command is broadcasted to all the field devices, the devices then start broadcasting arbitrarily; when the vehicle encounters the broadcast its preprogrammed program kicks in to control the vehicle. This paper describes the precision vehicle using IoT and compares the vehicle system with the existing and proposed one.

***Keywords*:** Internet of Things (IOT), sensors, user interface, vehicle control

12.1 Introduction

The IoT is a worldwide network of interconnected objects and equipment that exchange data. IoT was one of the most recent of the technology innovations that are gaining traction across a wide range of businesses. According to estimates, the number of IoT-based equipment will reach 21 billion by

**Corresponding author*: srinivasanrajkumar28@gmail.com

A. Chitra, V. Indragandhi and W. Razia Sultana (eds.) Intelligent and Soft Computing Systems for Green Energy, (163–172) © 2023 Scrivener Publishing LLC

2020. IoT systems include dependable device-to-device and human-to-device interactions. Business and data analysis, control and tracking, and cooperation and data exchange are the three main deployment domains.

With economic development and population rise in cities, vehicle count also increases in tandem. The higher the number of vehicles, the higher the possibility of accidents, and safety measures and countermeasures are to be implemented at all levels. Reckless driving can be controlled through an autonomous vehicle. The concept of a smart city improves the sustainability and smartness of the city and the quality of life of the people. Evolvement in the Internet of Things (IoT) reduces the technical drawbacks in the smart city development. The physical infrastructure of the smart city is equipped with smart devices to produce a reliable environment for people with fewer risks. From that smart device, data is shared to centralized intelligence to improve the performance of the infrastructure. In this development there is a backlog in structuring the automobile vehicle traffic to overcome this issue. Smart city developers use the smart traffic infrastructure concepts to automate the vehicles and road signals to achieve efficiency of the model and reduce fuel consumption. Rapid changes in traffic infrastructure, such as increasing autonomous insertion will increase the tradeoff between vehicles. To overcome the issue, increasing the crossing delay time will reduce the traffic duration of the vehicles. Figure 12.1 shows the various applications of transport in IoT.

Despite the fact that traffic crossroads make up a minor portion of the road network, they are responsible for a substantial number of traffic accidents and delays. As a result, traffic engineers are continually concerned about safe and effective intersection management. Traffic lights have been utilised as a key technique of intersection management since their introduction towards the end of the nineteenth century. When traffic signals automatically adjust to real-time traffic circumstances, the efficiency and capacity of a traffic network skyrockets. As a result, static traffic light switching patterns gave way to dynamic traffic signaling, which makes extensive use of communication, computation, and sensing technologies to manage traffic flow. To perceive traffic conditions in real time, state-of-the-art sensing systems such as RFID, microwave radar, and video image processor are deployed either on the road or at the roadside.

12.2 Related Works

This is an overview of the literature. M. Sarrab *et al.* proposed the IoT-based model to improve the smart cities. The authors use the physical

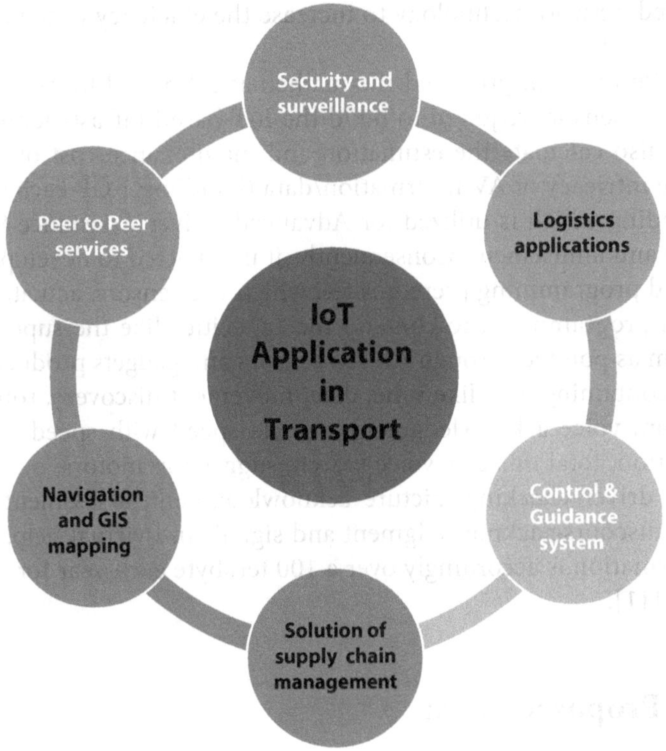

Figure 12.1 IoT application in transport.

infrastructure of the city with smart devices to achieve the smartness of the city. From these model data is stored in different spaces and data is used to improve the intelligence of the city. The model improves the environmental impact to control pollution of the city. M. Liyanage *et al.* proposed the multi-accessing Edge computing model to reduce the computational offloading. The authors use MEC structure instead of IoT structure to improve the performance and sustainability of the system. This model also improves the cloud-based offloading. But the system fails to achieve the computational efficiency compared to IoT-based smart devices [2]. X. Krasniqi *et al.* proposed the model to improve the low-power device efficiency in targeting mobile devices. The authors use mobile devices to structure the smart city environment to extrapolate the other mobility devices [3]. Existing systems lead to conjunction between vehicles in real-time environments [5]. A. Kassu *et al.* proposed the model to sense the modern vehicles which are connected through the cabled network technology to increase the efficiency in the network controller [7].

B. V. Philip *et al.* proposed the mathematical model to calculate the number of sensors required to build the IoT-based infrastructures. The authors also calculate the estimation and maintenance cost of the system. The intricacy of AV information/data (handling 1 GB each second) is expanding which is utilized for Advanced Driver Assistance Systems (ADAS) and amusement. Consequently, it is expected to develop equipment and programming prerequisites, which use sensors, actuators gadgets and programming, to contend the capacities like the superhuman cerebrum as pointed through AI. AV sensors and gadgets produce information containing data like time, date, movement discovery, route, fuel utilization, voice acknowledgment, vehicle speed with speed increase, deceleration, total mileage, voice search, suggestion motors, eye following and driver checking, picture acknowledgment, assessment examination, discourse acknowledgment and signal, and virtual help. The all out information is accordingly over a 100 terabyte each year for 100,000 vehicles [11].

12.3 Proposed Work

The information recovered from IoT gadgets is variable in term of construction and is frequently accomplished progressively. Nonetheless, dealing with this continuous information is an extensive issue as the entire daily practice of associated applications is seriously dependent on the properties of this information the executives administration. The network of gadgets is now and again inaccessible while the vehicle is on a far-off interstate (for example, away from an IoT framework or in a non-network zone). Critical measure of use and venture is required in keeping up with the workers to manage data trade. Construction of organization convention is certainly not a simple assignment as it needs to fulfill the prerequisites of cost, convenience and execution of the entire framework. Organization geography and convention ought to be chosen cautiously, relying on a few variables. Each organization geography and conveying convention has its own qualities and limits. Dependability in conveying IoT-based information from a legitimate and precise source is vital, explicitly with regard to the field of crisis control-based applications having numerous tough prerequisites [3].

12.4 Existing System

In existing models, maps in vehicles have been utilized primarily for route purposes, alongside applications around focal points. The goal of these guides isn't exact enough for independent driving. Moreover, current guides don't meet the necessity of continuous data (i.e., live guides) and don't give adequate data for independent driving. Particularly in metropolitan conditions with high traffic thickness, the prerequisites for safe, completely mechanized driving are huge—for vehicle details as well as for framework determinations. A decent outline of European advancement in this field of exploration gives the European guide for shrewd frameworks for robotized driving [12]. Figure 12.2 shows the existing system of vehicle control system with sensors. In this system the analyzer is used to send the signal to information sensor. Sensors are used for collecting the data or information about the vehicle control.

12.4.1 Advantages

IoT is a network of smart devices, capable to organize the system by automation techniques, responsible for protecting the environment and sharing information to intelligence system through the internet to improve the growth of the IoT-based devices. IoT applications are rapidly improving

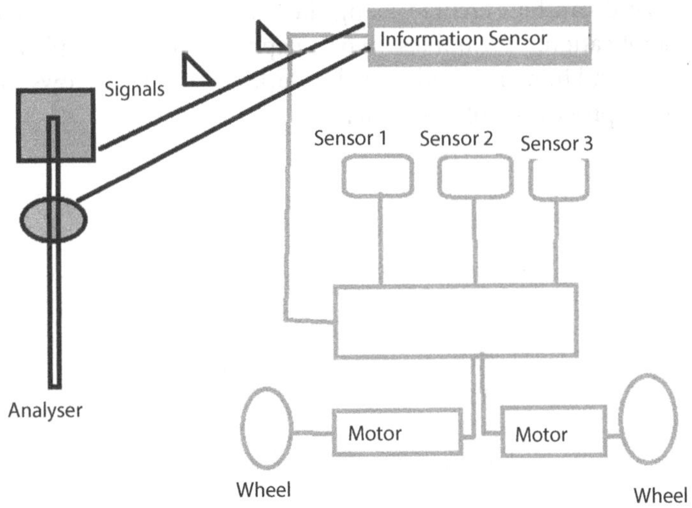

Figure 12.2 Existing architecture diagram.

their performance in a virtual environment. IoT devices are connected in a single environment and perform multiple tasks through the internet. It improves the smartness of the system.

12.4.2 Disadvantages

In the technical world all the technology has some limitation, and IoT has some limitation in choosing the sensors, protocols and communicating interface. IoT systems also have some security-related issues and latency. The selected sensors may affect the human life cycle, privacy issues, etc. It leads to attacking activity and leakage of the secure data. IoT technology needs rapid improvement and changes in network topology to reduce the hacking and security breaches. It is challenging to retain connection between nodes and allocate resources for exchanging data in real time.

12.4.3 Applications

- Once the instructions are passed from the registered and connected device and if the vehicle is connected to the Internet of Things device, then commands can be executed.
- IoT air-conditioners are one of the connected devices which are available in the market for public use. Once the device is connected to the network in any way, then the desired command can be executed using the registered and connected application. Wearables have experienced an explosive demand in markets all over the world. Figure 12.3 shows the data processing of IoT cycle.

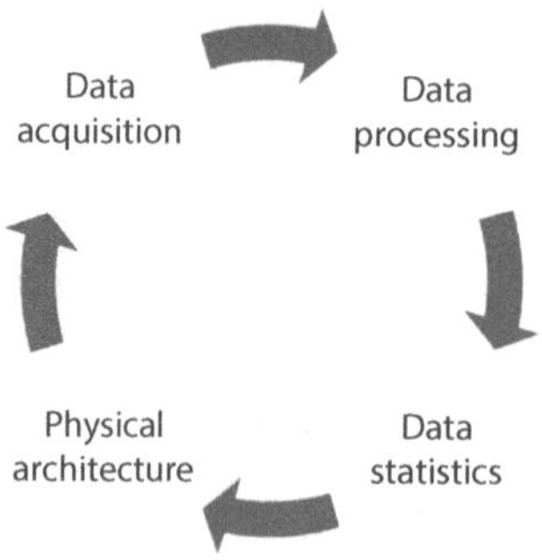

Figure 12.3 Data process in IoT.

- Sensors and software are included into wearable devices to collect data and information about their users. This information is then pre-processed to derive key user insights. These gadgets address a wide range of exercise, health, and entertainment needs.
- Wearable apps must be highly energy efficient or ultra-low power and tiny in size, according to Internet of Things technologies.
- Smart machines are more accurate and reliable in communicating through data than humans, according to the IoT's guiding principle. Furthermore, this information might assist businesses in identifying inefficiencies and problems earlier.

12.5 Proposed System

With the help of the Internet of Things we try to reduce the number of road accidents and vehicle theft. Here we use multiple connected devices and connected car to achieve the safety.

An IoT device sends signals to the vehicles in a deterministic or random way depending upon the configuration. To control the vehicle, a command is issued from the authorized person through the universal or region-specific interface; the given command is verified and broadcasted to the devices located all over the field. Depending on the priority of the issue the command will be issued by the authorized person and general configurations also can be made to stop the vehicles at the signals to adhere to the statutory warnings and the red/yellow signals.

Once the command is broadcasted to all the field devices, the devices then start broadcasting arbitrarily; when the vehicle encounters the broadcast its pre-programmed program kicks in to control the vehicle. Once the vehicle receives the commands from the broadcast or signal device, the received command will be validated and executed. Once the command starts to execute the Engine control unit will be controlled by the sensor to complete the instructions. Depending upon the instruction provided to the sensor the response is given accordingly. The response is processed in the broadcaster device and further it is sent to the command center as response to the command executed previously. The command center validates the response received from the connected vehicle. Multiple grids can be connected together to form a centralized grid to expand the network.

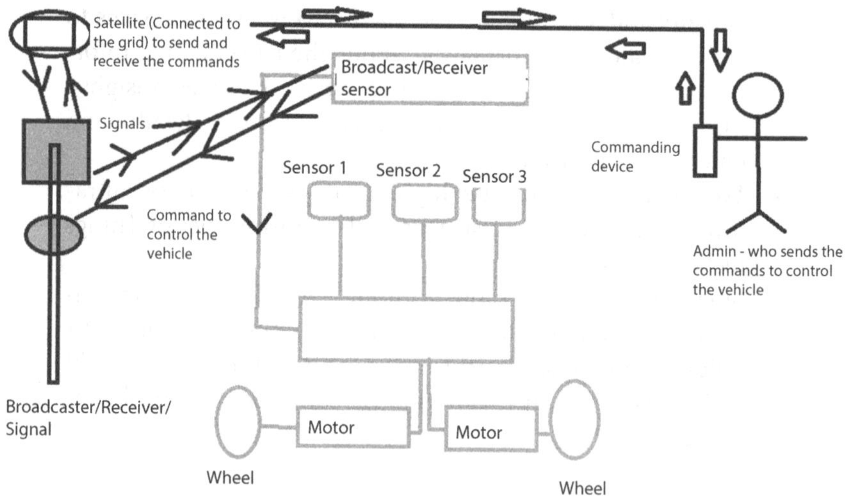

Figure 12.4 Proposed architecture diagram.

Nodes can be configured in such a way that each node can emit the different signals as required by the command center or as required. Each can be accessed individually or can access all nodes at a single instance; there connectivity provides the interoperability that makes the system redundant and always accessible. Two-way communication is enabled to achieve the high success rate; in two-way communications we can verify the result once the command is executed and the response is communicated. Figure 12.4 represents proposed architecture of Precision vehicle control, the function of IoT using sensors.

12.6 Conclusion and Future Enhancement

We have proposed a Precision vehicle control using Internet of Things to reduce the numbers of accidents and vehicle theft on the roads. This effectively safeguards passengers and pedestrians on the road. Future work aims to simplify the user interaction with the application connected to the centralised grid and by avoiding the admin, multiple redundant ways to connectivity can be implemented. By connecting to the database of stolen vehicles, vehicles can be protected.

References

1. M. Sarrab, S. Pulparambil, and M. Awadalla, "Development of an IoT based real-time traffic monitoring system for city governance", *Global Transitions*, vol. 2, pp. 230–245, 2020.
2. M. Liyanage, P. Porambage, A. Y. Ding, and A. Kalla, "Driving forces for Multi-Access Edge Computing (MEC) IoT integration in 5G", *ICT Express*, vol. 7, Issues 2, pp. 127–137, 2021.
3. J. Santa, L. Bernal-escobedo, R. Sanchez-iborra, J. Santa, L. Bernal-escobedo, and R. Sanchez-iborra, "On-board unit to connect personal mobility vehicles to the IoT", *Procedia Computer Science*, vol. 175, pp. 173–180, 2020.
4. M. Simsek, A. Boukerche, B. Kantarci, and S. Khan, "AI driven autonomous vehicles as COVID-19 assessment centers: A novel crowd sensing-enabled strategy", *Pervasive and Mobile Computing*, vol. 75, pp. 101426, 2021.
5. X. Krasniqi and E. Hajrizi, "Use of IoT Technology to Drive the Automotive Industry from Connected to Full Autonomous Vehicles", *IFAC-Papers OnLine*, vol. 49, issue 29, pp. 269–274, 2016.
6. H. G. Seif and X. Hu, "Autonomous Driving in the iCity — HD Maps as a Key Challenge of the Automotive Industry", *Engineering*, vol. 2, issue 2, pp. 159–162, 2016.
7. A. Kassu and M. Hasan," Factors associated with traffic crashes on urban freeways", *Transportation Engineering*, vol. 2, pp. 100014, 2020.
8. N. Yousefnezhad, A. Malhi, and K. Framling, "Security in product lifecycle of IoT devices: A survey", *Journal of Network and Computer Applications*, vol. 171, p. 102779, 2020.
9. R. Ramapriya, P. Mp, G. Ap, A. Kamath, A. Srinivas, and M. Rajasekar, "IoT Green Corridor", *Procedia Computer Science*, vol. 151, pp. 953–958, 2019.
10. S. Sachdev, J. Macwan, C. Patel, and N. Doshi, "Voice Controlled Autonomous Vehicle Using IoT", Procedia Computer *Science*, vol. 160, pp. 712–717, 2019.
11. B. V. Philip, S. Member, T. Alpcan, S. Member, and J. Jin, "Distributed Real-Time IoT for Autonomous Vehicles", *IEEE Transactions on Industrial Informatics*, vol. 15, issue 2, pp. 1131–1140, 2019.
12. Z. Karami and R. Kashef, "Smart transportation planning: Data, models, and algorithms", *Transportation Engineering*, vol. 2, pp. 100013, 2020.

13

A Process of Analyzing Soil Moisture with the Integration of Internet of Things and Wireless Sensor Network

R. Srinivasan[1]*, Kavitha R[1], V. Murugananthan[2] and T. Mylsami[3]

[1]*Department of Computer Science and Engineering, Vel Tech Rangarajan Dr Sagunthala R&D Institute of Science and Technology, Chennai, India*
[2]*Faculty of Information Technology, INTI International University, Nilai, Negeri Sembilan, Malaysia*
[3]*Dr. G R. Damodaran College of Science, Coimbatore, Tamil Nadu, India*

Abstract

Hydrological investigations, environmental tracking, and agricultural productivity are considered significant in soil moisture data. Agricultural yield is influenced by many factors such as pH, moisture, relative humidity, and temperature. Farmers have been increasingly using technology to enhance the quality and quantity of their agricultural output as time has passed. This study presents a multi-parameter tracking system that will keep farmers and users informed via the internet. The growing demand for organic farming has necessitated continual plant health tracking in recent times. Temperature, humidity, weather report, soil moisture, and air pressure are all important factors to consider when cultivating. Additional devices are equipped to keep track of crop conditions and infestations. Steps can be taken to avert harm or improve crop output depending on the data gathered. Scientists can use the information acquired to predict crop yield or make policy decisions. This becomes even more important to ensure quantity and quality. As a result, the goal of this study is to create a remote-tracking method that consistently analyzes the plant's moisture levels. To achieve the stated goal, the WSN (Wireless Sensor Network) is combined with the IoT (Internet of Things). In addition, the suggested study uses the Exponential Weighted Moving Average (EWMA) incident detection technique to extend the lifetime of the network.

**Corresponding author*: srinivasanrajkumar28@gmail.com

A. Chitra, V. Indragandhi and W. Razia Sultana (eds.) Intelligent and Soft Computing Systems for Green Energy, (173–184) © 2023 Scrivener Publishing LLC

Keywords: IoT, soil moisture sensor, Wi-Fi, exponential weighted moving average, WSN

13.1 Introduction

Because agriculture was such an essential part of many developing nations' economies, it is critical to concentrate on improving farming practices. These nations not only meet their own needs, but they also help to deliver high-quality food to the rest of the globe. Irrigation and unbalanced wastewater treatment for crops have an impact on agricultural output as well as the potential for water waste, which was a valuable resource. Temperature, humidity, weather report, soil moisture, and air pressure are all important factors to consider when cultivating. Additional devices are equipped to keep track of crop conditions and infestations. Steps can be taken to avert harm or improve crop output depending on the data gathered. Scientists can use the information acquired to predict crop yield or make policy decisions.

13.1.1 WSN

A WSN is a collection of sensor networks that are deployed in different locations and communicate with one another to create a network. Detector, processor, antenna, battery system, transceiver, analog to digital converter, and data memory make up a sensor node. WSNs are favored because of their minimal price, low voltage, and ease of maintenance, and they play an important role in smart home, agricultural, health care, and military sectors.

13.1.2 IoT

The IoT is a worldwide network of interconnected objects and equipment that exchange data. IoT is one of the most recent technology innovations that are gaining traction across a wide range of businesses. According to estimates, the number of IoT-based equipment will reach 21 billion by 2020. IoT systems include dependable device-to-device and human-to-device interactions. Business and data analysis, control and tracking, and cooperation, and data exchange are the three main deployment domains [2]. The major goal of this study is to create a plant-soil humidity tracking system that allows the user to virtually check plant growth. Zigbee software was used to capture information for wireless transmission, which was then sent to a server. The EDA (Energy Driven Architecture) is used to extend the WSN lifetime.

13.2 Literature Study

Farmers face a huge difficulty with rain in general and fresh water in particular. This is because the climates are so different. As a result, to properly utilize limited water sources without hurting production, it is necessary to use appropriate water for the plant. In this case, Cloud-based IoT technologies can be utilized to successfully feed water to the crops until the land around it reaches the desired humidity, and then shut off the water source by turning off the pump from a remote location. A mobile phone can detect the soil moisture and turn off the motor. Several studies have been published in the literature that describe the use of various processors and detectors to monitor and measure agricultural data. The following are some of the studies that have been done on soil moisture estimation by some scientists: In [3] a smart sensor network web technology-based device for in situ moisture detection was developed. The EC-5 soil humidity detector and the XBee pro unit are part of the device. Wireless sensing technology provides 3D soil moist data as a factor of time for areas [4]. It is proposed that the environmental parameters acquired from the sensors be recorded in a system. The data is processed and analyzed using a smartphone app (software). TDR100 moisture detector, DTH22 humidity sensor [14], ATMega2560 microcontroller, and LM35 temperature sensor, are all part of the system. In 2012, a method was designed that uses embedded sensors in a surveillance area to assess humidity, soil moisture, and temperature. The data from the sensors is sent to an Atmel ATmega328P microprocessor through BLE. The actual data was monitored using cloud technologies. The system's goals are to reduce energy and water use.

[5] A GPRS-based device was created to monitor and measure the farming landscape in Indian agricultural fields. The moisture and temperature sensors are installed in the plant's root zone. The data from the detectors is collected using the PIC24FJ64GA004 microcontroller. The information obtained is represented graphically using the GPRS unit. Photovoltaic solar panels are used to operate the entire system. The created system's goals are to reduce water and electricity usage while also improving the quality of crop and food grains production.

[6] A system is being designed that uses the ECHERP protocol and WSN to effectively distribute water in agricultural areas. The system collects and analyses information from various sensors before supplying water to the crop field [7]. A method is being developed to monitor and measure the moisture levels in agricultural areas. Moisture is determined by the resistance change between two sites in the soil, according to the moisture sensor's theory. XBee

data transmission technology is used to send sensor information. A moisture sensor and a PIC16F876A microcontroller were used to create the hardware. Using GUI software, the moisture information is represented graphically on a PC. In 2015, a technology was created that uses wireless subsurface sensor networks and Spatial-Temporal connections to estimate moisture levels [8].

13.3 Proposed Work

This section lists the proposed study, which includes the following components [4]:

1. Transmitter and Sensing Module
2. Receiver Unit
3. IoT activation
4. Event recognition method. The suggested soil moisture tracking system is [18] depicted in Figure 13.1.

13.3.1 Sensing and Transmitter Module

Figure 13.2 displays the sensing unit, which includes a soil moisture monitoring probe and a sensor panel with an LM393 microcontroller. For moisture monitoring, the sensing probe was inserted into the plant. Analog output, VCC, and GND are the three pins that link the sensor panel to the microprocessor [10].

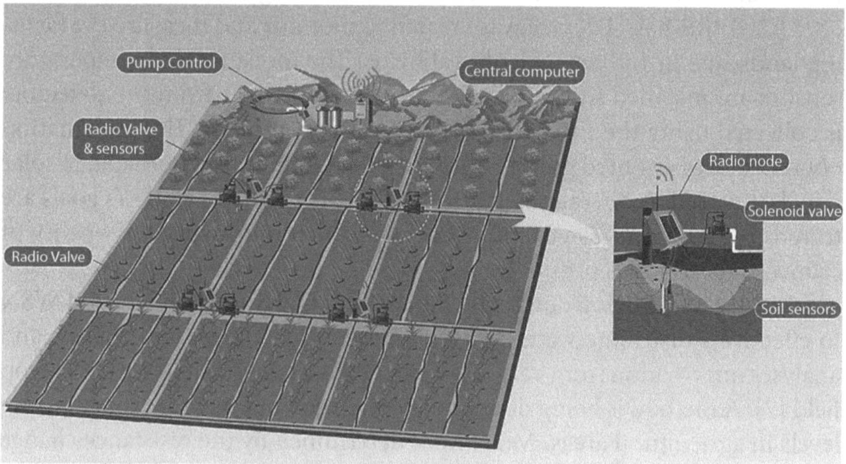

Figure 13.1 Proposed soil moisture tracking system [9].

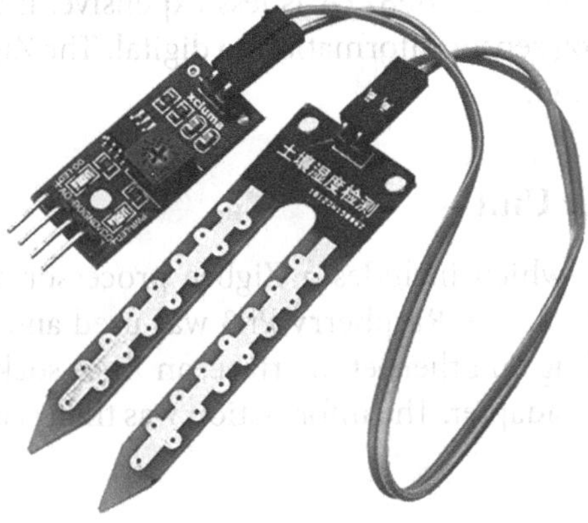

Figure 13.2 Soil moisture sensor.

In the proposed study, a CC2500 Zigbee pair was used, which has a highly sensitive and cost-effective transmitter. It is made up of three Light Emitting Diodes that serve as indicators for energy, reception, and communication. The range of frequencies is 2108MHz to 2541.9MHz. Battery power is used to power the whole transmitter and sensing unit. The interaction between the soil humidity sensor and the PIC microprocessor is shown in Figure 13.3. The suggested research employs a PIC microcontroller since it is highly programmable. When contrasted to other

Figure 13.3 Sensor interfaced with PIC microcontroller and Zigbee.

microcontrollers, the PIC 16F877A is less expensive. It includes an ADC that converts analog sensor information to digital. The ZigbeeTx transmits digitized information.

13.3.2 Receiver Unit

The receiver unit, which includes a Zigbee processor and a receiver, is shown in Figure 13.4. The Raspberry Pi 3 was used and it includes many capabilities including an Ethernet interface, an Xbee socket, a CPU, RAM, and a power source adapter. The information was then sent to the cloud via a wired connection.

13.3.3 IoT Activation

IoT is a type of inter-networking that allows a variety of physical devices or objects to be connected to the internet. These gadgets are capable of collecting, exchanging, and storing data. IoT cloud is a data storage service that allows information from these sensors to be saved and viewed remotely. Dropbox was the cloud file system used in this study, as indicated in Figure 13.5. Dropbox provides 2GB of free data storage. The presence of daily events is recorded in a file.

The EWMA method is used to extend the lifetime of the WSN. If the value drops below or beyond the two predetermined criteria, the data is transferred to the cloud regularly.

Figure 13.4 Receiver Module [11].

Figure 13.5 Cloud enablement [12].

13.3.4 Event Recognition Algorithm [4]

The transmission among sensor networks consumes a lot of energy, which was a major downside of WSN. To address the aforementioned problem EDA is used [13]. EWMA was one of the threshold-based EDAs that was utilized in this study to set the moisture control parameters. For recognizing tiny shifts in position, the EWMA chart approach is a better and more effective tool [5, 6]. The following is the EWMA mathematical prototype.

$$EWMA_t = Z_t = \lambda x_t + (1 - \lambda) Z_{t-1} \tag{13.1}$$

for t=1,2,3,4……,n

Where t denotes the number of observations made at regular intervals, and w denotes the weighted sum of the preceding values and consists of a number between 0 and 1, x = the value acquired.

Zt-1 = Zt's former value

Equation (13.2) is used to calculate the Upper Control Limit (UCL) and Equation (13.3) is used to calculate the Lower Control Limit (LCL).

$$UCL = \lambda_0 + LQ\sqrt{\frac{\lambda}{(2-\lambda)[1-(1-\lambda)^{2t}]}} \tag{13.2}$$

$$LCL = \lambda_0 + LQ\sqrt{\frac{\lambda}{(2-\lambda)[1-(1-\lambda)^{2t}]}} \qquad (13.3)$$

Where μ0 denotes the mean, L is the breadth of the control boundaries, and = denotes the standard variance from the dynamic range graph.

13.4 Result and Discussion

In real time, an IoT-enabled soil moisture tracking system employing WSN [15] has been deployed. Figures 13.7 and 13.8 illustrate a soil moisture detector that is submerged in the soil and indicates the amount of water existing on an LCD [16]. The information from the detectors was then communicated to the Raspberry Pi via a Zigbee transceiver, from whence it is posted to the internet. When a soil moisture detector is submerged in dry soil, it reads zero percent. The reading rises as the amount of water injected into the soil rises [17].

The moisture level of the soil is shown in Figure 13.6, which is presented on the LCD. When data is transferred by Zigbee, an LED indicates this.

The 3 LEDs in the Zigbee Rx flicker to represent data reception, as seen in Figure 13.7.

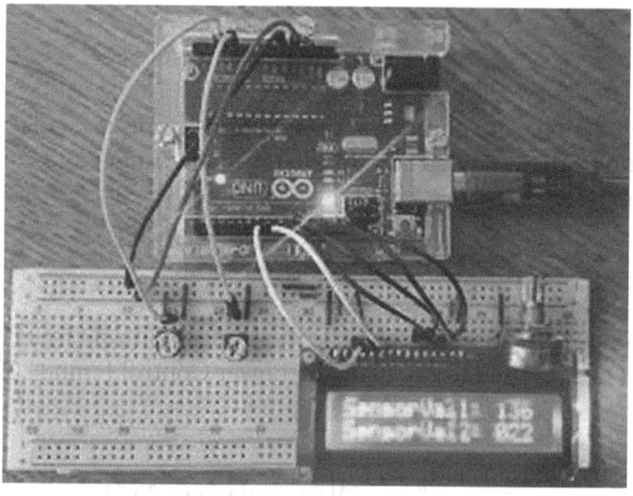

Figure 13.6 Moisture level indicated on LCD.

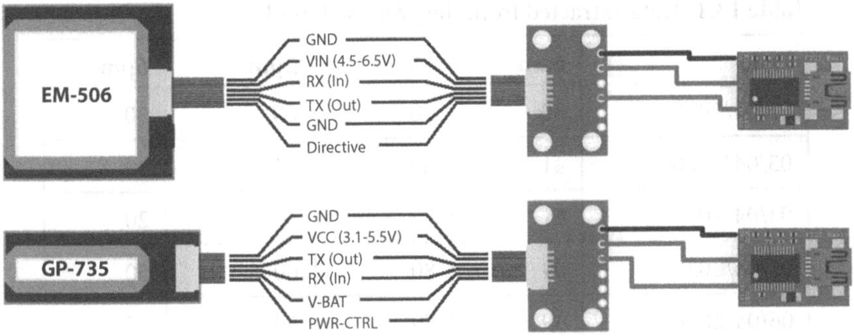

Figure 13.7 Data receiving image at the receiver unit.

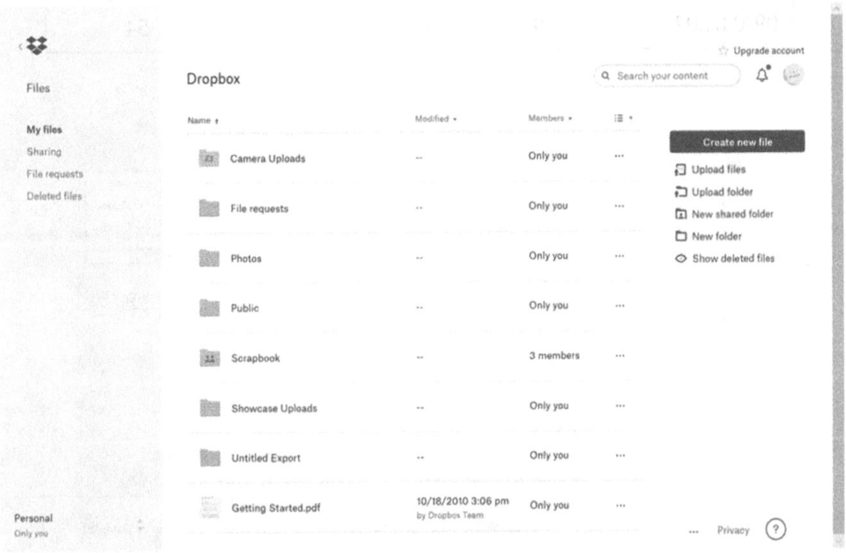

Figure 13.8 Drop box-created files.

The files generated daily are depicted in Figure 13.8. Each file contains four soil moisture level readings collected at periodic intervals. Table 13.1 displays the data collected from the document.

The control graph for the EWMA event recognition method is shown in Figure 13.9. The information from the detector is fed into the EWMA method as input. This method generates the upper and lower limits that are used as WSN input parameters [19]. LCL and UCL can be 40 and 50, correspondingly, based on the plot.

Table 13.1 Data extracted from the proposed model.

Date	6am	10am	2pm	6pm
02/04/2020	80	70	60	50
03/04/2020	41	69	52	49
04/04/2020	41	35	19	20
05/04/2020	09	80	60	50
06/04/2020	39	79	59	50
07/04/2020	45	39	29	35
08/04/2020	20	15	7	8
09/04/2020	79	69	59	54

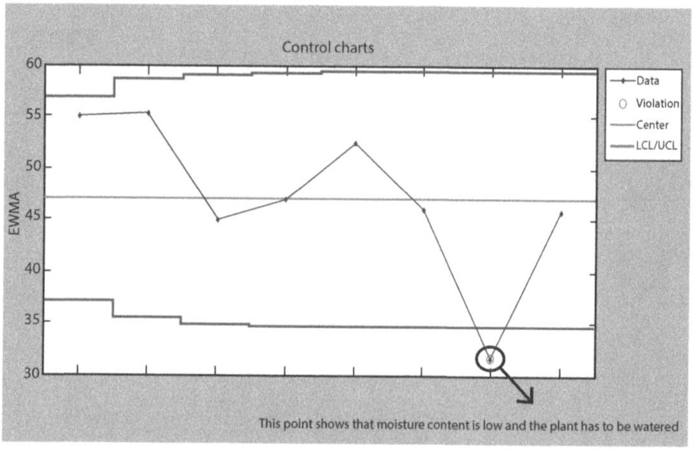

Figure 13.9 EWMA's control chart.

13.5 Conclusion

Real-time monitoring and control of agricultural factors are critical for agriculture development. Researchers created and built a method for measuring and monitoring soil moisture utilizing WSN and the Cloud IoT system in this study. Although Indian farmers face numerous challenges, the advent of wireless communication in agriculture could be beneficial in the current climate of water shortages and uncertain weather. The deployment of a WSN-based soil moisture tracking system was described in this work.

Because WSN is battery-powered, it has a longer lifespan. The EWMA object tracking method is utilized, and occurrences are only generated when certain threshold criteria are met. The networks are in a rest condition most of the time, which allows them to conserve energy. This research can be expanded upon by examining more than one detector.

References

1. *Inventive Computation Technologies*, Springer Science and Business Media LLC, 2020.
2. Boobalan, J., V. Jacintha, J. Nagarajan, K. Thangayogesh, and S. Tamilarasu. "An IoT based agriculture monitoring system." In *2018 International Conference on Communication and Signal Processing (ICCSP)*, pp. 0594–0598. IEEE, 2018.
3. Dutta, Arjun, Poulami Paul, Swarnabha Roy, Rahul Agarwala, Abhisikta Chakraborty, Asesh Ghosh, HimadriNathSaha, "GSM Based Irrigation System." In *2019 IEEE 10th Annual Ubiquitous Computing, Electronics & Mobile Communication Conference (UEMCON)*, pp. 0867–0872. IEEE, 2019.
4. Ezhilazhahi. A, P.T.V. Bhuvaneswari, "IoT enabled plant soil moisture monitoring using wireless sensor networks", *Third International Conference on Sensing, Signal Processing and Security (ICSSS)*, pp. 345–349, 2017.
5. Hamouda, Yousef, and Mohammed Msallam, "Variable sampling interval for energy-efficient heterogeneous precision agriculture using Wireless Sensor Networks", *Journal of King Saud University-Computer and Information Sciences*, pp. 88–98, 2020.
6. Jawad, Haider Mahmood, Rosdiadee Nordin, Sadik Kamel Gharghan, Aqeel Mahmood Jawad, and Mahamod Ismail, "Energy-efficient wireless sensor networks for precision agriculture: A review", *Sensors*, 1781, 2017.
7. Math, Rajinder Kumar M., and Nagaraj V. Dharwadkar, "IoT Based low-cost weather station and monitoring system for precision agriculture in India", *International Conference on I-SMAC (I-SMAC)*, pp. 81–86. IEEE, 2018.
8. Math, Rajinder Kumar, and Nagaraj V. Dharwadkar, "A wireless sensor network based low cost and energy efficient frame work for precision agriculture", *International Conference on Nascent Technologies in Engineering (ICNTE)*, pp. 1–6, 2017.
9. Nandal, Vijay, and Sunita Dahiya, "IoT Based Energy-Efficient Data Aggregation Wireless Sensor Network in Agriculture: A Review", *Psychology and Education Journal*, pp. 2985–3007, 2021.
10. Nisar Ahmad, Ali Hussain, Ihsan Ullah, Bizzat Hussain Zaidi, "IOT based Wireless Sensor Network for Precision Agriculture", *International Electrical Engineering Congress (iEECON)*, 2019.

11. *Proceedings of International Conference on Recent Trends in Machine Learning, IoT, Smart Cities and Applications*, Springer Science and Business Media LLC, 2021.
12. Saraf, Shweta B., and Dhanashri H. Gawali, "IoT based smart irrigation monitoring and controlling system" *IEEE International Conference on Recent Trends in Electronics, Information & Communication Technology (RTEICT)*, pp. 815–819, 2017.
13. Swathi, NS and K. Deeba, "Automated Control of Irrigation in Agriculture Using IOT and Wireless Sensor Networks", *International Journal for Advance Research and Development*, pp. 28–32, 2018.
14. Thakur, Divyansh, Yugal Kumar, Arvind Kumar, and Pradeep Kumar Singh, "Applicability of wireless sensor networks in precision agriculture: A review", *Wireless Personal Communications*, pp. 471–512, 2019.
15. Vippon Preet Kour, Sakshi Arora, "Recent developments of the Internet of Things in Agriculture: A Survey", *IEEE Access*, 2020.
16. Vuran, Mehmet C., Abdul Salam, Rigoberto Wong, and Suat Irmak, "Internet of underground things in precision agriculture: Architecture and technology aspects", *Ad Hoc Networks*, pp. 160–173, 2018
17. www.hindawi.com
18. Xing, Xijun, Jiancheng Song, Lingyan Lin, Muqin Tian, and Zhipeng Lei, "Development of intelligent information monitoring system in greenhouse based on wireless sensor network", *4th International Conference on Information Science and Control Engineering (ICISCE)*, pp. 970–974, 2017.
19. T. Mylsami, V. Murugananthan, "Enhancing Agriculture Growth through Internet of Things", *Journal of Critical Reviews*, Vol. 7, Issue 18, 2020.
20. P. Visu, V. Murugananthan, "A dependable smart security system for dominant attacks on agriculture crop fields", *International Journal of Engineering & Technology*, pp. 195–197, 7 (1.3), 2018.

14

Automatic Angular Position Stabilization of Ambulance Stretcher in Real Time

Vedant Joshi, Maheshwari S. and Kathirvelan J.*

Master of Technology in "IoT and Sensor Systems", Department of Sensors & Bio-Medical Technology, School of Electronics Engineering, Vellore Institute of Technology, Vellore, India

Abstract
Ambulances are widely used to transport patients injured in road accidents, and other emergency cases, etc. Due to rough road conditions these patients experience discomfort as the angle of the stretcher deviates from its original position, which makes the condition of the patient even worse. In this paper, an ambulance stretcher mechanism has been proposed which reduces the angular deviation. The idea is to use an Inertial Measurement Unit sensor which will detect the change in stretcher angle in two major axes, pitch and roll, and actuate the servo motors to bring the stretcher back to the desired position in real time. This makes the system an open-loop controlled system which stabilizes the angular tilt of the ambulance stretcher in real time.

***Keywords*:** Ambulance, stretcher, inertial measurement unit, servo motor, stabilize, pitch, roll

14.1 Introduction

Ambulance transport is uncomfortable for a patient with moderate to severe injuries, considering two scenarios: linear vibrations and angular displacement of the stretcher. Along with the linear vibrations that the patient experiences, in some rare road conditions the ambulance is tilted for a brief amount of time, which tilts the stretcher. Some causes for this

**Corresponding author*: j.kathirvelan@vit.ac.in

A. Chitra, V. Indragandhi and W. Razia Sultana (eds.) Intelligent and Soft Computing Systems for Green Energy, (185–198) © 2023 Scrivener Publishing LLC

angular tilt are shown in Figure 14.1. They are, first, an overbridge situation where the ambulance goes uphill and downhill, tilting the stretcher. The second cause is during breaking and acceleration of the ambulance where the patient experiences an inertial force in backward and forward motion. The third cause is a situation where there is an array of bumps on the road which leads to misalignment of the stretcher. The effects the patient experiences increase the intracranial pressure inside the brain, resulting in an increase in pain of patients in accident cases, discomfort for sensitive newborn babies, etc. [1]. This angular tilt, which affects the patients in several ways discussed earlier, cannot be eliminated fully with the use of mechanical suspension systems. To eliminate this angular tilt, we need an active system which will detect the change in angular tilt with respect to the desired position of the stretcher and counteract it with the use of active devices such as electrical actuators, Inertial Measurement Unit sensors and microcontrollers.

In study [3] the authors utilized an exclusive purpose equivalent robot parallel to the stretcher. Furthermore, a potential control structure has been deployed to compensate for the unevenness of the road. [5] proposed a pitch-roll-interconnected hydro-pneumatic suspension which assets by seven DOF dynamic model, in which the behaviours of this are clearly formalized using hydrodynamic condition derivation, which is capable to achieve the obstacle control for pitch, roll, and bounce strategies for ambulances to improve the quality and narrow the vibrations for the patients lying on the stretcher. Cohort [6] depicts the improvement of a suspension system for the stretcher with hydro-pneumatics, so as to decrease the vibration-actuated uneasiness. Elastic, hydro-pneumatic, slow-dynamic and dynamic isolators system are introduced to help the patient,

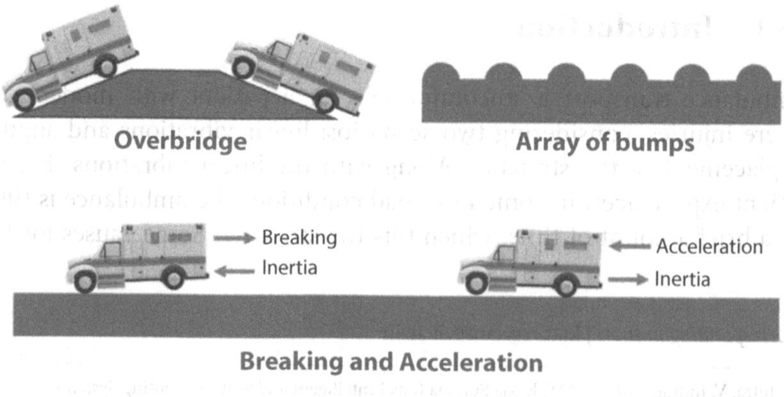

Figure 14.1 Causes for the angular tilt of the stretcher.

permitting movement comparative with the vehicle. The dynamic qualities and dependability of a mechanical plan for a solution to lessen problematic and destructive road actuated vibrations experienced in the patient-care compartment of ambulance using universal-joint were determined in [7], while in study [4], the system is designed with the help of a mechanical component called as "Inerter", placed between the ambulance floor and the stretcher in both horizontal and vertical directions to reduce the vibrations. In [8] the authors proposed a method to stabilize the stretcher using IMU sensors and AVR/ARM controller architecture which stabilizes the ambulance stretcher using complementary filter algorithm; the results were demonstrated with the help of a prototype. [12] manages the investigation of the Inverse Kinematics (IK) and the reachable workspace of a three-degrees-of-freedom (3-DOF) parallel manipulator, proposing various changes and enhancements to improve on its utilization with Motion Cueing Algorithms (MCA) for self-movement generation in VR simulators. It also incorporates objective measures (safe zones) on the workspace volume that can give a basic but productive method of contrasting the kinematic capacities of various types of movement stages for this specific application. The overshoot rate, rise time, and information arrangement perfection of Sensor Fusion (Complementary and Kalman channel) and Moving Average channel reaction in IMU information securing from step contribution of 20-degree rotation. Moving normal channel brought about the smallest overshoot level of 0% yet produced the slowest reaction with 0.42 second rise time [11].

As the patients are in a much more sensitive state than a healthy person, they experience the effects of these angular tilt displacements even more. Hence to reduce their discomfort and provide them a safe ride, in this paper, an actively controlled system is proposed which will detect as well as eliminate the angular displacement of the stretcher.

14.2 Materials and Methods

14.2.1 Interior and Flaws

The interior of any typical ambulance is enriched with many facilities such as advanced medical equipment, safety equipment, comfortable seats for an attendant and a stretcher for the patient to rest on.

The stretcher is kept on a bed (Figure 14.2) or on the base of the ambulance interior. Due to this, the angular and linear vibrations that the ambulance floor experiences are transferred to the stretcher where the patient rests.

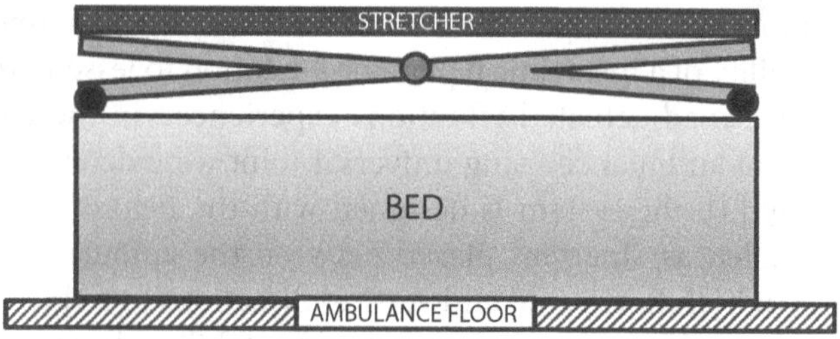

Figure 14.2 Typical position of ambulance stretcher.

The basic desire of any transported patient is to minimize his/her pain, but the current type of typical stretcher arrangement shown in Figure 14.2 is not suitable for the same.

14.2.2 Proposed Position of the Stretcher

The proposed system is shown in Figure 14.3. This system will eliminate the angular displacements transferred from the base of the ambulance to the stretcher.

14.2.3 Hardware and Software

The components for hardware and electronics used for designing the prototype are listed below.

- Acrylic body.
- Three ball joints and two revolute joints.
- Metal links.

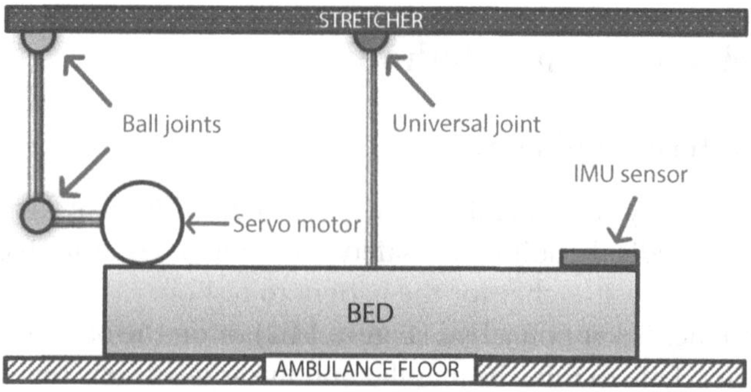

Figure 14.3 Proposed design of stretcher.

- Servo motor.
- Arduino Uno microcontroller board.
- MPU6050 IMU Sensor.

In the prototype, for data acquisition, the MPU6050 Inertial Measurement Unit sensor which has a 3-axis accelerometer and 3-axis gyroscope embedded on a chip is used in order to measure the angular displacement. To process the data and to compute the output of the prototype model Arduino UNO microcontroller board is used. The electrical actuators used to rotate the stretcher in order to stabilize it are two DC angular rotation servo motors each for Pitch axis and Roll axis. To increase the precision in the proposed system, servo motors are used instead of Permanent Magnet DC motors [8].

14.2.4 Methodology

The angular tilt is measured by the MPU6050 sensor placed at the interior base of the ambulance and is provided as input to the Arduino UNO microcontroller board. The Arduino UNO is programmed to compute the required commands that should be given to the servo motors such that the stretcher rotates and goes back to the initial position. This action happens in real time so the patient experiences much less angular displacement (Figure 14.4).

As shown in Figure 14.5, with the help of an open-source library the offset values for the MPU6050 sensor used are calculated. In order to reduce the

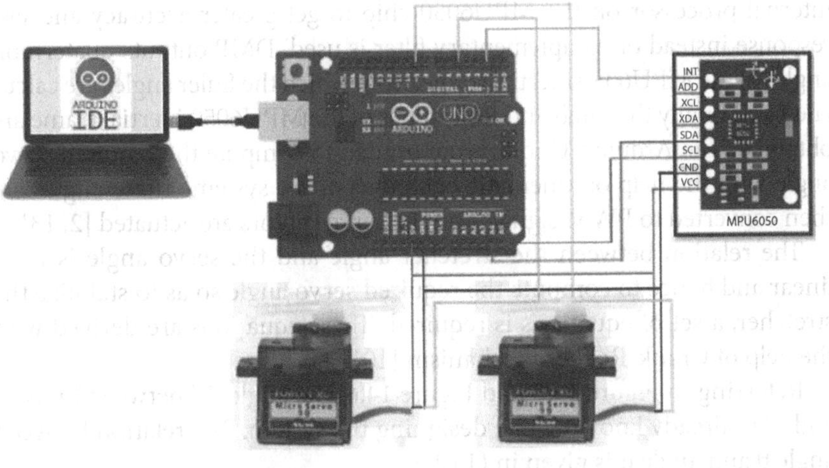

Figure 14.4 Schematic of the system.

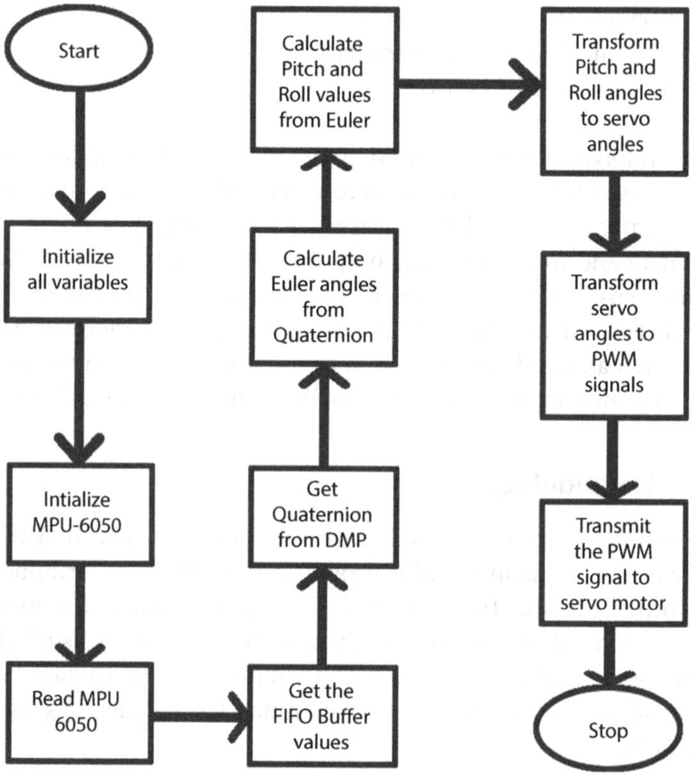

Figure 14.5 Flow chart of the system.

effect of noise and drift, the DMP (Digital Motion Processor) which is the internal processor on the MPU6050 chip to get greater accuracy and fast response instead of complementary filter is used. DMP outputs quaternion angles of the MPU6050 inertial frame, from which the Euler angles are calculated and finally the Pitch and Roll values of the MPU6050 inertial frame are obtained. The Arduino Uno is programmed to compute the required servo angle with the help of kinematic equations of the system. These angles are then converted to PWM signals and the servo motors are actuated [2, 13].

The relation between the stretcher angle and the servo angle is non-linear and hence to compute the required servo angle so as to stabilize the stretcher, a set of equations is required. These equations are derived with the help of Crank-Rocker mechanism [10].

Referring to Figure 14.3 and Figure 14.6, the angle 'Y' between links 'd' and 'c' is already known while designing the system. The relation between angle θ and angle α is given in (14.1):

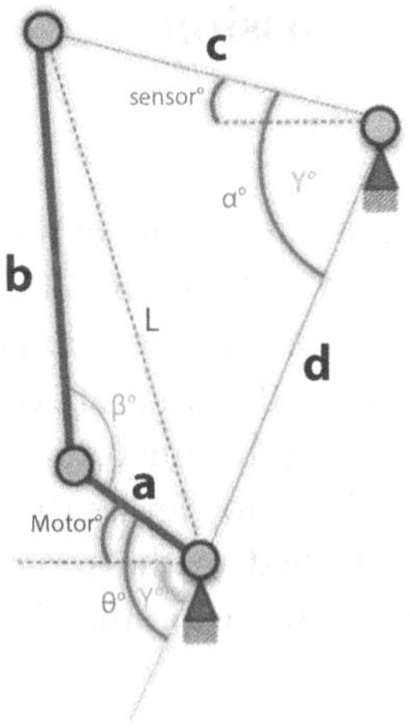

Figure 14.6 Modeling of links.

$$\theta = 2 * \tan^{-1} * \frac{\left(c * \sin(\alpha) - b * \sin(\beta)\right)}{\left(c * \cos(\alpha) + a - d - b * \cos(\beta)\right)} \tag{14.1}$$

The angle between the links a and b, i.e., 'β' is computed with the help of diagonal length 'L', as in (14.2):

$$\beta = \cos^{-1} * \frac{a^2 + b^2 - L^2}{2 * a * b} \tag{14.2}$$

The diagonal length 'L' is calculated using the angle 'α', lengths of the links 'd', 'c' and the sensor reading which gives the tilt of the stretcher as in (14.3).

$$L = \sqrt{d^2 + c^2 - 2 * d * c * c\cos(\alpha)} \tag{14.3}$$

By calculating the angle 'θ' the servo motor angle for every sensor reading is calculated using (14.4):

$$\text{Motor} = Y - \theta \tag{14.4}$$

14.3 Results and Discussion

14.3.1 Results

To demonstrate the working and feasibility of the proposed system, a prototype model has been developed and shown in Figure 14.7. For construction of the prototype, two sheets of acrylic with dimensions 15 cm × 30 cm × 0.5 cm each, two acrylic links – 10.5 cm each, one metal link – 15 cm with 0.25-inch diameter, universal joint having 0.25-inch diameter are used. Two wooden platforms 3.5 cm × 2 cm to elevate the servo motors required for movement of servo arms attached to the servo shaft are attached below the servo motors. The MPU6050 sensor and the Arduino UNO board are embedded in the electronics box.

The MPU6050 sensor is placed in the electronics box which senses the tilt of the ambulance base. The Arduino UNO then calculates the servo angles based on the sensor reading and the kinematic equations. The servo motors are then driven by the PWM pulses which are converted from the servo angles calculated by the Arduino UNO. Figure 14.8 shows the angle of the motor (Motor_Roll) for stabilizing the Roll motion of the stretcher based on the IMU sensor reading (Roll).

We verified the calculations using an online tool which calculates the interior angles of irregular quadrilaterals based on any one angle as an input [9]. We can observe in Figure 14.9 that if we increase the angle (Roll) by 10 degrees, then the angle (Motor_Roll) is increased by 12 degrees.

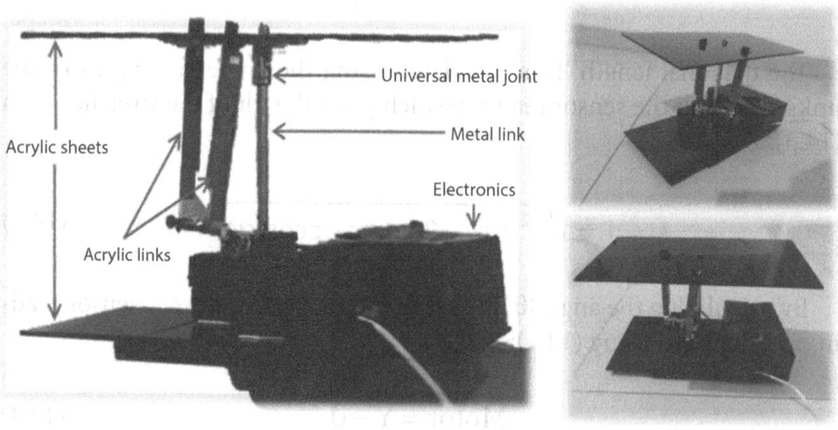

Figure 14.7 Prototype images.

19:46:58.920 -> Roll=8.19	Alpha2=92.77	L2=11.89	Beta2=100.35	Theta2=94.84	Motor_Roll=-10.25
19:46:58.920 -> Roll=8.33	Alpha2=92.92	L2=11.90	Beta2=100.53	Theta2=95.01	Motor_Roll=-10.43
19:46:58.920 -> Roll=8.49	Alpha2=93.07	L2=11.91	Beta2=100.73	Theta2=95.21	Motor_Roll=-10.62
19:46:58.966 -> Roll=8.63	Alpha2=93.22	L2=11.92	Beta2=100.92	Theta2=95.39	Motor_Roll=-10.81
19:46:58.966 -> Roll=8.79	Alpha2=93.37	L2=11.93	Beta2=101.11	Theta2=95.59	Motor_Roll=-11.01
19:46:58.966 -> Roll=8.95	Alpha2=93.53	L2=11.94	Beta2=101.32	Theta2=95.79	Motor_Roll=-11.21
19:46:58.966 -> Roll=9.11	Alpha2=93.70	L2=11.96	Beta2=101.53	Theta2=96.00	Motor_Roll=-11.41
19:46:58.966 -> Roll=9.29	Alpha2=93.87	L2=11.97	Beta2=101.75	Theta2=96.22	Motor_Roll=-11.63
19:46:59.013 -> Roll=9.47	Alpha2=94.05	L2=11.98	Beta2=101.98	Theta2=96.45	Motor_Roll=-11.86
19:46:59.013 -> Roll=9.67	Alpha2=94.25	L2=12.00	Beta2=102.24	Theta2=96.70	Motor_Roll=-12.11
19:46:59.013 -> Roll=9.86	Alpha2=94.45	L2=12.01	Beta2=102.49	Theta2=96.94	Motor_Roll=-12.36
19:46:59.013 -> Roll=10.07	Alpha2=94.65	L2=12.03	Beta2=102.75	Theta2=97.20	Motor_Roll=-12.62
19:46:59.059 -> Roll=10.27	Alpha2=94.86	L2=12.05	Beta2=103.01	Theta2=97.46	Motor_Roll=-12.87
19:46:59.059 -> Roll=10.48	Alpha2=95.07	L2=12.06	Beta2=103.28	Theta2=97.73	Motor_Roll=-13.14
19:46:59.059 -> Roll=10.69	Alpha2=95.27	L2=12.08	Beta2=103.55	Theta2=97.98	Motor_Roll=-13.40

Figure 14.8 Dimensions of bottom acrylic sheet of the prototype.

Hence to make the angle to its original value, angle needs to be rotated by 12 degrees in negative direction. Comparing Figure 14.6, Figure 14.8 and Figure 14.9 we can also verify the other angles used for calculating the "Motor_Roll" which are α, θ and β.

In Figure 14.10 we can observe the plot of the Roll (stretcher angle) versus the Motor_Roll (servo motor angle) which shows that when the "Roll" of the stretcher is positive the motors are rotated with "Motor_Roll" angle in negative direction and vice versa to stabilize the stretcher.

In Figure 14.11 the results from the prototype model designed are shown. We can observe how the top acrylic sheet (imagined as a stretcher) stabilizes in the horizontal position against the tilted bottom acrylic sheet (imagined as the ambulance floor).

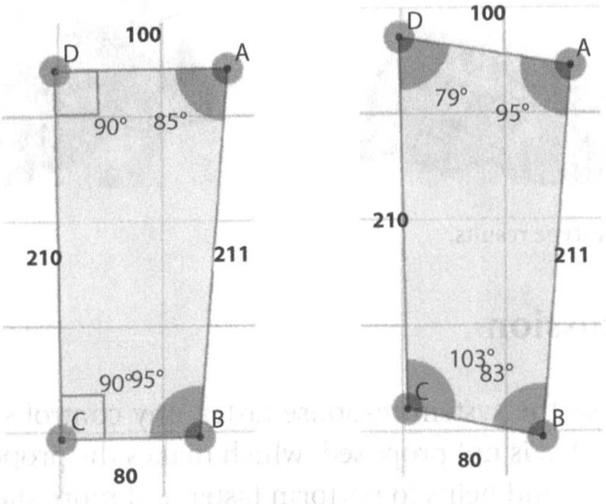

Figure 14.9 Verification of the calculation.

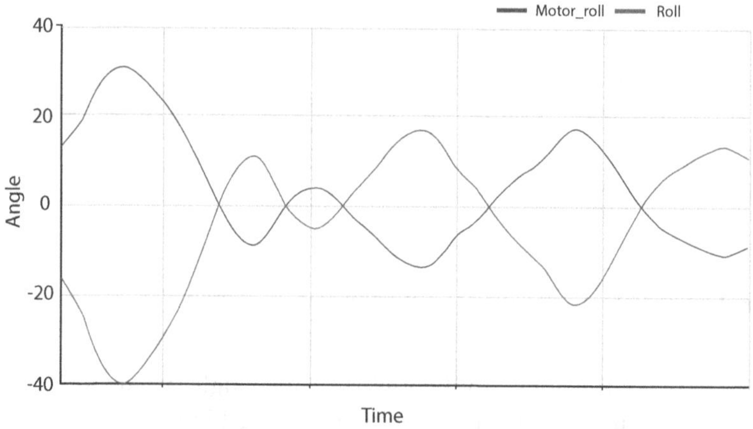

Figure 14.10 Roll (Stretcher angle) vs. Motor_Roll (Servo angle).

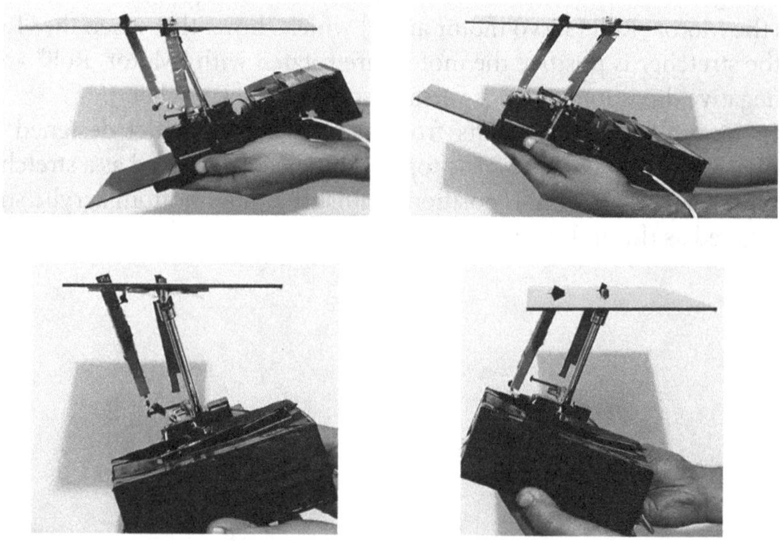

Figure 14.11 Prototype results.

14.4 Discussion

In order to make the system response faster, any control strategy such as PID controller, etc., is not proposed, which makes the proposed system an open loop system and helps to perform faster and more stable than [6, 14, 17]. Moreover, in contrast to [5, 6, 14–16], use of only electric actuators (servo motors) is proposed due to the drawbacks of hydraulic, pneumatic, electro-hydraulic and electro-pneumatic systems which are, bulkiness, complex, dangerous in certain circumstances, high operating costs, high maintenance costs, less accuracy and slow response compared to electric actuators [18]. We also observed that the position of the servo motors is affecting the feasibility of the design in [8] even though only electric actuators are been used. So, the design of the system was optimized and is now more feasible and reliable. The aim of the proposed system is to stabilize the stretcher when there is a continuous tilt in the pitch and roll orientation. Considering this, the systems proposed in [4, 6, 7, 17] are more focused on reducing the vibration depending on the acceleration of the ambulance while braking and accelerating the vehicle. But when there is no acceleration and only tilt in the case of overbridged roads, these systems fail to stabilize the stretcher.

14.5 Conclusion

An open-loop system which can stabilize the angular tilt of the ambulance stretcher in real time is proposed. The system stabilizes the stretcher by measuring the tilt of the base of the ambulance and then calculating the servo angles with the help of the kinematics equations derived with the help of crank-rocker mechanism. The servo motors are then driven by PWM pulses and the stretcher is stabilized immediately. The prototype of the system was designed using SG90 servo motors, IMU sensor (MPU6050), Arduino UNO microcontroller board, metal and acrylic links, universal joint, acrylic sheets and wooden base for the motors. The results were observed, verified and discussed. We believe that the system is feasible and will be an added advantage to the existing suspension systems for ambulances. Our future work will include the elimination of the vertical vibrations and designing a bigger prototype.

References

1. C. Chumchan and K. Tontiwattanakul, Health risk and ride comfort assessment by ISO2631 ofan ambulance. *International Conference on Engineering, Applied Sciences and Technology (ICEAST), 1, 2019.*
2. V. Popelka., A self stabilizing platform. International Carpathian Control Conference (ICCC), 458, 2014.
3. L.E. Bruzzone, M.P. Lazzarotto, R.M. Molfino, M. Zoppi, Parallel robot for ambulance stretcher active suspension: mechanical modelling and

simulation. Proc. of the IASTED International Conference Modelling, Identification and Control (MIC2002), 474, 2002.

4. R. Koyanagi and M. Takahashi, Design of Vibration-Isolating Bed for Ambulances Using Inerter. *Asian Control Conference (ASCC)*, 567, 2019.
5. Bohuan Tan, Yang Wu1, Nong Zhang, Bangji Zhang and Yuanchang Chen, Improvement of ride quality for patient lying in ambulance with a new hydro-pneumatic suspension. Advances in Mechanical Engineering, 11(4), 1, 2019.
6. Ali M. Abd-El-Tawwab, Ambulance Stretcher with Active Control Isolator System. *Journal of Low Frequency Noise, Vibration and active Control, 20(4), 217, 2001.*
7. Rinto K Anto, Varun Krishna T V, Nikesh V M, Rahuldas M, Manu Joseph, Nithin P N., Stabilization of Stretcher in an Emergency Ambulance. *International Research Journal of Engineering and Technology (IRJET)*, 5, 2018.
8. Tripathi, V., Gupta, R., Self-stabilizing platform using mpu 6050-a boon for the society to reduce accidental death. *International Journal of Engineering and Advanced Technology*, 8, 254, 2019.
9. Mathsisfun.com, *Interactive Quadrilaterals*. https://www.mathsisfun.com/geometry/quadrilaterals-interactive.html, 2021.
10. Huseyin Mutlu, Design of the crank–rocker mechanism for various design cases based on the closed-form solution. *Sadhana*, 46, 4, 2021.
11. G. G. Redhyka, D. Setiawan and D. Soetraprawata, Embedded sensor fusion and moving-average filter for Inertial Measurement Unit (IMU) on the microcontroller-based stabilized platform. International Conference on Automation, Cognitive Science, Optics, Micro Electro-Mechanical System, and Information Technology (ICACOMIT), 72, 2015.
12. Casas-Yrurzum, Sergio & Coma, Inmaculada & Portalés, Cristina & Fernández, Marcos, Optimization of 3-DOF parallel motion devices for low-cost vehicle simulators. *Journal of Advanced Mechanical Design, Systems, and Manufacturing*, 11(2), 2017.
13. G. Madhumitha, R. Srividhya, J. Johnson and D. Annamalai, Physical modeling and control of self-balancing platform on a cart. *International Conference on Robotics: Current Trends and Future Challenges (RCTFC)*, 1, 2016.
14. L. Hou, S. Liu, Q. Sun, N. Niu and L. Sun, Optimal control of active ambulance stretcher suspension based on genetic algorithm. *Chinese Automation Congress (CAC)*, 6771, 2017.
15. Tan, Bohuan & Wu, Yang & Zhang, Nong & Zhang, Bangji & Chen, Yuanchang, Improvement of Ride Quality for Patient Lying in Ambulance with a New Hydro-Pneumatic Suspension. *Advances in Mechanical Engineering*, 11, 1, 2019.
16. *Shaer, Bassel & Kenné, Jean-Pierre & Kaddissi, Claude & Mintsa, Honorine. Real-time hybrid control of electrohydraulic active suspension: Electrohydraulic*

Active Suspension Hybrid Control. Int. J. Robust. Nonlinear Control, 27, 4968, 2017.

17. Sagawa, K., & Inooka, H., Ride quality evaluation of an actively-controlled stretcher for an ambulance. *Proc. Instn. Mech. Engrs. Part H: J Engineering in Medicine*, 216(4), 247, 2002.
18. TiMOTION Technology, Advantages and Drawbacks of Pneumatic, Hydraulic, and Electric Linear Actuators - TiMOTION, https://www.timotion.com/en/news/news_content/news-and-articles/general/advantages-and-drawbacks-of-pneumatic%2C-hydraulic%2C-and-electric-linear-actuators?upcls=1481189409&guid=1499762723, 2021.

15

Automated Ploughing Seeding with Water Management System

Anto Sheeba J., Shyam D., Sivamani D.*, Sangari A., Jayashree K. and Nazar Ali A.

Department of Electrical and Electronics Engineering, Rajalakshmi Engineering College, Chennai, India

Abstract
Ploughing is a method which is used to separate the soil on the surface and bring fresh nutrients to the surface. Before each crop is planted in the agricultural field, the farmers will plough their field using tractors. But this needs a lot of labor and more manual efforts required for seed feeding. To reduce this manual effort, automation in a ploughing, seeding and water management system is proposed. This employs ultrasonic sensors, a high torque low speed gear DC motor, and relay circuit, connected to two microcontrollers, Arduino and ATmega16. The sensor mechanism will help the microcontroller to control the movement of the machine in such a way that it will operate in the desired field without crossing the borders. This machine is specially designed to do ploughing and seeding for paddy fields. The water pump motor is controlled by a microcontroller using ultrasonic sensor during ploughing and a soil moisture sensor further. It is a pollution-free vehicle as it uses electrical power.

Keywords: Ploughing, seeding, automation, high torque DC motor, ultrasonic sensor, soil moisture sensor

15.1 Introduction

India is a predominantly agricultural country. In south India, paddy is the most important crop. Working in the field for longer periods of time is difficult and requires more effort. Ploughing the field area, sowing, water management, cutting, and other agricultural procedures are only a few examples. In the past, animals were used to perform all of these tasks.

**Corresponding author*: sivamani.d@rajalakshmi.edu.in

A. Chitra, V. Indragandhi and W. Razia Sultana (eds.) Intelligent and Soft Computing Systems for Green Energy, (199–210) © 2023 Scrivener Publishing LLC

Later, machines took the place of animals, resulting in a significant rise in agricultural productivity [1–3]. Humans must, however, operate the machines. The use of equipment such as tractors, for example, has increased pollution. Agricultural systems are operated automatically in several countries. Because agriculture is such a big deal in India, farmers will benefit if automation is applied. As a result, our effort focuses on automating basic working procedures such as ploughing and seeding [4–6].

The project's goal is to use automation to modernize farm technologies. It ploughs any type of field, regardless of size or shape, automatically. Existing systems have problems such as automated systems not being created, remote control systems requiring an operator with operational experience, and computer operated systems requiring internet access [7].

The goal of this study is to automate agriculture for better results and lower overall costs by combining ploughing and sowing into a single pack, as each of these tasks is now performed by separate machines. Additionally, to reduce pollution by minimizing the use of petroleum-based automobiles in favor of battery-powered vehicles [8–10].

15.2 Block Diagram

The description of the various blocks and how they work is explained in this section. The many blocks involved in the design and implementation of an automated ploughing, sowing, and water management system are depicted in this diagram. The system's block diagram is shown in Figure 15.1. The Arduino UNO, ATmega 16, DC gear motor, ultrasonic sensor, driver circuit, relay, soil moisture sensor, battery, and seeder make up the system design [11].

The motor driver circuit named IC L293D can drive a variety of motors like DC geared motors and even bi-polar stepper motors [12–16]. It acts as a switch between a low voltage circuit and a high voltage circuit. The motor terminals and the input are connected to the pins of L293D according to the logic in it. A set of similar DC geared motors are used for prototype of our project. The DC motor takes an input voltage of 12V, which can be used in variety of robotics applications. The motor can run at a speed 60 RPM.

The block diagram contains two separate systems. One is a ploughing, seeding system which uses ATmega 328p and the other is a water management system which uses ATmega 16.

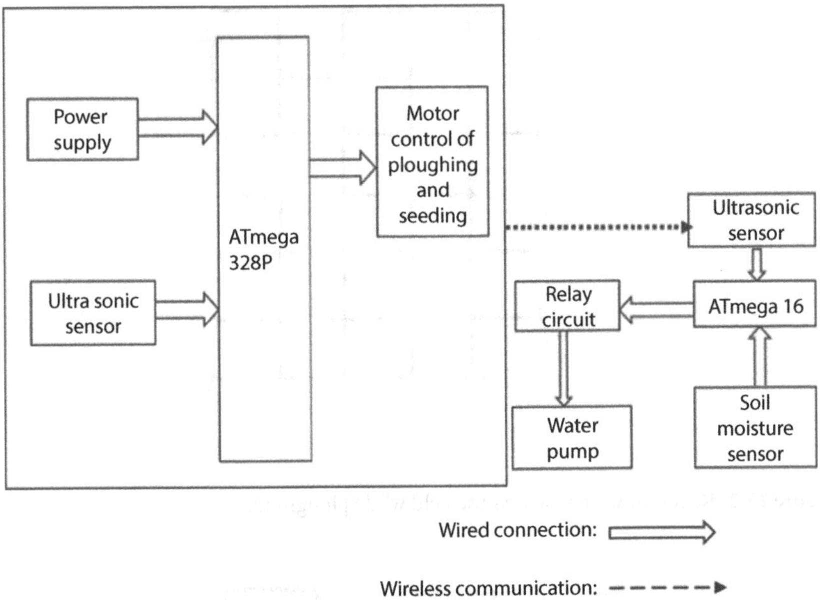

Figure 15.1 Block diagram.

15.3 Working Methodology

The ultrasonic sensor is used to determine the distance between two points. This will be connected to the ATmega 328p microcontroller. When the rover approaches the end of the field, the sensor detects it and sends a signal to the controller. According to the coding, the controller will cause the rover to turn in the desired direction [17]. When the rover reaches another field corner, it will do a U-turn and repeat the process until it reaches another field corner. When the rover is first at the first corner another ultrasonic sensor will sense its presence; it will give a signal to ATmega16 which will turn on the water pump. A soil moisture sensor will be placed in the field and when the field is sufficiently moisturized then the water pump will be turned off automatically. This mechanism can be used further [18]. A box with holes will be placed under the rover, through which the seeds will come out when it rotates [19, 20]. A normal DC motor will be used for this purpose. Figure 15.2 shows the rover movement across the field while ploughing. Separate commands will be given for ploughing and seeding. The movement that is used for ploughing and seeding is shown below. Figure 15.2 shows rover movement across the field while ploughing; Figure 15.3 shows the rover movement across the field while seeding [21–26].

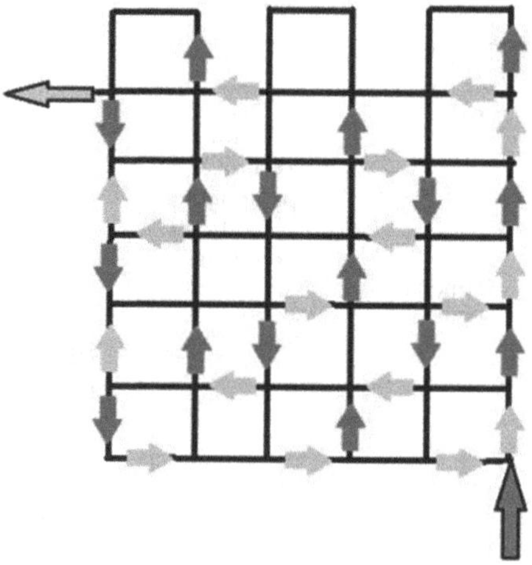

Figure 15.2 Rover movement across the field while ploughing.

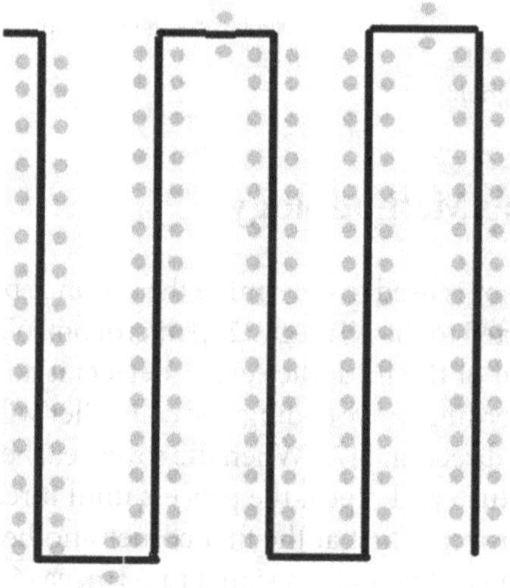

Figure 15.3 Rover movement across the field while seeding.

15.4 Design Calculation

The movement of the rover depends upon the output of the ultrasonic sensor. So proper distance calculation should be done for accurate movement of the rover. The distance is calculated using the formula,

$$\text{Distance} = ((N * T) / 20000) \text{ in cm} \tag{15.1}$$

Where N is the speed of sound in air, T is the time it takes the receiver to read the echo from the obstacle in seconds. When the plougher is attached, the machine is 32 cm in length. As a result, the distance input for the ploughing operation is set to 35 cm in the coding. When the seeder is connected, the machine measures 16 cm in length. As a result, the distance input for seeding operation in the code is 18 cm.

15.5 Simulation

Proteus Design Suite software is used to create the working model. It combines computing, visualization, and a programming environment in one package. Proteus is also a modern simulation programme with advanced data structures and built-in editing and debugging capabilities that support many types of sensors and simulation connections. Proteus is a test software that simulates several types of microcontrollers, including the Arduino series. A stable version of Proteus such as V 8.0 is used to develop our working model [27]. The components required can be added by clicking on "Pick devices" button on the left of the schematic capture. After adding all the components it will be connected to form the complete circuit. The software circuit for the robot is shown below in Figure 15.4.

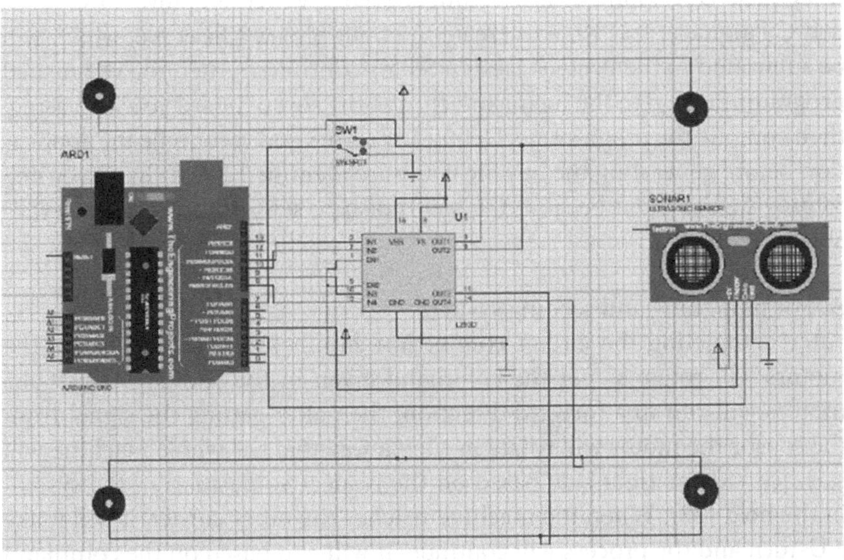

Figure 15.4 Proteus design circuit for plougher system.

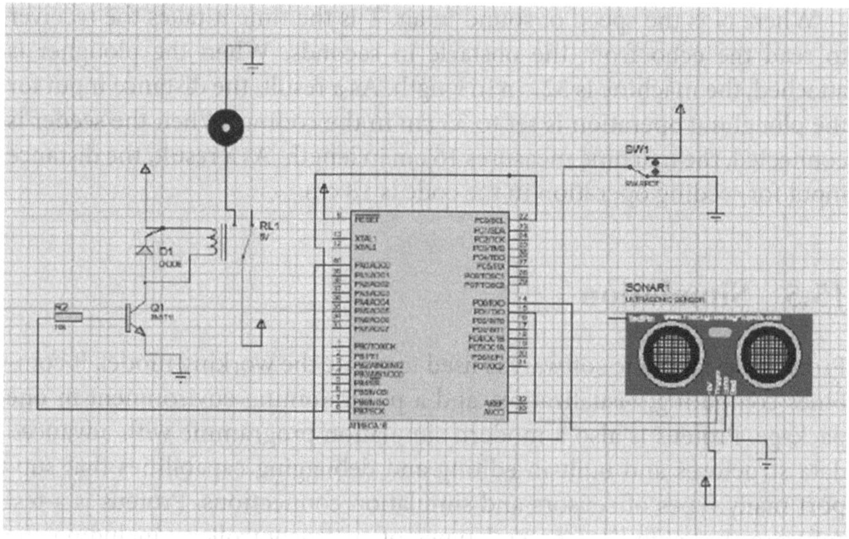

Figure 15.5 Proteus design circuit for water management system.

The controller will function based on the coding that has been created. The ATmega 328p controller is used in this robotic system. Ultrasonic sensor, DC motors, and motor driver kit are all included. The Proteus design circuit for the plougher system and water management system is shown in Figures 15.4 and 15.5. The logic is programmed in the Arduino Integrated Development Environment (IDE), which employs an embedded C language. It is time to create the built binary (.hex) file, which will be submitted to the Proteus Design Suite's controller. After you've finished programming, go to the menu and choose the sort of board you'll be using, then compile the programme. Debug any errors and recompile the application before generating the hex file if any are identified. We utilized a switch to provide the input because the soil moisture sensor library is not available in Proteus.

A. For ploughing and seeding system

After completing the circuit connection simulation will be started. The motors will rotate as per the command given in the coding. At first the motors will run in a forward direction, then after getting the signal from the sensor the motor will rotate in such a way that the whole machine will turn left side or right side based on the respective position of the system in the field [28]. When the machine reaches the other corner, it will make a U-turn and the process will continue. It will end based on the command given in the coding.

During the initial condition, the supply will be OFF and the switch will be open so that there will be no rotation left and right. When the supply is ON and switching is open the left and right wheel perform forward rotation. When supply is ON and the switch is closed the left wheel performs forward rotation and the right wheel performs backward rotation.

B. For water management system
This system uses soil moisture sensor and relay mechanism to operate the water pump motor. When supply is ON and switch is open the water pump will become ON and the relay open. When supply is ON and switch is closed the water pump will become ON and the relay closed. When the controller gets the input from the soil moisture sensor the water pump motor will be turned on. When it gets a second input from the sensor the motor will be turned off.

15.6 Hardware Implementation

The hardware implementation of an automated ploughing, sowing, and water management system, as well as a description of the circuit's main components, are presented. The outcomes are examined and debated. Table 15.1 shows the component description. The Arduino has 14 digital input/output pins (six of which can be used as PWM outputs), six analogue inputs, a 16 MHz quartz crystal, a USB connection, a power jack, an ICSP header, and a reset button on the hardware side.

It comes with everything needed to get started with the microcontroller; it simply has to be plugged into a computer with a USB cable or power it with an AC-to-DC adapter or battery. The Atmel AVR 8-bit microcontroller ATmega 16 is a high-performance, low-power microcontroller based on sophisticated RISC architecture. It has 131 effective instructions. It can operate at a frequency of up to 16 MHz. It contains a customizable flash memory of 16 KB. It features 16 KB of programmable flash memory, 1 KB of static RAM, and 512 bytes of EEPROM. The endurance cycles of flash memory and EEPROM are respectively 10000 and 100000. It is a microcontroller with 40 pins.

A set of similar DC geared motors are used for the movement of the robot. The DC motor takes an input voltage of 12V, which can be used in a variety of robotics applications. The motor has a speed range of 10 to 500 RPM. The Ultrasonic sensor offers information on the absolute position of a target or moving object. Ultrasonic methods are typically the

Table 15.1 Components used in hardware setup.

Components	Quantity	Rating
Arduino (ATMEGA 328P)	1	5V
ATMEGA 16	1	5V
DC motor	4	12V
DC motor	2	6V
Soil moisture sensor	1	-
Ultrasonic sensor	2	-
Relay	1	12V
Battery	1	12V,1.2A
IC L293D	1	-

sole alternative to mechanical probing for glossy surfaces, transparent objects, or environments with a high level of dust and humidity. In this project the sensor gives the input to ATmega 16 and ATmega 328p. It is used for distance detection. The IC L293D can drive a variety of motors like DC geared motors and even bi-polar stepper motors. It acts as a switch between a low voltage circuit and a high voltage circuit. The motor terminals and the inputs are connected to the pins of L293D according to the logic in it.

The volumetric water content in soil is measured by a soil moisture sensor. Because direct gravimetric measurement of free soil necessitates the removal, drying, and weighting of a sample, soil moisture sensors use features such as electrical resistance and neutron interaction to determine the volumetric water content. It is utilized to water the agricultural area by operating the pump in this project. A 12V, 1.2A battery is utilized to supply the circuit with the power it requires. It is a rechargeable battery that uses a 12V converter to recharge.

The ultrasonic sensor is used to measure distance. This will be connected to the microcontroller ATmega 328p. The sensor will sense when the rover reaches the end of the field and give a signal to the controller. The controller will make the rover turn in a desired manner according to the coding. When the rover reaches another corner of the field then it will do a U-turn and the process goes on until it reaches another corner. When the rover is first at the first corner another ultrasonic sensor will sense its presence; it will give a signal to ATmega 16 which will turn on the water pump. Soil moisture sensor will be placed in the field and when the field is sufficiently moisturized then the water pump will be turned off automatically. This mechanism can be used further. A box with holes will be placed under the rover, through which the seeds will come out when it rotates. A normal DC motor will be used for this purpose. Separate commands will be given for ploughing and seeding. Figure 15.6 shows the hardware deployed for ploughing and seeding system. The hardware for water management system is shown in Figure 15.7.

Figure 15.6 Hardware circuit of ploughing and seeding system.

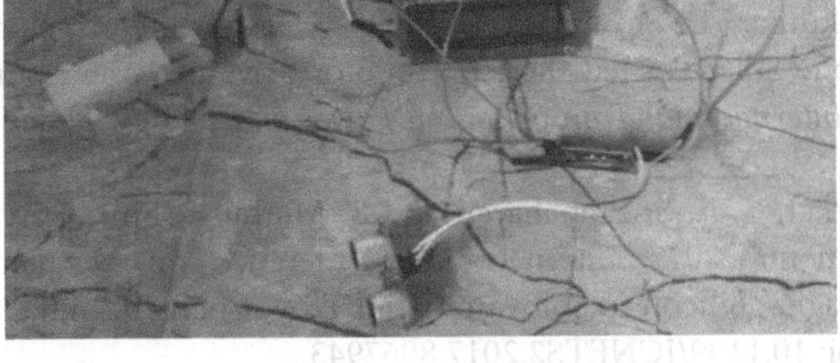

Figure 15.7 Hardware circuit of water management system.

15.7 Conclusion

We may conclude from this research that the system executes the majority of agriculture labour automatically without the need for human intervention. Water management systems deliver precise water to specific crops while reducing water waste that occurs with other methods. Overall, we can state that the system is totally automated, saving money and time while also improving manufacturing quality. If properly implemented in every farm, it will result in green evolution. It should be noted that the movement utilized for seeding can also be used for harvesting. As a result, the cutting system can be installed in the same rover in the future, allowing for the development of a fully automated irrigation system.

References

1. H. H. Kadar, S. S. Sameon and P. A. Rafee, "Sustainable Water Resource Management Using IOT Solution for Agriculture," *2019 9th IEEE International Conference on Control System, Computing and Engineering (ICCSCE)*, Penang, Malaysia, pp. 121–125, 2019. doi: 10.1109/ICCSCE47578.2019.9068592.
2. A. K. Mishra and B. Singh, "Stage Solar PV Powered Water Pump with a Storage System," *2018 8th IEEE India International Conference on Power Electronics (IICPE)*, Jaipur, India, pp. 1–6, 2018. doi: 10.1109/IICPE.2018.8709497.
3. A. P. Barros, "Water for Food Production - Opportunities for Sustainable Land-Water Management using Remote Sensing," *IGARSS 2008 - 2008 IEEE International Geoscience and Remote Sensing Symposium*, Boston, MA, USA, pp. IV - 271-IV - 274, 2008. doi: 10.1109/IGARSS.2008.4779710.
4. S. A. Amrita, E. Abirami, A. Ankita, R. Praveena and R. Srimeena, "Agricultural Robot for automatic ploughing and seeding," *2015 IEEE Technological Innovation in ICT for Agriculture and Rural Development (TIAR)*, pp. 17–23, 2015. doi: 10.1109/TIAR.2015.7358525.
5. Chandika, "Automation and Emerging Technology Development of Seed Sowing Robot", *Journal of Agricultural Science*, Vol. 1, Issue No. 3, June 2009.
6. K Premkumar, D Shyam, D Sivamani, "Design and Implementation of Standalone Solar PV fed Induction motor drive for Water pumping application", *International Transaction on Power and Energy Systems*, pp. 59–69, 2021.
7. C. Rajurkar, S. R. S. Prabaharan and S. Muthulakshmi, "IoT based water management," 2017 International Conference on Nextgen Electronic Technologies: Silicon to Software (ICNETS2), Chennai, India, pp. 255–259, 2017. doi: 10.1109/ICNETS2.2017.8067943.

8. Gholap Dipak Dattatraya, "Robotic Agriculture Machine", *International Journal of Innovative Research in Science, Engineering and Technology*, Vol. 3, Issue No. 4, April 2014.
9. K. Ramesh, K. T. Prajwal, C. Roopini, M. Gowda M.H. and V. V. S. N. S. Gupta, "Design and Development of an Agri-bot for Automatic Seeding and Watering Applications," *2020 2nd International Conference on Innovative Mechanisms for Industry Applications (ICIMIA)*, pp. 686–691, 2020. doi: 10.1109/ICIMIA48430.2020.9074856.
10. Ejiofor Virginia, "Microcontroller based Automatic Water level Control System", *International Journal of Innovative Research in Computer and Communication Engineering*, Vol. 1, Issue No. 6, August 2013.
11. A. Yalavarthi and B. Singh, "Grid-Integrated Solar PV Fed SRM Water Pump Drive for Small-Scale Irrigation and Household Supply," *2020 IEEE Energy Conversion Congress and Exposition (ECCE)*, Detroit, MI, USA, pp. 3073–3078, 2020. doi: 10.1109/ECCE44975.2020.9236299.
12. A. Varshney, U. Sharma and B. Singh, "A Grid Interactive Sensorless Synchronous Reluctance Motor Drive for Solar Powered Water Pump for Agriculture and Residential Applications," *2020 IEEE International Conference on Power Electronics, Drives and Energy Systems (PEDES)*, Jaipur, India, pp. 1–6, 2020. doi: 10.1109/PEDES49360.2020.9379885.
13. A. Nazar Ali, D. Sivamani, R. Jaiganesh, M. Pradeep, "Solar powered air conditioner using BLDC motor." *IOP Conference Series: Material and Science Engineering*, Vol. 623, Oct. 2019.
14. V.Venkatesh, A.NazarAli, R.Jaiganesh, V.Indragandhi "Extraction and conversion of exhaust heat from automobile engine in to electrical energy." *IOP Conference Series: Material and Science Engineering*, Vol. 623, Oct. 2019.
15. Shyam D, Premkumar K, Thamizhselvan T, Nazar Ali A, Vishnu Priya M. "Symmetrically Modified Laddered H-Bridge Multilevel Inverter with Reduced Configurational Parameters," *International journal of engineering and advanced technology*, Vol. 9, issue 1, Oct. 2019.
16. A. Nazar Ali, R. Jayabharath, and M. D. Udayakumar. "An ANFIS Based Advanced MPPT Control of a Wind-Solar Hybrid Power Generation System." *International Review on Modelling and Simulations (IREMOS)* 7.4, pp. 638–643, 2014.
17. A. Nazar ali, Dr. R. Jayabharath, R. Shanthi Priyadharshini. "A Single phase highly efficient transformer less inverter for PV Grid connected power system using ISPWM technique." *International Journal of Applied Engineering Research (IJAER)* Vol. 10, No. 9, pp. 7489–7496, 2019.
18. A. Nazar Ali, and R. Jeyabharath. "Ride through Strategy for a Three-Level Dual Z-Source Inverter Using TRIAC." *Circuits and Systems* 7.11, pp. 3911, 2016.
19. A. Nazar Ali and R. Jayabharath. "Performance Enhancement of Hybrid Wind/Photo Voltaic System Using Z Source Inverter with Cuk-sepic Fused

Converter." *Research Journal of Applied Sciences, Engineering and Technology,* 7.19, pp. 3964–3970, 2014.

20. A. Nazar Ali, N. Gowri, and P. G. Scholor. "Power factor correction based bridgeless single switch SEPIC converter fed BLDC motor." *Advances in Natural and Applied Sciences* 10.3 pp. 190–198, 2016.
21. A.T. Sankara Subramanian, P.Sabarish and A. Nazar Ali, "A Power factor correction based canonical switching cell converter for VSI fed BLDC motor by using voltage follower technique." *IEEE xplore Digital Library*, pp. 1–8, 2017.
22. R. Jai Ganesh, A. Nazar Ali, S .Kodeeswaran, B. Karthikeyan, L. Nagarajan, "Iot based Water Management System for Highly Populated Residential Buildings." *International Journal of Disaster Recovery and Business Continuity,* Vol. 11, issue 01, pp. 4018–22, 2020.
23. A. Nazar Ali, R. Jai Ganesh, D. Sivamani, D.Shyam, "Solar powered highly efficient Seven-level inverter with switched Capacitors." *IOP Conference Series: Material and Science Engineering,* Vol. 906. Aug 2020, 17578981,1757899X.
24. Vijayalakshmi, S., Sivaraman, P.R., Karthick, R., Nazar Ali, A. "Implementation of a new Bi-Directional Switch multilevel Inverter for the reduction of harmonics." *IOP Conference Series: Materials Science and Engineering,* 2020, 937 (1), 012026.
25. Sankara Subramanian, A.T., Priyadharsini, S., Palaniyappan, S., Nazar Ali, A. "A novel IOT based domestic automation system for load monitoring and efficient control." *IOP Conference Series: Materials Science and Engineering,* 2020, 937(1), 012028.
26. Anton Amala Praveen, A., Muthu Kumaran, M. Nazar Ali, A., "Minimization of Power Factor Penalty Charges for Non-Linear Domestic Loads with IOT Technology." *IOP Conference Series: Materials Science and Engineering,* 2020, 937(1), 012011.
27. D.Sivamani, R.Ramkumar, A.Nazarali, D.Shyam, "Design and implementation of highly efficient UPS charging system with single stage converter." *Materials Today: Proceedings,* 10 Oct. 2020.
28. S.R. Paveethra, C. Kalavalli, S. Vijayalakshmi, Dr. A. Nazar Ali, D. Shyam (2020), Evaluation of Voltage Stability of Transmission Line with Contingency Analysis. *International Journal of Scientific & Technology Research*, Vol. 9, issue 02, pp. 4018–22.

16

Detecting Fraudulent Data Using Stacked Auto-Encoding: A Three-Layer Approach

P. Saravanan[1], V. Indragandhi[2] and V. Subramaniyaswamy[1]*

[1]School of Computing, SASTRA Deemed University, Thanjavur, India
[2]School of Electrical Engineering, Vellore Institute of Technology, Vellore, India

Abstract
With the growing number of online transactions due to ease of use, the probability of fraud being involved in smart card transactions is increased. Online users are most vulnerable with a credit card, and banks and vendors employ debit card facilities for online banking systems. There are multiple middle-man transaction processing websites, and the user may not know whether he is being subjected to fraud or not. Auto-encoder is a deep learning technology used for many real-world applications, especially in fault detection and fault diagnosis. In this paper, a deep learning-based, stacked auto-encoder system has been designed for predicting the accuracy of normal and fraud clauses in credit card fraudulent datasets. The system is built using TensorFlow with a three-layer stacked auto-encoder and tested using a real-world credit card transaction set. The experimental results show that the proposed system outperforms in terms of accuracy of prediction.

Keywords: Smart card security, fraudulent data, classification, prediction, deep learning, stacked auto-encoder

16.1 Introduction

Banking systems have generally been built using expert systems running on outdated rules to identify fraud, and these systems usually are easy to bypass. An increase in the complexity of fraud detection algorithms is to be implemented to raise the standard of security in the financial services

**Corresponding author*: vsubramaniyaswamy@gmail.com

A. Chitra, V. Indragandhi and W. Razia Sultana (eds.) Intelligent and Soft Computing Systems for Green Energy, (211–222) © 2023 Scrivener Publishing LLC

industry. According to reports [1], credit card fraud in the USA for the year 2016 amounted to $9 billion. This was higher than the revenue of companies such as PayPal and MasterCard for 2017, which amounted to $10.8 billion each [2]. The advancement of the fraud detection algorithm would prove to be essential for such companies.

Machine Learning algorithms have proved to be highly efficient in solving problems of this sort, involving data analysis to predict when an occurrence is errant and to detect any similar occurrences as they are under process. Methods such as classifiers and Support Vector Machines have been employed to solve such problems. But the more advanced machine learning algorithms, such as deep neural networks and auto-encoders, have proved to be even more accurate in solving complex problems. Auto-encoders involve analyzing given input, identifying the input's core features, and reconstructing the input based on the core features. This process is unsupervised, and it encodes the input by determining the vital features of the input by itself. Thus, recreating the input proves to be useful for several applications.

16.1.1 Deep Learning

Deep learning helps with the modeling of complex relationships between data, helping advance the field of Machine Learning. The representation of deep learning models involves multiple layers to represent the model, employing supervised and unsupervised learning algorithms for successive levels. The lower levels one brings in high abstraction levels, as the higher levels are defined by modifying the output variables from lower levels of the data model. Deep neural networks such as the Deep Belief Network (DBN) and the Stacked Autoencoder (SAE) are often employed for extracting significant characteristics from input before prediction [3].

16.1.2 Auto-Encoders

Auto-encoders are widely employed in the field of deep learning and are used in the current solution. A feature extraction algorithm runs on unsupervised neural network architecture, choosing the best features from the given set to construct an encoded input set that produces the output set with the best accuracy. It is robust compared to Principle Component Analysis (PCA) as it provides generalization in a non-linear manner [4]. Figure 16.1 shows the block diagram of an auto-encoding system. The auto-encoding system comprises an encoder function (We) to encode the given input data (x) and decoder function (Wd), which decodes the encoded data.

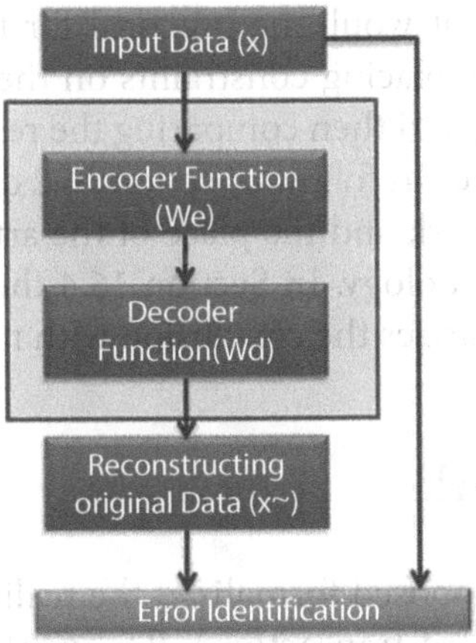

Figure 16.1 Block diagram of general auto-encoding system.

The reconstruction part brings the decoded data into its original format(x~). The performance of the auto-encoding system will be analysed based on the deviated value (error) from the original.

The basic ideology behind the model is to apply back-propagation and make the target values match the input. We are trying to make the model learn an approximation to the identity function.

An auto-encoder generally has an input layer, a hidden layer, and an output layer, similar to neural network architecture. The difference is the output layer having the same dimension as the input layer, as we are trying to reconstruct the input to make the algorithm efficient. The hidden layer has a smaller dimension compared to the input layer.

Figure 16.2 shows the typical architecture of an Auto-encoder system. Auto-encoders are generally used to modify input structures to create

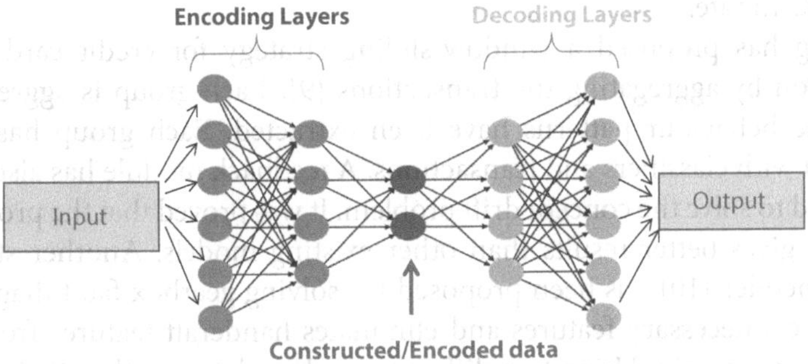

Figure 16.2 Architecture of auto-encoder system.

other data structures that would be practical for the problem statement at hand. This is done by placing constraints on the network, creating the alternate data structure, and then comparing the results with PCA.

This paper is structured as follows. Section 16.2 delivers related research on the corresponding work and the place of the article. Section 16.3 provides a proposed methodology. In Section 16.4 the results are discussed, and Section 16.5 summarizes the conclusion with future work.

16.2 Related Work

Dal. A *et al.* [5] have proposed formalized the realistic detection problem using everyday credit card transactions. Real-world data, which consists of more than 75 million card transactions, have been used, and a learning strategy has been designed for assessment. The system also addresses the class imbalance problem with concept drift. A Scalable Real-time Fraud Finder (SCARFF) was proposed by Carcillo. F *et al.* [6] is the machine learning approach for real-time fraud finding. This is the integration of Spark, Cassandra, and Kafka. The system has been tested with used imbalance, feedback, and nonstationarity parameters. The framework has used real-world card transactions.

Pure machine learning algorithm-based credit card detection is proposed by Randhawa *et al.* [7]. It is the hybrid method to evaluate the model efficacy using Adaboost and Majority voting algorithms. The proposed system has been tested with real-world data and shows that Adaboost and Majority voting outperform well. The results have also been compared with NB, DL, and SVM algorithms. V. Bhusari and S. Patil [8] have proposed a fraud detection model using Hidden Markov Model (HMM). The model uses a sequence of operations, while card transactions create HMM and are trained with cardholder behavior. The system has resulted in a low false alarm rate.

Jiang has proposed a window-sliding strategy for credit card fraud detection by aggregating the transactions [9]. Each group is aggregated, and the behaviour patterns have been extracted. Each group has been trained with classifiers and transactions. A feedback module has also been adopted to solve the concept drift problem. It was proved that the proposed system gives better results than other existing models. Another stacked auto-encoder [10] has been proposed for solving gearbox fault diagnosis. It extracts necessary features and eliminates handcraft features from the frequency domain. Using two different gearbox datasets, the efficiency of the system has been proved.

Randomization-based autoencoders (AE) are used in multi-layer (deep) neural networks to extract unsupervised features, which are then processed by an Extreme Learn Machine (ELM) classifier. Each randomization-based AE functions as a stand-alone feature extractor, and a deep network is built by stacking several of them [11]. An autoencoder can automatically learn deep hierarchical semantic features representations of data automatically, replacing operations extracted with hand-designed features [12].

16.3 Proposed Methodology

The paradigm put to use here is an encoder-decoder structure. The input is converted to a lower-dimensional hidden layer then reconstructed into a higher dimensional output layer that matches the dimensionality of the input layer [13]. The code layer present in the middle of the deep auto-encoder architecture uses classification by employing a compressed feature vector. It can also be combined within a stacked auto-encoder.

The conversion of a lower-dimensional layer in the form of the hidden layer forces the auto-encoder to select the features of the data that influence the output the most. In the most optimal solution possible, the output of the auto-encoder provides a better result and envisioning of the data entries compared to the raw data.

The non-linear hypothesis to be learned by the auto-encoder is defined as shown in equation (16.1)

$$hW,b\ (x) \approx x \tag{16.1}$$

where h represents the non-linear hypothesis to learn using the parameters W and b, which represent the weighting and bias to fit the given data. It tries to approximate itself to the identity of the function. The process employed is reconstruction error minimization, using a loss function to calculate the error and penalizing based on the error.

The loss function is defined as in equation (16.2).

$$L(x, d(f(x))) \tag{16.2}$$

L is the loss function which penalizes d(f(x)) for varying from x. In auto-encoder terms, d is the decoder, and f is the encoder function.

A deep auto-encoder structure involves two deep-belief network places symmetrically. Each consists of four or five shallow layers that encode and another set of four or five layers. Deep learning, as such, can be placed in

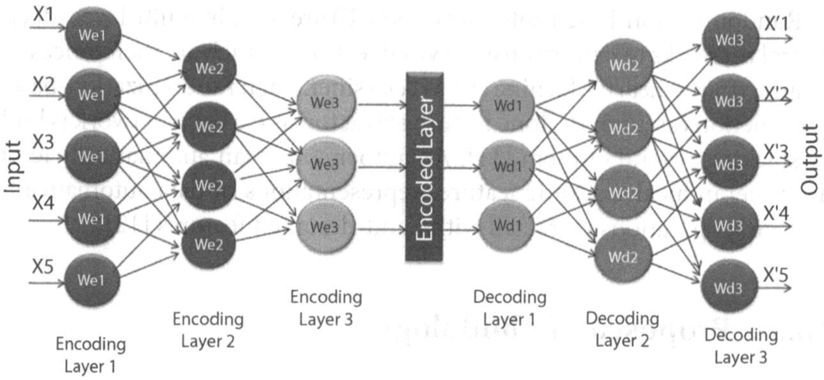

Figure 16.3 Structure of stacked auto-encoder with three layers [15].

auto-encoders, multiple hidden layers can be placed to provide complexity in a method called stacked auto-encoder [14]. The output from a hidden layer is sent as input to a higher level. Each layer progressively learns first-order, second-order features, and so on. Figure 16.3 shows the structure of a stacked auto-encoder.

Where xi is the input, and x`i is the output. Wei represents the encoding function with respect to 1st layer and Wdi denotes the decoding function with respect to 1st layer. The dataset is initially scaled to bring all the features to fit in similar scales of magnitude. This is done to ensure that one feature does not overwhelm the others just because the values of the feature are higher. As mentioned before, the PCA transform is fitted onto the dataset. The dataset is then split into the training and test datasets, where the training set is used to build the model, which is then fitted onto the test dataset. The tuples are then labeled as Normal and Fraud classes, based on the detection of fraud. Since the number of fraud cases is highly nominal compared to a Normal transaction, the frequency graph, for instance, of Fraud and Nominal classes, shows the frequency of the Fraud class to seem insignificant. Figure 16.4 shows the frequency graph for the class labels present in the dataset.

The auto-encoder structure involves symmetric encoding and decoding layers. In this specific case, the layers are dense, making sure the layers are fully connected. Designing an auto-encoder for this particular fraud detection is done using different models for the auto-encoder, varying number of layers in the auto-encoder, the activation function, and the threshold value.

The variation in the number of layers is shown using a two-layer stacked auto-encoder and a three-layered stacked auto-encoder to explore the accuracy differences shown with variation in the layers.

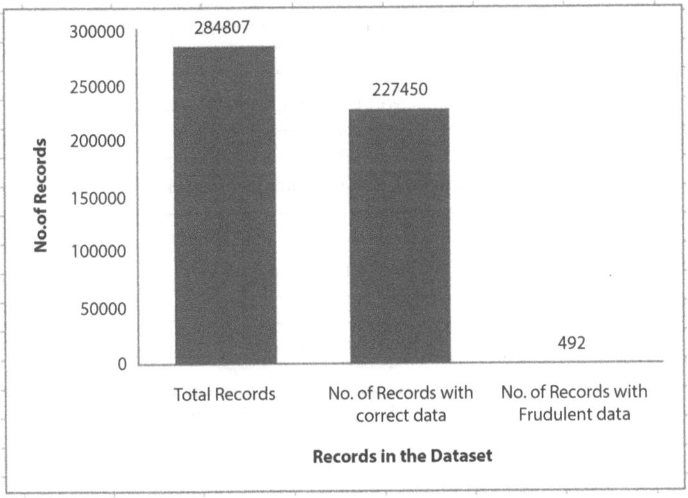

Figure 16.4 Frequency observation for normal and fraud classes.

The activation function is used in the auto-encoder to map a given node's output into the required range based on the input to be sent to the next node. This is done to ensure all the nodes get inputs of a similar range and make sure the desired solution is obtained.

The activation functions used are tanh and relu, where tanh brings data into the range (-1,1), and relu brings data into the range (0,x). The threshold value is used to validate whether a specific output value is to be taken into consideration or not. If the value does not exceed the threshold value, it can be deemed insignificant and need not be taken into account. The stacked auto-encoder is built using these parameters, the flow diagram of which is shown in Figure 16.5.

The dataset taken into consideration contains transaction details for credit cards used by European cardholders during September 2013, spanning two days. The dataset has 284,807 entries, of which 492 entries are labeled to be a fraud. The number of fraud accounts for only 0.172% of all transactions, making the dataset highly unbalanced. The dataset is subject to an initial PCA transformation, as all original details about cardholders and the transactions cannot be disclosed. The dataset contains 28 columns of PCA transformed features and two other features. One is 'Time' for indicating the elapsed time between the first and current transaction. The second additional feature is the 'Amount' which indicates the amount involved in that transaction. The feature 'Class' is the response variable; it takes the value 1 in case of fraud and 0 otherwise.

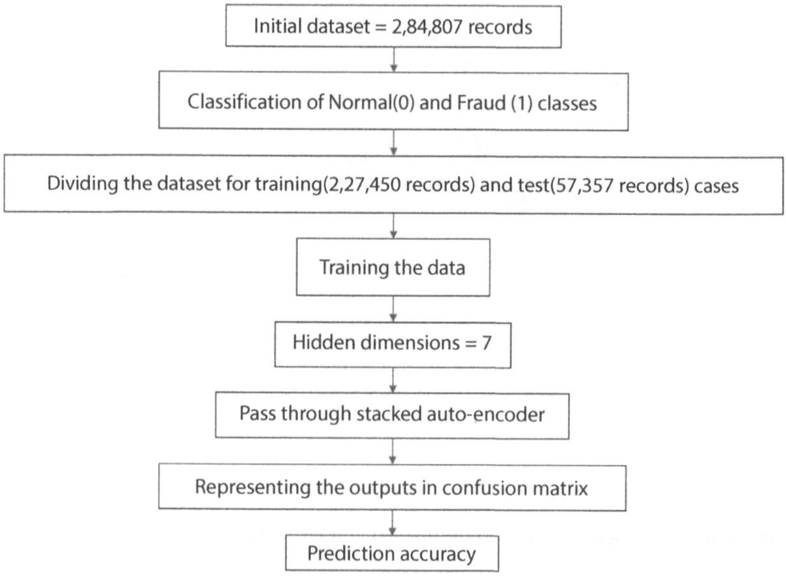

Figure 16.5 Flow diagram for the building of stacked auto-encoder model.

16.4 Results and Discussion

The proposed system is implemented in Keras; an API is used to build deep learning models for complex operations. It is used in research activities, fast prototyping, and production because of its advantages, as it is user-friendly, modular, composable, and easy to extend. The Keras API is imported and extended using the TensorFlow platform provided by Google. Deep learning models are designed, built, and trained using TensorFlow. To do numerical computations using data flow graphs, we used the TensorFlow framework. TF's is more advanced, especially when it comes to high-level processes such as threading and queues, and debugging. TensorFlow is valuable because it can scale issues indefinitely—graph nodes can run throughout a dispersed network. The logic in TF is unique in that it uses both the CPU and the GPU of the computer. TF now has a lot more power per machine because of the addition of the graphical processor unit.

Statistical data for the dataset is calculated, the mean, standard deviation, min, and max values for the original data and the reconstructed data after passing through the auto-encoder. The statistical data is shown in Table 16.1.

There is a general loss in data quality after passing it through the auto-encoder. Hence, we have designed an auto-encoder for predicting the

Table 16.1 Statistical output for the proposed system.

	Reconstruction error	True class
Count	56962.000000	56962.000000
Mean	1.094645	0.001703
Std. Deviation	3.999082	0.041231
Min	0.164545	0.000000
25%	0.430645	0.000000
50%	0.658607	0.000000
75%	1.003150	0.000000
Max	333.940175	1.000000

accuracy of normal and fraud clauses in the credit card fraudulent dataset. There are three different conclusions obtained with the variations in the accuracy of the prediction. All these values directly affect prediction accuracy.

Case 1: For the first case, the accuracy predicted is 87.19512%. Here there is no threshold value used, and those prediction list values that are more than 0.126 are used for the accurate prediction. It is a simple stacked auto-encoder with two layers of encoding and decoding. The activation functions "tanh" and "relu" are used in predicting a non-linear output.

Case 2: For the second case, the accuracy predicted is 74.39024%. Here a three-layer stacked auto-encoder is used with additional usage of the "sigmoid" activation function. By this experiment, there was a belief in the increase in the accuracy of the prediction. But with the threshold value fixed to 1, the accuracy reduced to an approximate value of 74%. Thus with the increase in layer and introduction of the threshold value, the accuracy seemed to reduce.

Case 3: The second case was extended to new conclusions by increasing the threshold values. When the value reached 5, the accuracy obtained was 98.17073%. This is the highest value of accuracy obtained, and with further change, the accuracy fluctuates around the same value.

Figure 16.6 shows the various accuracy achieved through the stacked Auto-encoder model under no threshold value, threshold value=1 and threshold value=5. Thus for the three-layer stacked auto-encoder with a given threshold value of 5, maximum accuracy is obtained. The confusion matrix for this scenario is shown in Figure 16.7.

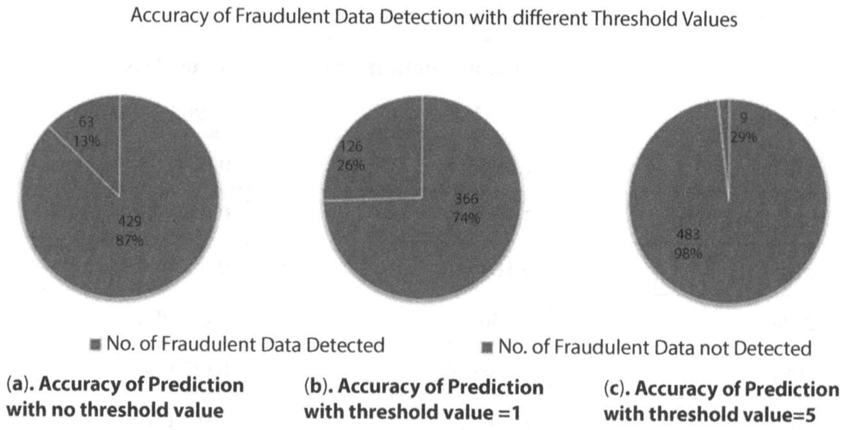

Figure 16.6 Accuracy of stacked auto-encoder system with different threshold values.

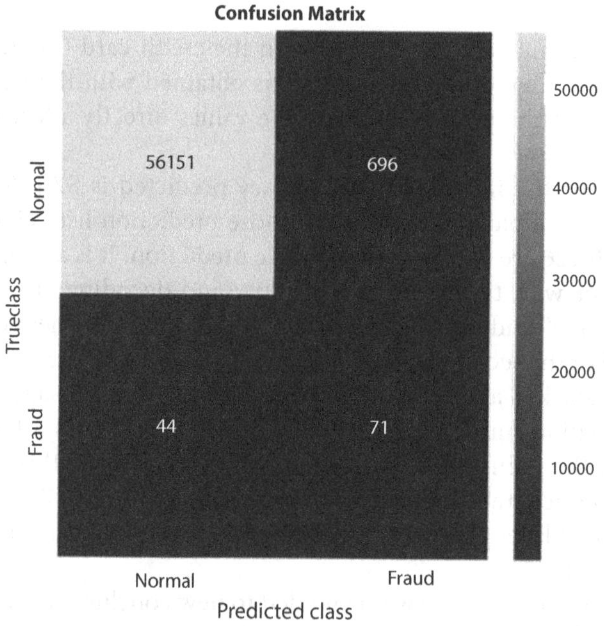

Figure 16.7 Confusion matrix is shown for case 3.

The key features for the change-in accuracy are due to different activation functions, the number of layers in stacked auto-encoders, and the threshold values. An activation function is one that is used for getting non-linear outputs. In the case of different encoding layers, the activation function helps prevent the collapse of layers of encoding. The increase in stack level is directly proportional to the efficiency of the prediction as it involves an additional layer of processing. The threshold value is used for the construction of a confusion matrix.

16.5 Conclusion

Online users are most vulnerable at the credit card and debit card facilities for online banking systems. A stacked auto-encoder with three layers has been designed for detecting credit card fraud in this proposed system. The deep learning-based system was tested with credit card transaction data set, and the accuracy has been measured. The statistical data and final result in all three different cases show that the system outperforms accuracy. In the future, the system can be extended for medical image processing applications.

Acknowledgment

The authors gratefully acknowledge the Science and Engineering Research Board (SERB), Department of Science & Technology, India, for the financial support through the Mathematical Research Impact Centric Support (MATRICS) scheme (MTR/2019/000542). The authors also acknowledge SASTRA Deemed University, Thanjavur, for extending infrastructural support to carry out this research work.

References

1. https://www.nasdaq.com/articles/credit-card-fraud-and-id-theft-statistics-2015-09-16.
2. Verizon Annual Report (2018), https://www.verizon.com/about/sites/default/files/2018-Verizon Annual-Report.pdf
3. Kinh, B. T., Anh, D. T., & Hieu, D. N. (2020). A Comparison Between Stacked Auto-Encoder and Deep Belief Network in River Run-Off Prediction. In *Context-Aware Systems and Applications, and Nature of Computation and Communication* (pp. 65–81). Springer, Cham.
4. Marivate, V. N., Nelwamodo, F. V., & Marwala, T. (2007). Autoencoder, principal component analysis and support vector regression for data imputation. *arXiv preprint arXiv:0709.2506.*
5. Dal, A. P., Boracchi, G., Caelen, O., Alippi, C., & Bontempi, G. (2018). Credit Card Fraud Detection: A Realistic Modeling and a Novel Learning

Strategy. *IEEE Transactions on Neural Networks and Learning Systems, 29*(8), 3784–3797.

6. Carcillo, F., Dal Pozzolo, A., Le Borgne, Y. A., Caelen, O., Mazzer, Y., & Bontempi, G. (2018). Scarff: a scalable framework for streaming credit card fraud detection with spark. *Information Fusion, 41*, 182–194.
7. Randhawa, K., Loo, C. K., Seera, M., Lim, C. P., & Nandi, A. K. (2018). Credit card fraud detection using AdaBoost and majority voting. *IEEE Access, 6*, 14277–14284.
8. Bhusari, V., &Patil, S. (2011). Application of Hidden Markov Model in credit card fraud detection. *International Journal of Distributed and Parallel Systems, 2*(6), 203.
9. Jiang, C., Song, J., Liu, G., Zheng, L., & Luan, W. (2018). Credit card fraud detection: A novel approach using aggregation strategy and feedback mechanism. *IEEE Internet of Things Journal, 5*(5), 3637–3647.
10. Liu, G., Bao, H., & Han, B. (2018). A stacked auto-encoder-based deep neural network for achieving gearbox fault diagnosis. *Mathematical Problems in Engineering, 2018*.
11. Katuwal, R., &Suganthan, P. N. (2019). Stacked autoencoder based deep random vector functional link neural network for classification. *Applied Soft Computing*, 85, 105854.
12. Yuan, C., Chen, X., Yu, P., Meng, R., Cheng, W., Wu, Q. J., & Sun, X. (2020). Semi-supervised stacked autoencoder-based deep hierarchical semantic feature for real-time fingerprint liveness detection. *Journal of Real-Time Image Processing*, 17(1), 55–71.
13. Muhammad, G., Hossain, M. S., & Garg, S. (2020). Stacked autoencoder-based intrusion detection system to combat financial fraudulent. *IEEE Internet of Things Journal.*
14. Vincent, P., Larochelle, H., Lajoie, I., Bengio, Y., &Manzagol, P. A. (2010). Stacked denoising auto-encoders: Learning useful representations in a deep network with a local denoising criterion. *Journal of Machine Learning Research, 11*(Dec), 3371–3408.
15. Prakash, P. K. S., & Rao, A. S. K. (2017). *R Deep Learning Cookbook*. Packt Publishing Ltd.

17

Artificial Intelligence-Based Ambulance

Dr. R. Sumathi*, R.M. Gokul, M. Gokulakrishnan, K. Ganesh Babu and S. Pavithra

Department of Electrical and Electronics Engineering, Sri Krishna College of Engineering and Technology, Coimbatore, India

Abstract

Street gridlock turns into a significant issue for exceptionally packed metropolitan urban areas. India is the second most populated country on the planet and is a quickly developing economy. It is confronting awful street blockage in the urban communities. As per *Times of India*, around 30% of deaths are caused because emergency vehicles experienced delays in reaching a clinic. In the proposed framework we are attempting to reduce the delay for the rescue vehicle. To smooth the development of emergency vehicles we think of "Keen Ambulance". We are attempting to give green signs to the path where the emergency vehicle needs to go by physically turning a switch on a specific path. We will utilize the innovation loved RF Module. This framework was planned so that it would be initiated when it got a signal from a rescue vehicle dependent on radio recurrence (RF) transmission, and we utilized Arduino to change the succession back to the typical arrangement before the crisis mode was enacted. In the second stage, we are building up a site for doing enrollment about clinical history, everything being equal. This information will assist with saving time in the clinic as it prepares to treat the incoming patient. This information can be recovered by utilizing exceptional id and unique finger impression confirmation. This produced information will be shipped off to the specific clinic before the arrival of the emergency vehicle. As this framework is completely computerized, it perceives the emergency vehicle and control traffic lights. This framework controls traffic signals and saves time in a crisis period. Accordingly, it counts as a lifeline project.

Keywords: RF module, traffic monitoring, fingerprint authentication

**Corresponding author*: sumathir@skcet.ac.in

A. Chitra, V. Indragandhi and W. Razia Sultana (eds.) Intelligent and Soft Computing Systems for Green Energy, (223–236) © 2023 Scrivener Publishing LLC

17.1 Introduction

17.1.1 Problem Statement

In today's world even with an increase of the number of vehicles, traffic signals are programmed and still running on fixed timers, which do not vary based on the volume of vehicle accumulation at junctions. Due to this scenario there will be a chance of increased waiting time. No provisions are available with present traffic monitoring system for getting any information about vehicles. Because of this, it will become very difficult to track a vehicle and to control signals. So, this creates complexities in emergency situations and may put lives at risk. Medical records containing all information connected with medical care of the patient is the most crucial information in terms of treatment of the patient [1–5]. When a patient enters the hospital in emergency conditions, delay occurs in starting the treatment and analyzing the patient's medical history and maintaining the patient medical records manually may result in loss of information of patients. Nowadays, policies and technology are rapidly moving towards the security of patient records [6–10].

17.1.2 Field of the Project

The project mainly focuses on sensing information with the help of human biometrics. This is a secure method for keeping records. Just by perceiving aunique mark we can acquire the data related to that patient. This is exceptionally advantageous in terms of keeping records. Biometric reacts quickly, ordinarily distinguishing a patient in just one second. Just by perceiving finger impression reality that biometric certifications are special for every persistent and can't be neglected or replicated. This innovation will serve to precisely follow a patient time. The following field of our undertaking centers on correspondence innovation. For correspondence we use RF module 434 MHz; this is utilized for single-way correspondence and works under Amplitude Shift Keying (ASK) Modulation, otherwise called Binary ASK. RF Transmitter is joined to the crisis vehicle and RF Receiver will be introduced closer to the flagging framework. The other significant zone of this venture is robotization, where the patient's clinical records can be recovered and shipped to the specific medical clinic. By utilizing the gathered information, the specialists or medical attendant can plan essential therapy or emergency treatment for the specific patient.

17.1.3 Objectives

- to provide basic first aid to every patient on time.
- to attend to the patient and send him or her to hospital as soon as possible amidst the heavy traffic zone.
- to change the traffic signal according to the direction of the ambulance.
- To achieve this condition, the direction switch in the ambulance must be pressed, then the signal is immediately transmitted from the ambulance to the receiver of the traffic control system via RF Module.
- To start the treatment immediately, previous medical records can be accessed with the help of biometric sensor and the patient health records can be analyzed prior.

17.2 Proposed System

17.2.1 Block Diagram of Traffic Signal Control System

Whenever a switch is pressed, a parallel data is sent to HT12E which encodes the parallel data into serial data and feeds it to RF transmitter. The RF receiver receives the serial data and feeds it to HT12D which decodes the serial data into parallel data, and it is sent to Arduino NANO. It then processes the received data and adjusts the traffic signal accordingly. The overall block diagram of traffic signal control system is shown in Figure 17.1. The biometric system logic is shown in Figure 17.2.

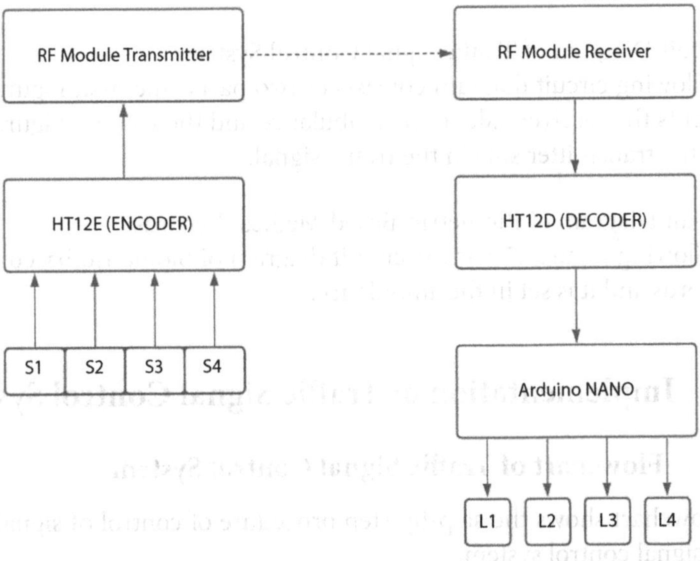

Figure 17.1 Traffic signal control system.

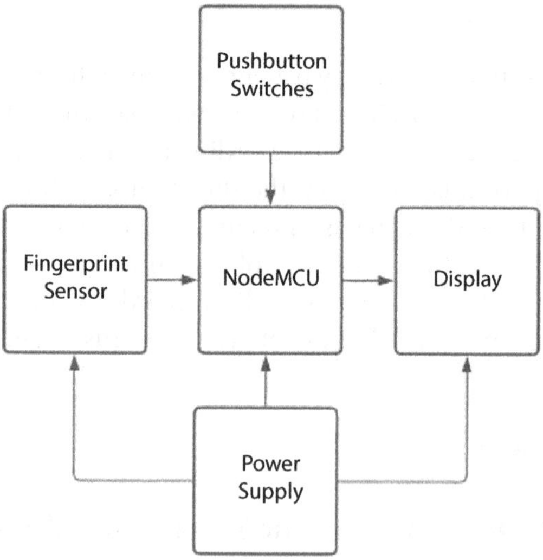

Figure 17.2 Biometric.

17.2.2 Block Diagram of Biometric-Based Medical Records

While escorting a patient to hospital, their medical records are accessed using their fingerprint and it will be sent to the hospital immediately via email. Another feature of this system is to send information about what type of patient is arriving to the hospital via email.

A. Circuit Diagram of Traffic Signal Control System:
The following circuit diagram consists of two parts. The first, Figure 17.3, represents the receiver side in the ambulance and the second, Figure 17.4, shows the transmitter side in the traffic signal.

B. Circuit Diagram of Biometric-Based Medical Records:
The following Figure 17.5 is the circuit diagram of biometric-based medical records and it is set in the ambulance.

17.3 Implementation of Traffic Signal Control System

17.3.1 Flowchart of Traffic Signal Control System

This flowchart shows the step-by-step procedure of control of signals in a traffic signal control system.

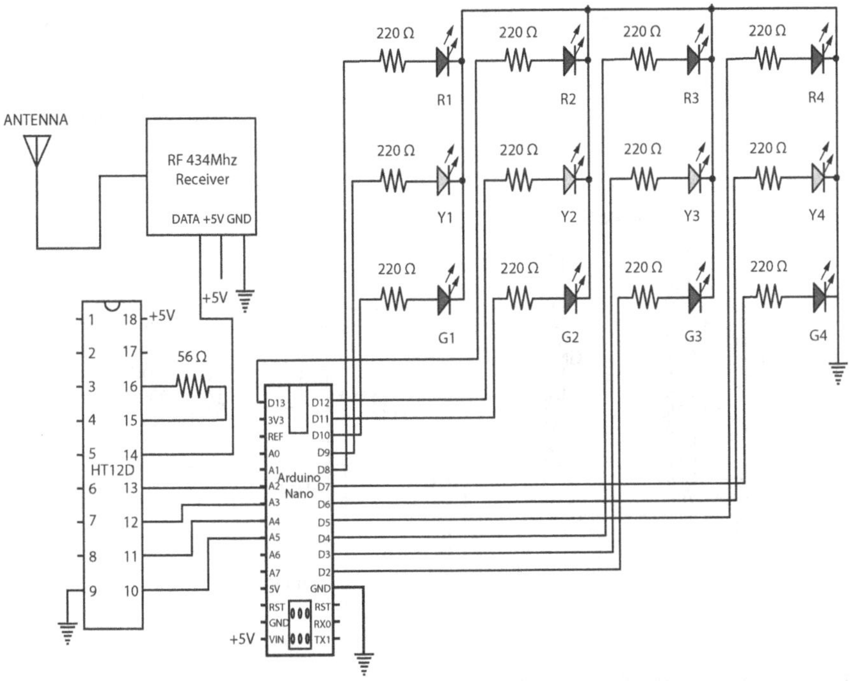

Figure 17.3 Receiver side.

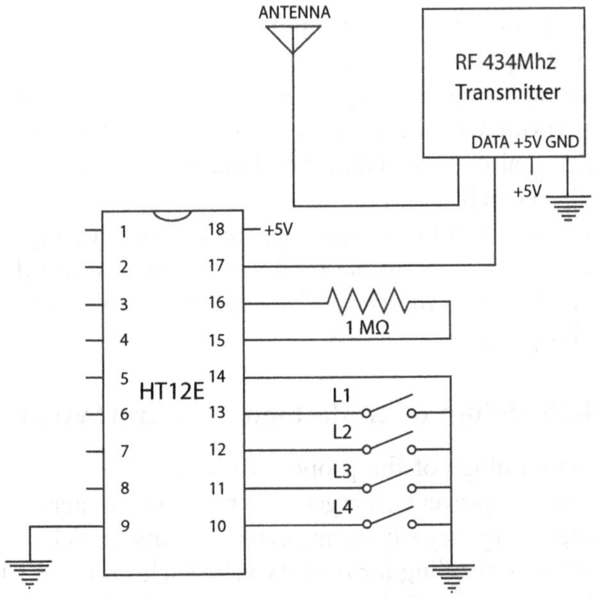

Figure 17.4 Transmitter side.

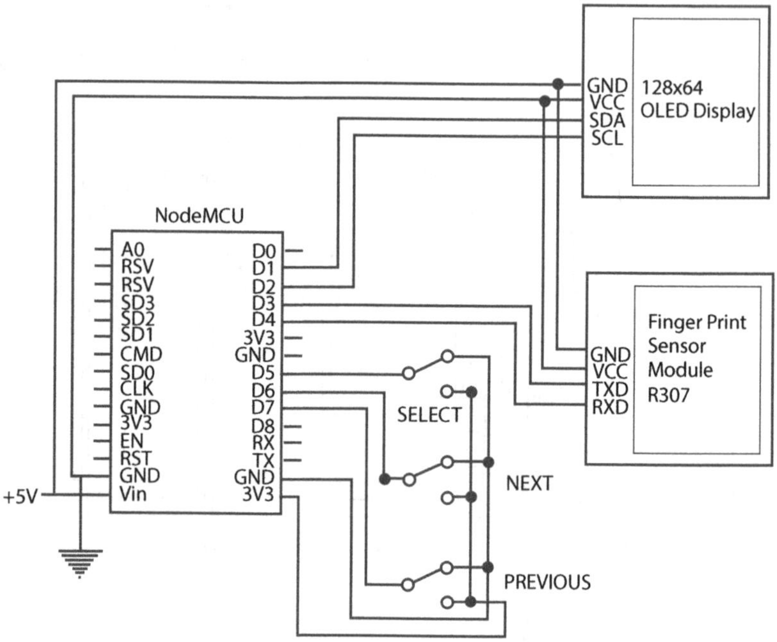

Figure 17.5 Biometric-based medical records.

17.3.2 Algorithm of Biometric-Based Medical Records System

- The system waits for user input.
- Then it captures the fingerprint using a sensor.
- After capturing the fingerprint, it analyses and matches the fingerprint with its medical data of the particular person which is already stored in it and that data is sent to the hospital via email.
- If the fingerprint captured does not match with any data, it again comes to its initial mode, where we can use the predefined options to notify what type of patient is arriving at the hospital.

17.3.3 Methodology of Traffic Signal Control System

The main methodology of the proposed model is to allow clear flow of vehicles in order to prevent emergency vehicles encountering traffic congestion during emergency situations, as the existing model is inefficient to solve congestion controlling for priority vehicle clearance. So, this project illustrates the "Intelligent Traffic Control System" using radio frequency wireless communication technology. The adopted flow is depicted in the

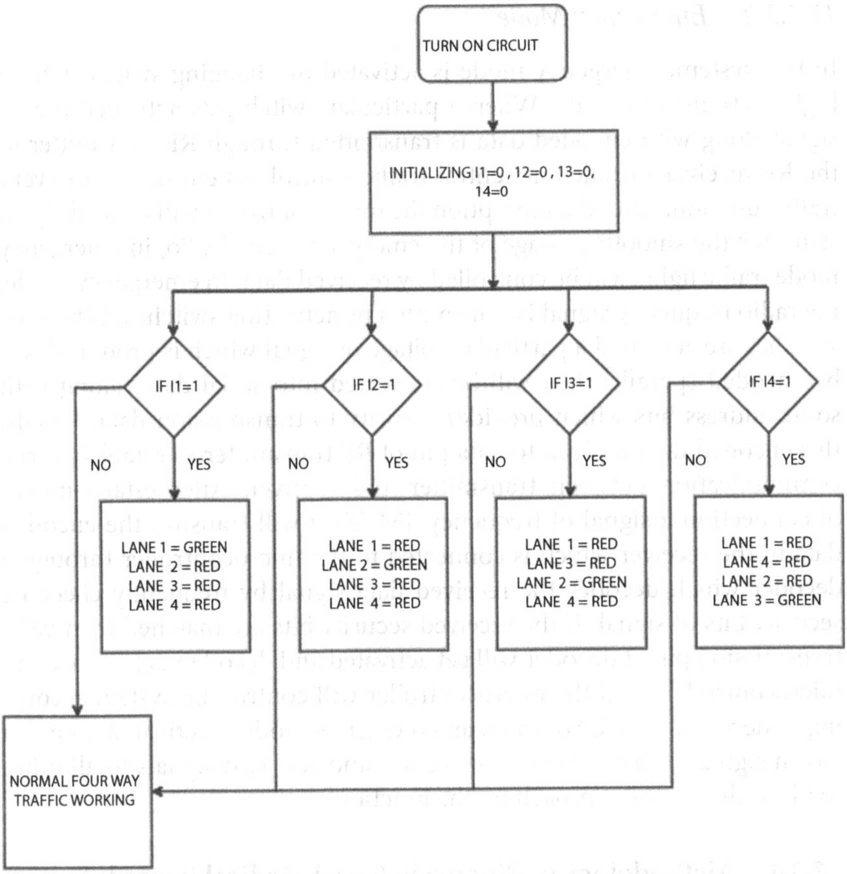

Figure 17.6 Flowchart.

flowchart shown in Figure 17.6. This system primarily consists of Arduino Uno (ATmega 328p) microcontroller, Encoder (HT12E), Decoder (HT12D), RF transmitter and RF receiver are mainly designed to work under two modes, normal mode and emergency mode.

17.3.3.1 Normal Mode

In normal mode the whole working of the system is based on operation of on-chip microcontroller which is programmed to control traffic signals with fixed predefined time intervals. So, based on predefined time intervals signals in different ways are getting altered at different intervals of time. Similarly, every 18 central traffic control system is programmed with different functions and methods according to the traffic congestion in a particular way.

17.3.3.2 Emergency Mode

In this system, emergency mode is activated by changing switch state to high on transmitter side. When a particular switch gets activated then a signal along with encoded data is transmitted through RF transmitter to the RF receiver installed at central traffic control system nearer to every traffic junction. After data reception the microcontroller will control signal states for the smooth passage of the emergency vehicle. So, in emergency mode traffic lights will be controlled by received data. In emergency mode, the radio frequency signal is transmitted by activating switches. When the switches are activated a particular voltage of signal which is produced will be encoded (parallel data will be converted into serial data) along with some address bits which provides security to transmission data. Finally, this encoded data is given to data pin of RF transmitter to establish serial communication between transmitter and receiver. After establishment of connection a signal of frequency 434 MHz will transmit the encoded data to the receiver which is connected to the microcontroller through a decoder which decodes the received data signal by frequently checking security bits of signal. If the received security bits are matched then valid transmission pin of decoder will get activated and decoded data is given to microcontroller. And the microcontroller will control the system according to decoded signal. So, when an emergency mode is activated it gives a Green signal to that particular direction and sets Red signals to all other roads or directions approaching the junction.

17.3.4 Methodology of Biometric-Based Medical Records System

This Biometric-based Medical Record System works in the following manner. At whatever point a patient gets into the emergency vehicle, his/her unique mark is filtered; their recently put away clinical information is recovered and it is shipped off the clinic by means of email. Something else about this framework is that it has predefined choices in it which is gotten to utilizing the given press catches and the ideal alternative is chosen which is additionally shipped off to the clinic through email. Those alternatives address what sort of patient will be showing up at the emergency clinic. The convention utilized here is SMTP (Simple Mail Transfer Protocol).

17.3.4.1 SMPT

Simple Mail Transfer Protocol is a TCP/IP convention utilized in sending and accepting email. Be that as it may, since it is restricted in its capacity to line messages at the accepting end, it is normally utilized with one of two different conventions, POP3 (Post Office Protocol) or IMAP (Internet Message Access Protocol), that let the client save messages in a worker letter box and download them occasionally from the worker. As such, clients ordinarily utilize a program that utilizes SMTP for sending email and either POP3 or IMAP for accepting email. SMTP fills in as a three-venture measure, utilizing a customer/worker model. Initial, an email worker utilizes SMTP to communicate something specific from an email customer, like Outlook or Gmail, to an email worker. Second, the email worker utilizes SMTP as a hand-off help to send the email to the accepting email worker. Third, the receiving worker utilizes an email customer to download approaching email by means of IMAP and place it in the inbox of the recipient.

17.4 Result and Discussion

17.4.1 Comparison of Results

The proposed traffic control system was implemented and the following Table 17.1 shows the output of the traffic signal in which the ambulance or any emergency vehicle passes through that particular lane.

The Biometric-Based Medical Records System is implemented when a patient gets into the ambulance; their fingerprint is scanned and it will find its match in the database. After finding its match, the acquired data will be sent to the hospital immediately via email.

17.4.2 Hardware Result

In this manner, the traffic light control framework was tried under various conditions and the yield was checked. The typical data transfer logic is shown in Figure 17.7a and the respective mail output is displayed in Figure 17.7b. Figure 17.8 below shows the hardware setup of Traffic Signal Control System.

Table 17.1 Output of the traffic signal in four lanes.

Switch	R1	Y1	G1	R2	Y2	G2	R3	Y3	G3	R4	Y4	G4
L1	0	0	1	1	0	0	1	0	0	0	0	0
L2	1	0	0	0	0	1	1	0	0	0	0	0
L3	1	0	0	1	0	0	0	0	0	1	0	0
L4	1	0	0	0	0	0	1	0	0	0	0	1

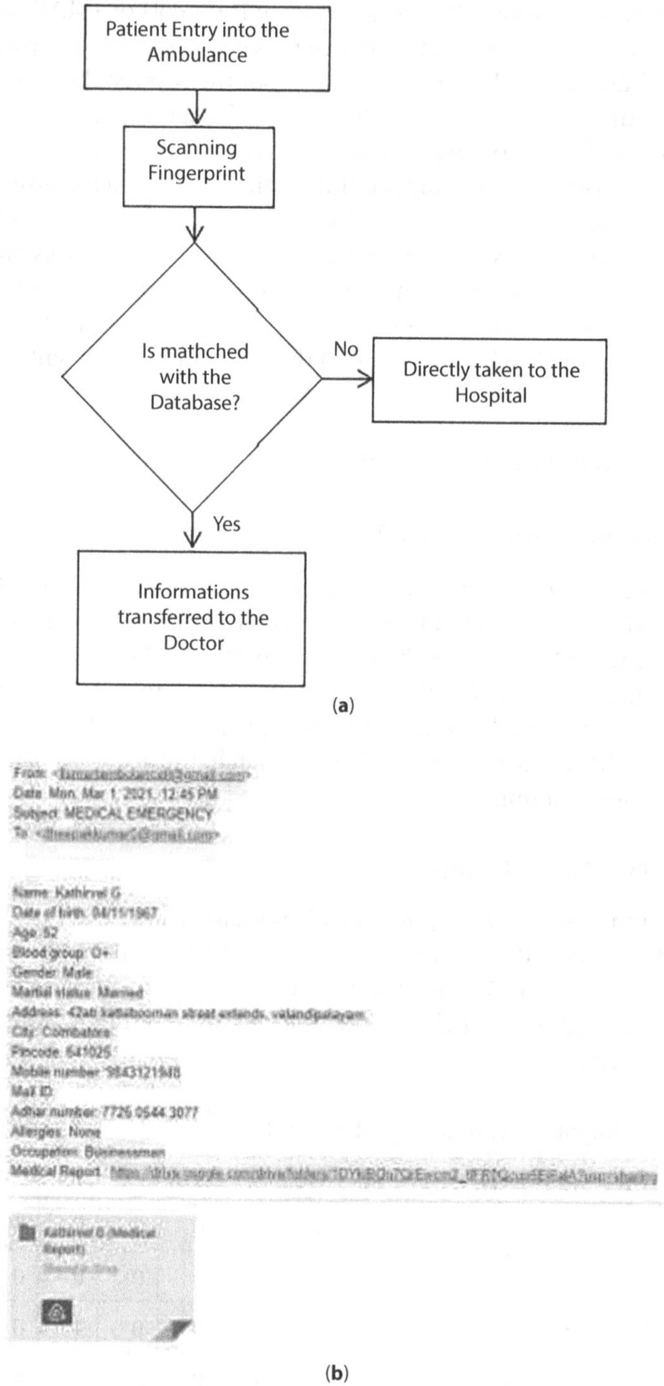

Figure 17.7 (a) Flowchart for data transfer. (b) Mail output.

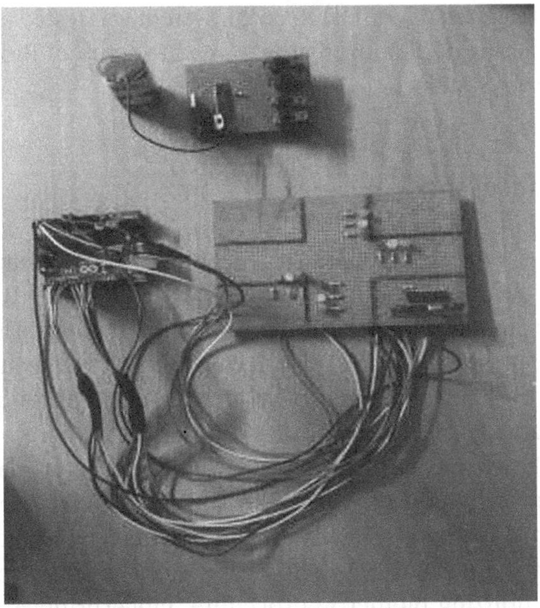

Figure 17.8 Hardware.

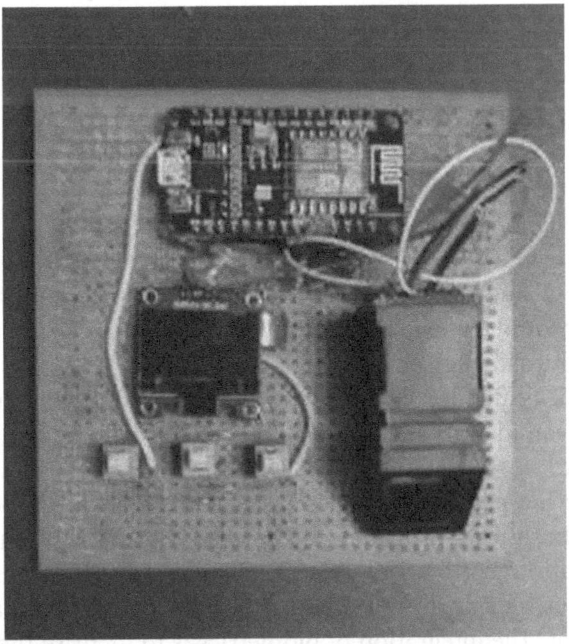

Figure 17.9 Hardware 1.

The Biometric-Based Medical Records System was tested and the output was verified. Figure 17.9 above is the hardware setup of Biometric-Based Medical Records System.

17.5 Conclusion

This venture, "Savvy Ambulance", has been effectively planned and tried. In this execution of traffic light control framework, we have utilized Radio Frequency Technology. It is created with reconciliation of all equipment segments. The presence of each module has been analyzed out and set cautiously along these lines adding to the best working of the unit. Besides, with the advantage of growing innovation utilizing exceptionally progressed IC's the task has been effectively executed. The primary advantage of this Biometric-Based Medical Record System is online increase of the patient information base. Quite possibly the main application is that it very well may be utilized during a crisis situation. Patient record errors can be limited by utilizing biometric procedure. Fingerprints are permanent. Our fundamental objective is to make a maintainable, helpful and secure arrangement thatpermits the specialists to all the more successfully utilize patient information to improve generally well-being, quality and effectiveness of care.

17.6 Future Scope

- Further upgrades should be possible to the model by testing it with longer reach RF modules.
- As of now, we have carried out a framework by thinking about one street of the traffic intersection. It tends to be improved by extending it to every one of the streets in a multiroad intersection.
- Basic health details can be included by government. Further, we can identify the address and contact number by using a card.
- The Biometric-Based Medical Records system can be used by healthcare providers to keep records and secure patient health records. The system is expected to enhance the effectiveness and the overall efficiency of hospital management. The integration of biometrics is to increase the user's confidence in this system.

References

1. Inam Ullah Khan, Muhammad Umar Khan, Hafiz Muhammmad, Syed Bilal Hussain Shah, "Traffic Density Based Light Control System", 2017.
2. R. Sureshkumar, R. Balaji, G. Manikandan and Appanaboina Masthan, "Advanced Traffic Clearance System for Ambulance Clearance Using Rf-434 Module", 2016.
3. Bilal, J. M., & Jacob, D. Intelligent Traffic Controller System. *IEEE International Conference on Signal Processing and Communications*, 2007.
4. R. Puviarasi, Mritha Ramalingam, Elanchezhian Chinnavan," Design of Intelligent Traffic Controlling System Using RF Transponder", 2018.
5. Neerav Bhatia, Abhishek Bedi, RavinderNath Rajotiya, "Smart Ambulance using RFID Technology", 2018.
6. https://www.elprocus.com/rf-module-transmitter-receiver
7. Kardas G, Tunali ET, 2006, Design and implementation of smart card based health care system for patient record Management, *Comput Methods Programs Biomed*, Jan., 81(1):66–78; doi: 0.1016/j.cmpb.2005.10.006
8. Yang Qinfeng, Mu Li, 2007, Out-patient-centered information system design, *New England Journal of Medicine*, pp. 209–304.
9. Jay Lohokare, Reshul Dani, Sumedh Sontakke, Ameya Apte, Rishabh Sahni, "Emergency services platform for smart cities", 2017.
10. Veeramuthu Venkatesh, V Vaithyanathan, B.Manikandan & Pethuru Raj, "A Smart Ambulance for the Synchronized Health Care –A service Oriented Device Architecture-Based", 2012.

18

LoRa-Based Flaw Location Detection in HT Line Using GSM

Dr. M. Senthilkumar, Abisheck D., Gnana Prakash K.*, Hari Babu S. and Hariharan R.

Department of EEE, Sri Krishna College of Engineering and Technology, Coimbatore, India

Abstract

In the cutting-edge situation, because of expansion sought after and supply through the world, the force framework field is going ahead and headways were being acquainted to our day-to-day routine. The Electric Power System is divided into an extensive range of sectors, one of which is transmission framework, where force is sent from creating stations and substations through transmission lines into shoppers. The HT lines flow between power transformer and distribution transformer. The voltage that flows through HT line is 11KV, 22KV and 33KV. Flaw is characterized as various unwanted yet unavoidable episodes that can incidentally upset the steady state of the force framework that happens when the protection of the framework fizzles anytime. A brilliant GSM-based flaw recognition and area frameworks was utilized to precisely show and find the deficiency confinement that had happened. LoRa is utilized for the long reach correspondence and the LoRa correspondence is utilized to impart in the middle of the transmitter and the recipient. The UNO gives the essential data the whole circuit and the potential transformer is utilized the action the voltage of the HT line and the LCD is utilized to show the situation with the yield like voltage and address of the post. The Relay is utilized to guarantee the stumbling conduction under any deficiency conductions and the heap is associated with guarantee the stumbling of whole circuit. Finally, the issue data is communicated to the control room and the time needed to find an issue is radically diminished, as the framework consequently and precisely gives exact deficiency area data. GSM is utilized to send SMS to determine versatile number. The proposed technology will help to sense and rectify the faults within a short time period. This system will help us to save the power system from damage and disasters.

***Keywords*:** LoRa, GSM, Arduino uno, relay, transformer

**Corresponding author*: 17euee038@skcet.ac.in

A. Chitra, V. Indragandhi and W. Razia Sultana (eds.) Intelligent and Soft Computing Systems for Green Energy, (237–254) © 2023 Scrivener Publishing LLC

18.1 Introduction

The power system has different operating elements, and there may be the possibility of disturbances and faults at all times. HT lines are more vulnerable to faults and they are exposed to atmosphere and there is a chance of occurring faults, the fault to be sensed and cleared as soon as possible to ensure the system consistency. The framework imperfections are naturally recognized and afterward convey the information, for example, location and transmission fault to the control room. The GSM network gives solid correspondence quality cross country inclusion. Message system has now suited the most broadly utilized help dependent on GSM standard. Simultaneously the diminishing expense of GSM gadgets, for example, cell phones and the GSM SMS give a remarkable location to the controller unit and orders can be sent in the remote correspondence organization. There are numerous courses of deficiencies in power transmission prompting blackouts, if not appropriately overseen.

Notable among these deficiencies are:

- Faults because of inappropriate earth to HT posts.
- Damage HT lines (tree falling on lines or birds in the middle of lines).
- Faults because of inappropriate support of circles/pins and mistake at the appropriation transformer.
- Lightning strike.

18.1.1 Different Types of Transmission Line Fault

18.1.1.1 Single Line-to-Ground Fault

The most well-known sort of shunt inadequacies is Single Line-to-Ground flaws. This sort of shortcoming happens when one conveyor tumbles to the ground or gets into contact with the unbiased wire. It could likewise be the consequence of the falling of tree branches in a storm. The Single Line-to-Ground fault type is shown in Figure 18.1.

18.1.1.2 Line-to-Line Fault

This fault happens mostly due to shunt shortcomings in the Line-to-Line deficiency (LL). This is declared when two of the HT lines are short circuited, caused for example by a tree falling on the HT lines or by contact of birds in between two HT lines, as shown in Line-to-Line fault at Figure 18.2.

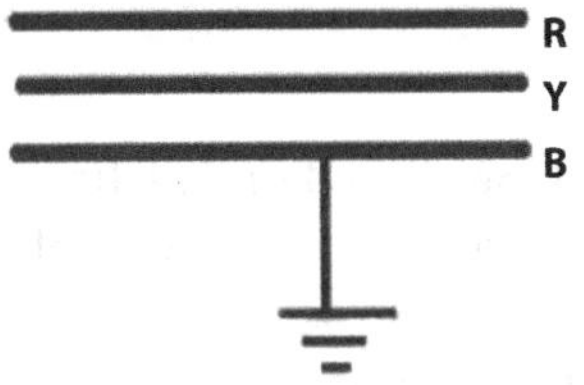

Figure 18.1 Single line-to-ground fault.

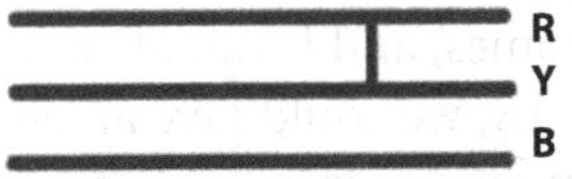

Figure 18.2 Line-to-line fault.

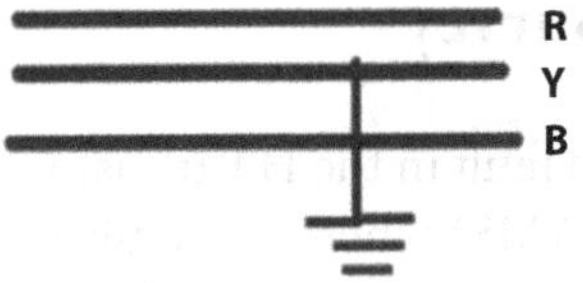

Figure 18.3 Double line-to-ground fault.

18.1.1.3 Double Line-to-Ground Fault

The Double Line-to-Ground deficiency (DLG) is shown in Figure 18.3. This could be a consequence of birds in between two electrical cables, or from different causes. This causes a short in two stages where the twofold line gets short during any shortcoming conditions.

18.1.1.4 Balance Three-Phase Fault

The three phases come in contact with each other and get shorted due to any critical situations. The power supply of HT lines gets interrupted by this type of fault. Balance three-phase fault is shown in Figure 18.4.

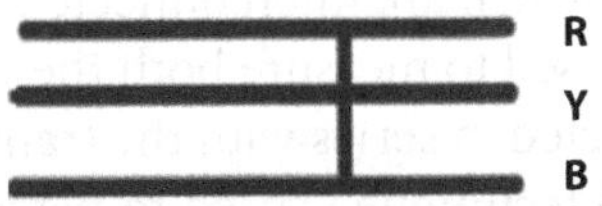

Figure 18.4 Balance three phase fault.

18.2 Objective

This work deals with the GSM based on HT line checking and sign framework that sends data of the equivalent to control room through messaging system. The carried-out framework arrangement mainly focuses on the appropriation framework. The main venture has continuous monitoring framework organizing ARDUINO microcontroller with GSM-based communication. The execution of the framework will save a huge measure of power and subsequently power will be available for a greater number of consumers, residential homes, and hospitals in a heavily populated country like India. Consequently, we could give an electrical inventory with no aggravations and can amend when the issue happens. For this situation, there can be large savings in labor and time.

18.3 Literature Survey

A system is used to find a fault in the HT transmission lines using the Wide Area Measurements (WAMS) where the fault is detected with their locations that uses the voltage to measure [1–3]. Here different techniques are used for the finding, location and organization of several faults in a transmission network [5, 6]. The transmission losses are also included that estimation factors like to affect the physical losses to various technical losses [7]. There are different faults on the transmission lines, where the voltage is measured and a signal is sent to the microcontroller, where the fault condition message to display the LCD [4]. Using Discrete Wavelet Transform (DWT) the fault condition to be detected on the three-phase transmission line, whereas the detailed coefficient of fault signals is obtained using software coding to classify the fault in transmission line [8].

The fault to be detected on HT transmission lines which are monitored in terms of Temperature, Voltage, Current which are transmitted by using GSM modem and also used to represent the hardware structure and software flow, also could save a large amount of electricity [9]. Detecting failure in the network of electric grid and also the electrical theft can be identified, various sensors are provided to gain symptoms that lead to network failure [11]. A new intelligent technique is used to sense and classify the transmission line faults which are interconnected to Phasor Measurement Units (PMUs) which is used to measure both the voltage and current [12].

Capacitors are connected in series with the transmission lines; Artificial Neural Network (ANN) technique is used to detect the fault and improve the performance transmission system, and also it is used to sense and classify the different faults power system [14]. GSM algorithm and IOT technique use a thermopile sensor and microcontroller which will help in better functioning of the fault controlling in the transmission lines [15]. The flaw can be quickly identified and the reactive control shutdown will not contribute to any harm to the load. Also the time taken to restore power to the affected region decreases significantly [16]. A fault indicator that comes with the GPS coordinated to find fault location, LoRa transceiver is used for fault indicator for increasing range [10].

18.4 Proposed System

In this system, considering three numbers of poles for the execution of line fault detection. The HT pole consists of only three phases, R, Y and B. The HT lines are placed in a 9.14-meter pole due to flow of high voltage of about 11KV, 22KV and 33KV. Also, they have a high induction, so they are placed by 9.14-meter pole. In our system, in this technique used Potential Transformers, LCD, LoRa, Power Supply Unit and the Microcontroller Unit as shown in block diagram of HT pole in Figure 18.5.

The Power Supply Unit provides necessary electrical power backup to the entire circuit placed in the pole. This paper consists of three Potential Transformer for three phases, R, Y and B. The main motivation of this system to behind this voltage level monitoring with end-to-end transmission

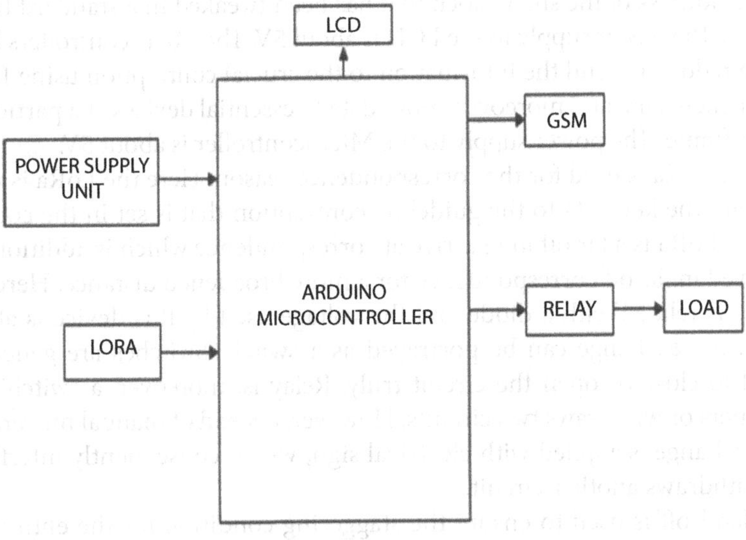

Figure 18.5 Block diagram of HT Pole.

line estimation. These are associated across or corresponding to the line which is to be checked. The power supply to the Potential Transformer is about 12V. LCD full forms liquid-crystal display. Here the LCD is used to display the output such as voltage of each phase, address of the pole, timer that has been programmed in a regular interval of time. The power supply to the LCD is about 5V. A Microcontroller is a small chip which is made up of a silicon semiconductor chip.

A Microcontroller has been programmed to send the information to the main device using LoRa as a transmitter, and it also transfers data to main devices at a particular interval of time. The power supply to the Microcontroller is about 5V. The LoRa is used for the communication purpose. Here the LoRa is used to transmit the essential data to the main device that is placed in the control room. LoRa is a wireless long-range communication which is more advanced in wireless communication for a longer distance [13]. Here the LoRa is connected with a diode and the voltage used by this device is about 3.3V. Similarly pole 1, pole 2 and pole 3 have the same block diagram.

Main Device

The portions that are related in the standard contraptions are Power Supply Unit, LCD, LoRa, Relay, Load, GSM and the Microcontroller Unit. The Power Supply Unit gives significant electrical power support to the entire circuit put in the Main Device. LCD full constructions liquid valuable stone grandstand. Here the LCD is used to show the yield like voltage of each stage, address of the shaft, clock that has been tweaked in a standard time-frame. The power supply to the LCD is about 5V. The Microcontrollers have been redone to send the information to the crucial contraption using LoRa as a beneficiary, and moreover, move data to essential devices at a particular time-frame. The power supply to the Microcontroller is about 5V.

The LoRa is used for the correspondence reason. Here the LoRa is used to send the key data to the guideline contraption that is set in the control room. LoRa is a far off long arrive at correspondence which is additionally evolved in far off correspondence for a more broadened distance. Here the LoRa is related with a diode and the voltage used by this device is about 3.3V. An exchange can be portrayed as a switch. Switches are generally used to close or open the circuit truly. Relay is, moreover, a switch that partners or withdraws two circuits. However, instead of manual movement an exchange is applied with electrical sign, which consequently interfaces or withdraws another circuit.

Hand-off is used to ensure the staggering condition for the entire circuit during the lack circumstances. GSM is a compact device; it addresses overall system for adaptable correspondence (GSM). GSM is an open and progressed cell development used for conveying convenient voice and data organizations work at different frequency ranges with repeat gatherings. GSM was made as an electronic system using the time division distinctive access (TDMA) procedure for correspondence purposes. Here the GSM is used to send the nuances to a foreordained compact number practically all the information, for instance, address of the post that is gone lack, voltage level. The Relay is connected in the main device which ensures the tripping condition of the entire circuit when any fault condition occurs. The main device which is connected to the sub-power station will act as a receiver, where the GSM is connected to the receiver (Main Device). When the HT feeder gets tripped, it automatically trips the entire feeder. Then fault location is located through GSM and the service line technical team sent to rectify the fault. After the fault is cleared the feeder is charged.

The voltage transmitted through the HT lines is about 11KV, 22KV and 33KV. In any fault condition the Address of the pole and the voltage level of the pole is transmitted by LoRa in the pole and this data is received in the main device, displayed in the LCD. After fault condition the message is forwarded to the specified mobile number through GSM. The Relay in main device Trips the entire circuit after fault. The technical crew is sent to the fault location to service it. After the rectification of the fault, the HT feeder line is charged. The block diagram for LoRa based flaw location detection is shown in Figure 18.6.

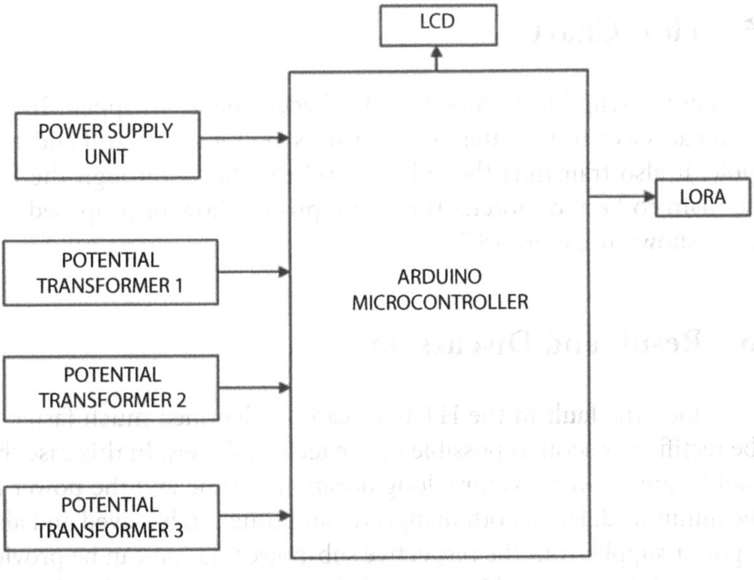

Figure 18.6 Block diagram of LoRa – based flaw –Location detection in HT line using GSM.

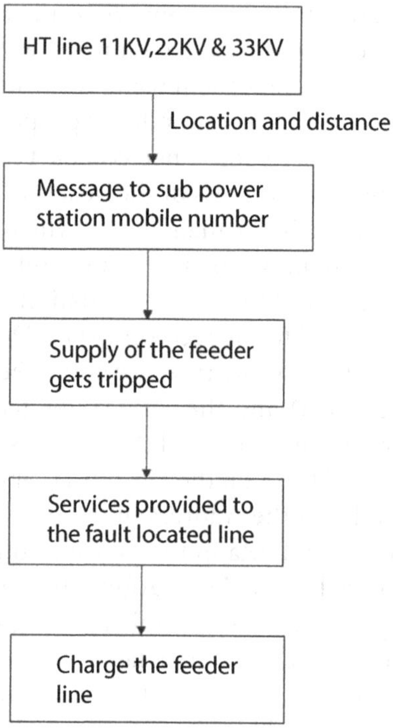

Figure 18.7 Process flow of proposed methodology.

18.5 Flow Chart

When any interruption occurs, the HT feeder line gets tripped. In pole, the LoRa acts as a transmitter and transmits fault location and address of the pole. It also transmits the voltage level that flows through the LoRa to the main to be monitored. The entire process flow of proposed techniques is shown in Figure 18.7.

18.6 Result and Discussion

In this paper, the fault in the HT lines can be identified much faster and can be rectified as soon as possible by the technical crew. In this case there will not be any power loss for a long duration of time and the power loss can be minimized. Here, both manpower and time can be saved and also a clear power supply from the respective sub-power station can be provided without any interruption. Here LoRa is being used as a mode of wireless communication which makes it much simpler. If the LoRa is used by gaining license the information that is communicated can be transmitted and received from any part of the word.

The Hardware setup of pole-1 is shown in Figure 18.8, which represents a real-time application of the pole. The circuit diagram consists of potential transformer, LoRa, Microcontroller Arduino, LCD. The switch box is to ensure the low voltage and could make fault at any phase. LoRa is used to communicate data through the LoRa antenna. The poles are connected in series as a real-time application.

The Hardware setup of pole-2 that is shown in Figure 18.9 represents a real-time application of the pole. The circuit diagram may consist of potential transformer, LoRa, Microcontroller Arduino, LCD. LoRa is used to communicate data through the LoRa antenna. The switch box is to ensure the low voltage and could make fault at any phase either at R, Y or B.

The Hardware setup of pole-3 that is shown in Figure 18.10 represents a real-time application of the pole. The circuit diagram consists of potential transformer, LoRa, Microcontroller Arduino, LCD. LoRa is used to communicate data through the LoRa antenna. The switch box is to ensure the

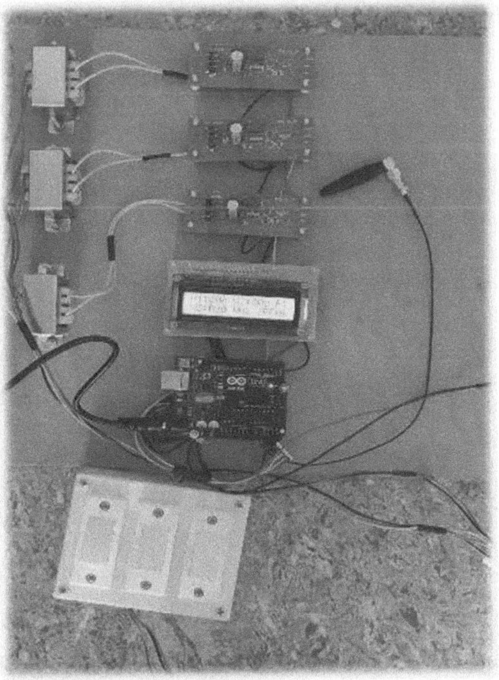

Figure 18.8 Hardware setup of Pole-1.

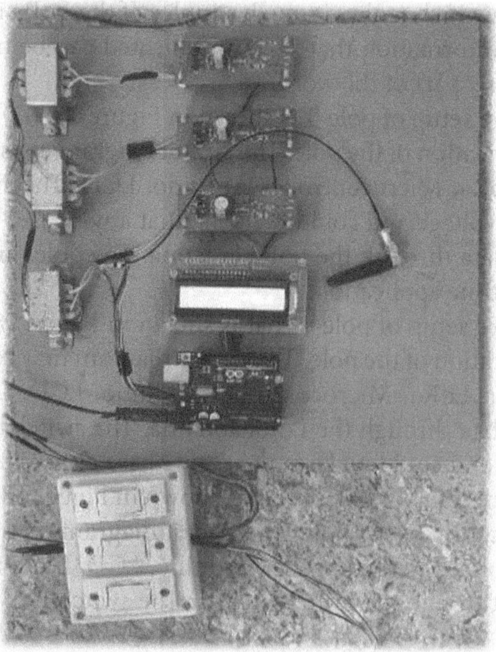

Figure 18.9 Hardware setup of Pole-2.

Figure 18.10 Hardware setup of Pole-3.

low voltage and could make fault at any phase either at R, Y or B. The load is connected to pole-3 to ensure the tripping condition and each phase is connected with separate load.

The Hardware setup of main device shown in Figure 18.11 is connected in the control room at the sub-power station. The circuit diagram of the main device consists of LoRa, Microcontroller Arduino, LCD, Relay, GSM module. LCD displays output data of poles 1, 2 and 3 in a particular interval of time. Relay is used to ensure the tripping condition of the entire circuit. GSM module is used to send a message to a particular mobile number at any fault conditions. A manually operated switch is provided; at the time the fault condition is rectified the switch is used to charge the entire circuit and the switch is operated manually.

Figure 18.12 represents the entire circuit. The entire circuit of main device, pole 1, pole 2 and pole 3 is connected in series as a real-time application and connected to a load. In any fault condition the load connected to the circuit gets tripped, which ensures the tripping of the entire circuit. A manually operated switch is provided; at the time the fault condition is rectified the switch is used to charge the entire circuit and the switch is operated manually. The main device is connected in the control room of the sub-power station and the pole circuits are connected in poles.

Figure 18.11 Hardware setup of main device.

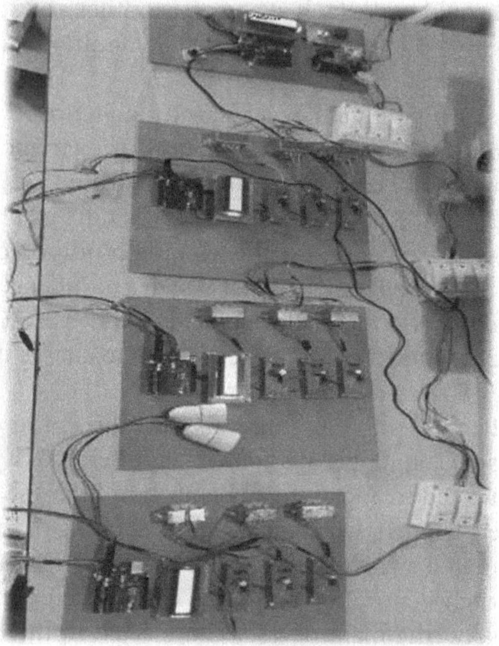

Figure 18.12 Entire circuit.

When the fault occurs in the line the data is transferred through LoRa. Data is received by the main device. Then the message is delivered to the specified mobile number as shown as the output in Figure 18.13. In the message the fault phase is indicated and the address of the pole is also indicated.

As mentioned, the message the fault pole addresses will be sent and the voltage level will be sent by the GSM module to the respective mobile number. Then the entire circuit gets tripped; that's no current through the entire circuit so these messages will be sent unless the fault condition is rectified; the output is represented as Figure 18.14.

18.7 Novelty of Work

Transmission lines (110kv, 230kv, 400kv) are electricity carrying conductors from the generator plant to the sub-power station. HT lines (11kv,22kv,33kv) are high tension line which carry electricity from sub-power station to distribution transformers (DTs). The IoT is used to

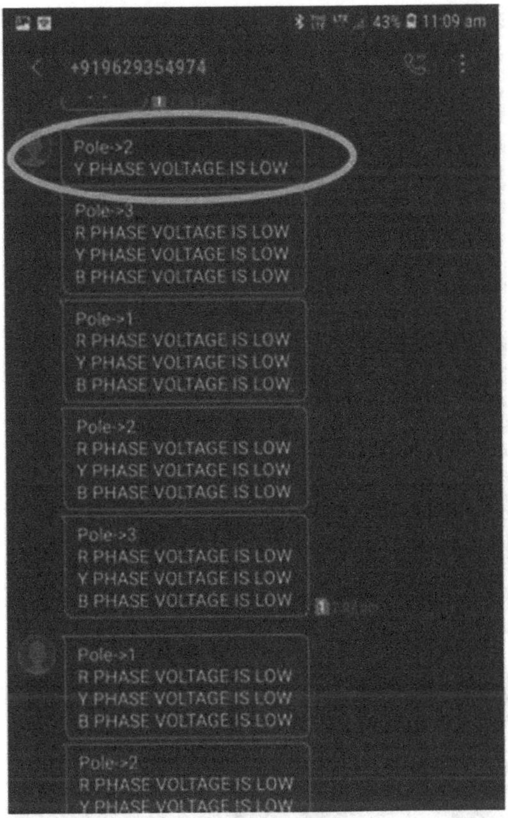

Figure 18.13 Output of GSM.

transmit the identified transmission line faults and the results are displayed in numerical value. By using LoRa and GSM, the exact location of the pole is identified and result is displayed through SMS. Normally the transmission lines are up to 50-100 km and when the fault occurs in the line, a large number of technical crews are required to rectify the faults.

In case of HT lines, the authorized person knows the exact position of the fault and hence fewer crew members are required to solve the problem and a lot of time can be saved. Each pole of HT lines is monitored separately and the final result will be viewed in the control room. The result will be as shown in the figure. Initially it displays the pole number and then it displays the voltage of R.Y. B phase. From this method, the entire HT line feeder can be monitored completely and if any fault occurs in the line, we can identify the exact position of the fault and can rectify it in a short interval of time rather than seeking help from line patrol.

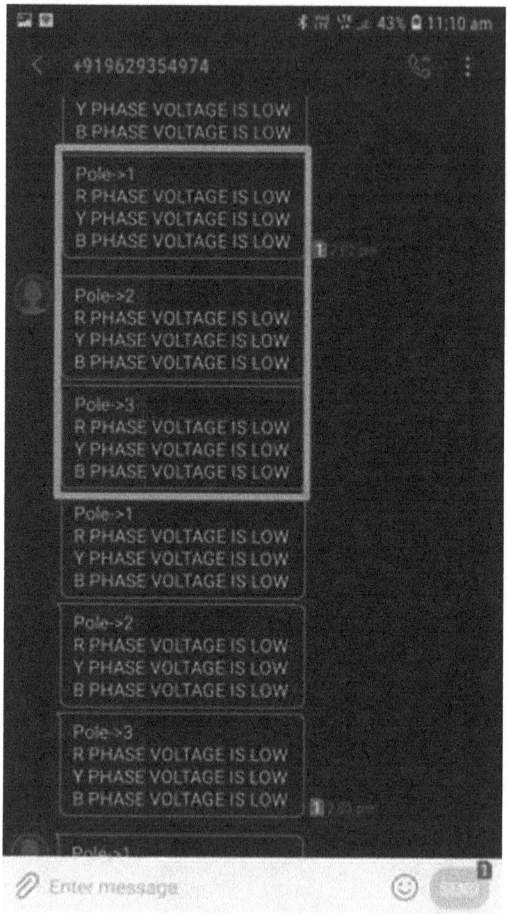

Figure 18.14 Output of GSM.

18.8 Conclusion

This paper presents an audit on the procedures utilized for flaw identification, order, and area in HT lines. Different sorts of changes are introduced. Here the flaw in the HT line is located through the LoRa. The goal of the project is success, in which uninterrupted power supply can be provided to consumers like the public, offices, hospitals, industries, etc. There comes a small search area rather than searching a vast area on- and off-territory. Also, wastage of power supply can be minimized and the fault can be rectified in the minimum of time by identifying the location. In older methods the high voltage transmission lines (where transmitted from generation plant to sub-power station) are provided with these types of features. Here this feature is provided to the HT lines and a non-interrupted power supply is provided directly to consumers.

18.9 Future Enhancement

A. Fault detection of a Generator and motor

A novel various leveled signal handling system is proposed for generator condition checking, also, flaw determination dependent on crude electrical waveform information empower organizations, which can regularly be estimated by determinedly tracking down the waveform sensors. The effect of generator hamper blames on deliberately found electrical waveform sensors in power networks are immediately examined and approved in MATLAB Simulink. In light of the huge arrangement of electrical waveform information created by MATLAB Simulink, a various leveled calculation is at that point intended to find deficiency site area and screen the state of generators in force organizations. At last, the proposed strategy is approved in 14-transport IEEE standard force network under various situations (e.g., one generator shortcoming, two-generator-flaw, different maturing levels, and so on). Our outcomes show that it could find the flaw site area and screen the maturing state of generators in force organizations. Contrasted with conventional condition observing, what's more, flaw analysis dependent on generator sensors, our proposed philosophy can screen countless generators based on a predetermined number of waveform sensors, which promises to decrease the expense of the upkeep and upgrade of the unwavering quality of the force matrix.

B. Improvements to human-machine interface

Visual showcase frameworks indicating the sidelong street position of a vehicle have been utilized in cutting-edge vehicle control and wellbeing frameworks for driver help with risky conditions, e.g., in snow-plowing tasks. A straightforward reproduction of parallel control soundness of the vehicle demonstrates the unfriendly impact of dead-time coming about because of the driver looking among windshield and show. An occasion driven inciting show is introduced to diminish the impact of the dead-time and improve the wellbeing of the driver help framework. The test brings about genuine driving gives an examination of the presentation of the new showcase versus the first showcase framework and demonstrates the improvement in sidelong security in driving a vehicle.

C. Improvements in computer-based protection of industry automation PC-based frameworks, by and large, alluded to as Programmable Electronic Systems (PESs), are being progressively utilized in the process business, additionally to perform security capacities. The interaction industry as we mean in this report incorporates, yet isn't restricted to, synthetics, oil and gas creation, oil refining, and force age. Beginning in the mid-1970s, the wide application prospects and the connected turn of event issues of such frameworks were perceived. From that point forward, numerous rules and principles have been created to coordinate and manage the utilization of PCs to perform wellbeing capacities (EWICS-TC7, EC, ISA).

Exercises learned over the most recent twenty years can be summed up as follows:

- Wellbeing is a social issue;
- Security is an administration issue;
- Security is a designing issue.

Specifically, security frameworks must be appropriately tended to in the general framework setting. No single technique can be viewed as adequate to accomplish wellbeing; highlights are needed in numerous security applications. Designing methodology needs to address not just equipment and programming issues in disengagement but also their interfaces and man- machine interface issues. At long last, the financial and mechanical parts of the security applications and improvement of PESs in measure plants are confirmed through all the Report. The extent of the Report is to add to the improvement of satisfactory attention to these issues and to represent specialized arrangements applied or being created.

References

1. Manoharsingh, Dr. B.K. Panigrahi and Dr. R.P. Maheshwari, Transmission Line Fault Detection and Classification, *Proceedings of ICETECT*, 2011.
2. Ravi V. Ghodchar and Dr. R.G. Karandikar, Fault Detection and Classification in Power Transmission Lines Based on Transient Signal Analysis: A Review of Different Methodologies, *International Journal of Engineering Research & Technology, Appl. Sci.* 2020, 10, 1312; doi:10.3390/app10041312, 2013.
3. Sayari Das, Shiv P. Singh, Bijaya K. Panigrahiba, Transmission line fault detection and location using Wide Area Measurements, *International Review of Electrical Engineering*, Vol. 14, No. 5, 2017.

4. Priya A. Gulbhile, Jitendra R. Rana and Bapu T. Deshmukh, Overhead Line Fault Detection Using GSM Technology, *IEEE Xplore Conference Proceedings*, 2017.
5. Debani Prasad Mishra and Papia Ray, Fault detection, location and classification of a transmission line, *Neural Computing and Applications*, Vol. 30, pp. 1377–1424, 2017.
6. S. Suresh and Nagarajan Ramalingam, Transmission Line Fault Monitoring and Identification System by Using Internet of Things, *International Journal of Advanced Engineering Research and Science*, Vol. 4, Issue 4, 2017.
7. Hui Hwang Goh, Syyi, Sim, Asad Shaykh, Md.Humayun Kabir, Chin Wan Ling, Qing Shi Chua and Kai Chen Goh, Transmission Line Fault Detection: A Review, *Indonesian Journal of Electrical Engineering and Computer* Science Vol. 8, No. 1, pp. 199–205, 2017.
8. Maliye Sri Pranav, Maliye Sri Pranav, Kanipakam Vishal, Jakkamsetti Tarun, Dr. D. Kavitha and Dr. V. Vanitha- Fault Detection and Classification in Three Phase Transmission Lines Using Signal Processing, *IEEE Xplore Conference Proceedings*, 2018.
9. Prof. Vikramsingh R. Parihar, Shivani Jijankar, Anand Dhore, Arti Sanganwar and Kapil Chalkhure, Automatic Fault Detection in Transmission Lines using GSM Technology, *International Journal of Innovative Research in Electrical, Electronics, Instrumentation and Control Engineering*, Vol. 6, Issue 4, 2018.
10. Patrick. S, Pouabe Eboule, Jan Harm C Pretorius, Nhlanhla Mbuli, and Collins Leke, Fault Detection and Location in Power Transmission Line Using Concurrent Neuro Fuzzy Technique, *IEEE Canada Electrical Power and Energy Conference*, 2018.
11. G.S. Nandakumar, V.Sudha and V. Sudha, Fault Detection in Overhead Power Transmission, *International Journal of Pure and Applied Mathematics*, Vol. 118 No. 8, pp. 377–381, 2018.
12. Abdul Qayyum Khan, Qudrat Ullah, Muhammad Sarwar, Sufi Tabassum Gul, Naeem Iqbal, Transmission Line Fault Detection and Identification in an Interconnected Power Network using Phasor Measurement Units, *ScienceDirect*, IFAC Papers on Line 51-24, pp. 1356–1363, 2018.
13. N Rosle, N F Fadzail, M I A Halim, M N K H Rohani, M I Fahmi, W Z Leo and N N A Baka, Fault detection and classification in three phase series compensated transmission line using ANN, Jamil *et al. Springer Plus Open Access Journal*, pp. 24:34, 2019.
14. K.Venkata Pavan, V. Srikanth, K.M. Prema, Dr. I. Chandra, Detection of Fault in Transmission Lines Using Various Technologies, *International Journal of Advance Research, Ideas and Innovations in Technology*, Vol. 5, Issue 3, 2019.
15. Sibisagar. B, Surya, V. R., Vignesh Vijayaraghavan, Dr. Suriya Krishnaan, Self-Regulating Line Fault Detection & Its Location in Transmission Lines, *International Journal of Scientific & Technology Research* Vol. 9, Issue 06, 2020.
16. Jignesh Gohil and Chintan Parmar, LoRa based Architecture for Fault Localization in Transmission Lines, *International Journal of Computer Applications*, Volume 177 - Number 39, 2020.

19

Classification Models for Breast Cancer Detection

Varsha B.*, Sneka P., Tanuja A. and Shana J.

Department of Artificial Intelligence and Machine Learning, Coimbatore Institute of Technology, Coimbatore, Tamil Nadu, India

Abstract

In the modern era, breast cancer is one of the most common cancers worldwide. In 2020, nearly 2.3 million women were diagnosed with breast cancer (one in every eight women). The goal of this study is to compare three machine learning models, namely, logistic regression, decision tree, and random forest classifier which have been implemented for breast cancer. The patient's dataset was collected; the dataset contained 569 rows of data, that is 569 patients' data and 33 columns which are the features based on the classification. The dataset consists of attributes of the nuclei measurements which consist of texture, radius, perimeter, area, concavity, etc. In this breast cancer classification, the cancer is mainly classified based on the type of cancer cells, that is either benign or malignant (already given in the dataset as patients having cancer or not). Benign are non-cancerous tumour cells which do not invade neighbouring cells, whereas malignant are the cancerous cells causing tumours which invade neighbouring cells. The split data consists of 25% of testing data and 75% of training data. Machine learning models such as logistic regression, random forest classifier and decision tree classifier models are implemented. The results showed that random forest classifier has a comparatively higher accuracy with 98%.

Keywords: Breast cancer classification, breast cancer prediction, benign, malignant, random forest, decision tree, logistic regression

19.1 Introduction

The goal and aim of this project is to compare the accuracy of the various classification models to detect whether the patient has cancer or not

**Corresponding author*: varshabalaji02@gmail.com

A. Chitra, V. Indragandhi and W. Razia Sultana (eds.) Intelligent and Soft Computing Systems for Green Energy, (255–264) © 2023 Scrivener Publishing LLC

given the relevant data. The classification of breast cancer divides carcinoma (cancer that occurs in skin cells) into categories based on how they have spread or whether they have spread. Classification algorithms predict and analyze one or more discrete variables while maintaining opposing attributes in a dataset. Classification and clustering are two of the most widely and often used methods of data processing. On the other hand, classification (supervised learning in machine learning) aims at classifying unknown situations supporting learning existing patterns and categories from the information set and then predicting future situations. The training set, which is used to create the classifying structure, and the test set, which tends to be evaluated by the classifier, are commonly referred to as classification task. Classification can be quite a complex optimization problem for this dataset. The researchers used a number of machine learning algorithm techniques to solve this classification problem. Women who are diagnosed with early carcinoma have a higher survival rate.

19.2 Related Work

Information from different papers based on breast cancer using machine learning with techniques like ultrasound, RBC and blood vessels, etc., was collected and analyzed. In [1] the authors used a novel feature selection technique that was used for diagnosing breast cancer survival. It is found that the data features were reduced dimensionally and the accuracy is improved. From [2] the authors used the Wisconsin diagnostic dataset for prediction. In [3] the authors have proposed classification and prediction using machine learning techniques. From [4] the authors have predicted the malignant tumour cells in the breasts. In [5] the authors have implemented the translation of the associated disease with gene structures across tissues. This proposed paper has attempted a new combination of feature selection approach for a dimensionally high data which uses independent component analysis. The papers [6] and [7] depict supervised machine learning techniques and prediction of diagnosis. In [8] the classification is using machine learning techniques. From papers [9] and [10] the authors have implemented various data mining techniques for prediction. In [11] applications of cancer diagnosis and prognosis were referred. From [12] a review of data mining and risk assessment techniques on the data is made. In [13] a medical screening on breast cancer is done. In paper [14], the authors have predicted breast cancer using the Decision tree algorithm. In paper [15] prediction, detection and classification on cancer data is performed. From [16] and [17] the authors have referred for factors that affect breast cancer. From [18] and [19] various deep learning techniques have been implemented by the authors. The authors of [20] discussed various machine learning classification approaches.

19.3 Research Objective

The main goal of the research is to build a model that will effectively identify or detect breast cancer among female patients. To achieve this objective, implementation of three classifiers, logistic regression classifier, random forest classifier and decision tree classifier has been made. With these models, the accuracy will be calculated for the training and testing data, respectively. The one with higher accuracy on testing data will be taken into account and will be compared with the original output dataset value, all under confusion matrix. This method does not take a long time and huge data analysis can be performed.

19.4 Methodology

Here, Figure 19.1 shows the steps involved in classification.

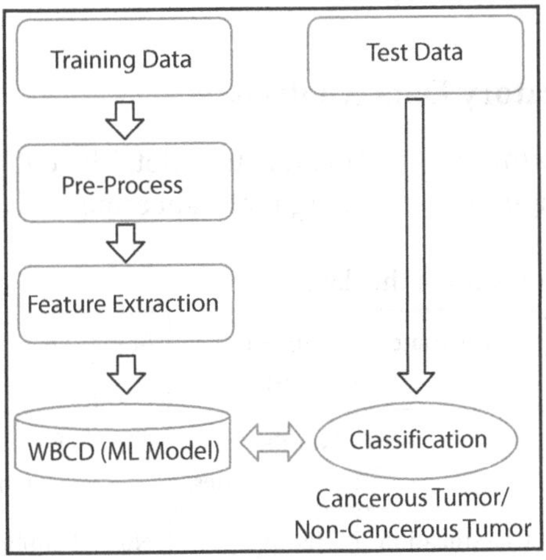

Figure 19.1 Classification model.

19.4.1 Dataset Description

The breast cancer dataset is taken from Wisconsin which is a diagnostic dataset of breast cancer. The dataset is collected from Kaggle in CSV format. The dataset contains attributes like patient's id and the cancer diagnosis whether it is benign or malignant. The dataset contains features and the measurements of the nuclei tumour cells like radius, the texture of the cells, perimeter, smoothness, concavity, symmetry and fractal dimension which is the coastline approximation of the invading tumour cells. There are 569 rows of data, representing patients and 33 columns which represent the 33 features or data points for each patient. Here the target class is mainly based on the diagnosis of cancer data.

From Table 19.1, it is found that the obtained dataset is a multivariate dataset and the attributes of the dataset taken are real entities of the patients. The number of instances and the number of attributes are found and checked for the missing and displayed in the above table.

19.4.2 Data Preprocessing

Data preprocessing is one of the major techniques that executes certain tasks like data cleaning, checking for the missing values, training and testing and makes it suitable for the machine learning models. The data preprocessing that has been applied here are the following:

1. Checking the missing or null values
2. Feature scaling of the dataset, here feature scaling is done using the standard scalar.

19.4.3 Exploratory Data Analysis

Visualizing the counts, by creating a count plot. The count plot depicts the count of the benign and the malignant cancerous cells. The blue colour

Table 19.1 Characteristics of the dataset.

Data Set Characteristics:	Multivariate	Number of Instances:	569	Area:	Life
Attribute Characteristics:	Real	Number of Attributes:	32	Date Donated	1995-11-01
Associated Task:	Classification	Missing Values?	No	Number of Web Hits:	1093558

indicates malignant cells and orange indicates benign tumour cells as shown in Figure 19.2.

From Figure 19.2, we infer that benign cells have a comparatively high ratio.

Representing the diagnosis of malignant (M) and benign (B) tumours in Boolean format as 1s and 0s as shown in Figure 19.3.

Creation of pair plot. A "pair plot" is also known as a scatter plot, where one variable in the same row of data is matched to the value of another variable. The pair plot for all the features has been plotted.

From Figure 19.4, the blue points represent Malignant cases and orange points depict the Benign cases.

A heat map is built to indicate the correlation between all features. Some of the features had higher numerical integer values and on the other hand had much lower numerical values as shown in Figure 19.5.

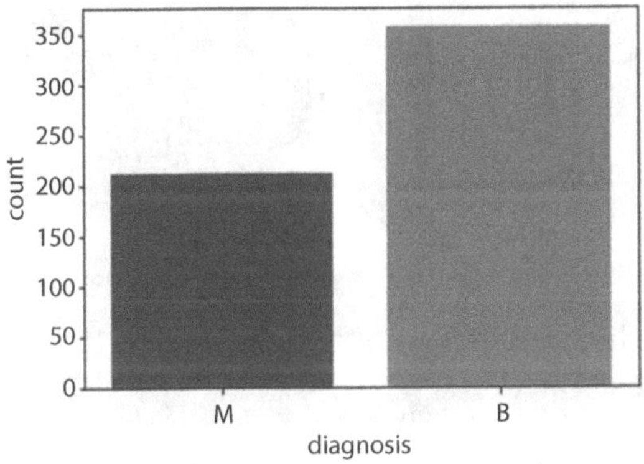

Figure 19.2 Count plot.

```
[1 1 1 1 1 1 1 1 1 1 1 1 1 1 1 1 1 1 1 0 0 0 1 1 1 1 1 1 1 1 1 1 1 1 1 1 1
 0 1 1 1 1 1 1 1 1 0 1 0 0 0 0 0 1 1 0 1 1 0 0 0 0 1 0 1 1 0 0 0 0 1 0 1 1
 0 1 0 1 1 0 0 0 1 1 0 1 1 1 0 0 0 1 0 0 1 1 0 0 0 1 1 0 0 0 0 1 0 0 1 0 0
 0 0 0 0 0 0 1 1 1 0 1 1 0 0 0 1 1 0 1 0 1 1 0 1 1 0 0 1 0 0 1 0 0 0 0 1 0
 0 0 0 0 0 0 0 0 1 0 0 0 0 1 1 0 1 0 0 1 1 0 0 1 1 0 0 0 0 1 0 0 1 1 1 0 1
 0 1 0 0 0 1 0 0 1 1 0 1 1 1 1 0 1 1 1 0 1 0 1 0 0 1 0 1 1 1 1 0 0 1 1 0 0
 0 1 0 0 0 0 0 1 1 0 0 1 0 0 1 1 0 1 0 0 0 0 1 0 0 0 0 0 1 0 1 1 1 1 1 1 1
 1 1 1 1 1 1 1 0 0 0 0 0 0 1 0 1 0 0 1 0 0 1 0 1 1 0 0 0 0 0 0 0 0 0 0 0 0
 0 1 0 0 1 0 1 0 0 0 0 0 0 0 0 0 0 0 0 0 0 1 0 0 0 1 0 1 0 0 0 0 1 1 1 0 0
 0 0 1 0 1 0 1 0 0 0 1 0 0 0 0 0 0 0 1 1 1 0 0 0 0 0 0 0 0 0 0 0 1 1 0 1 1
 1 0 1 1 0 0 0 0 0 1 0 0 0 0 0 1 0 0 0 1 0 0 1 1 0 0 0 0 0 0 1 0 0 0 0 0 0
 0 1 0 0 0 0 0 1 0 0 1 0 0 0 0 0 0 0 0 0 0 0 0 1 0 1 1 0 1 0 0 0 0 0 1 0 0
 1 0 1 0 0 1 0 1 0 0 0 0 0 0 0 0 1 1 0 0 0 0 0 0 1 0 0 0 0 0 0 0 0 0 0 1 0
 0 0 0 0 0 0 1 0 1 0 0 1 0 0 0 0 0 1 1 0 1 0 1 0 0 0 0 0 1 0 0 1 0 1 0 1 1
 0 0 0 1 0 0 0 0 0 0 0 0 0 0 0 1 0 1 1 0 0 0 0 0 0 0 0 0 0 0 0 0 0 0 0 0 0
 0 0 0 0 0 0 0 1 1 1 1 1 1 0]
```

Figure 19.3 Values after encoding the categorical data.

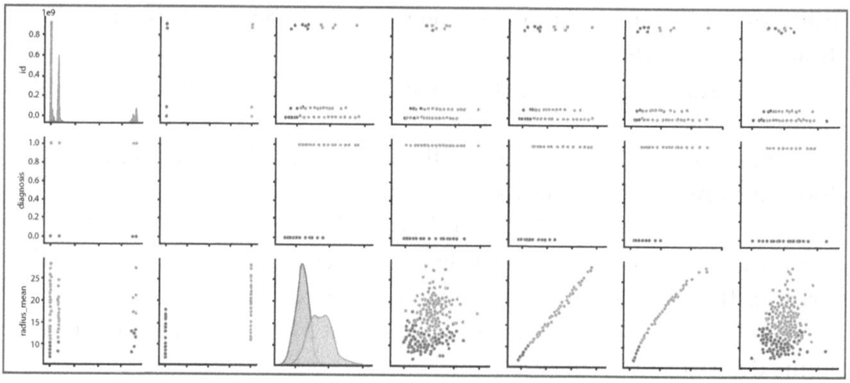

Figure 19.4 Pair plot.

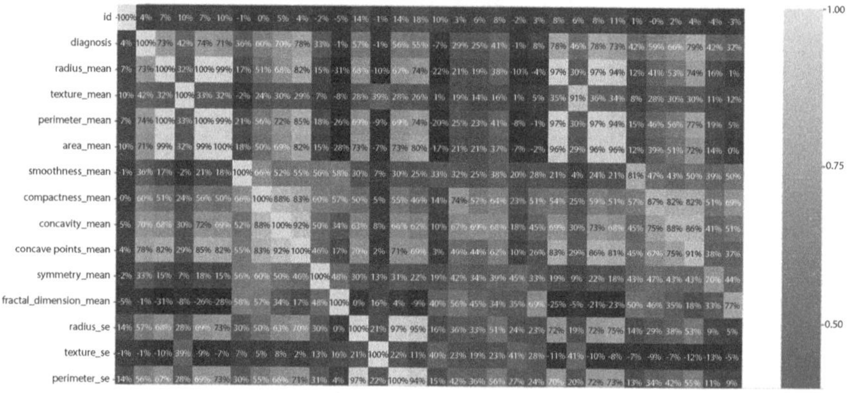

Figure 19.5 Heat map indicating correlation between all features.

19.5 Model Selection

To achieve the goal, the machine learning classification models are used to fit a function and to predict the class of the new input. Three classification models have been implemented for breast cancer classification.

19.5.1 Logistic Regression

Logistic regression is a form of supervised learning algorithm, which is used to predict a categorical dependent variable using a given set of independent variables. In medicine, frequent application is to find if variables have an influence on the disease. 0 could stand for ill, 1 for not ill. In this model, data is pre-processed and the criteria is chosen as 'entropy'. Fitting of the logistic regression to the training set. The formula for the logistic regression model is given in Equation (19.1). The accuracy of the training dataset in logistic regression classifier is given as 99.06%.

$$1/\ (1+\ e\wedge\text{-value}) \tag{19.1}$$

19.5.2 Decision Tree Classifier

Decision trees are a form of supervised machine learning technology that aids in determining a course of action by displaying various outcomes on each branch of the decision tree. Here, in this decision tree classifier the criterion "entropy" is used and random state is 0. In this classification model, the data is obtained and then the training and testing of data is done.

The decision tree generation is formulated for each step and then the model evaluation is done and finally the performance evaluation measures, i.e., accuracy and recall are obtained. The accuracy of the training dataset in random forest classifier is given as 100%.

19.5.3 Random Forest Classifier

Random forest model is a kind of non-parametric model, i.e., the population in it does not have a normal distribution. They operate by building a multitude of decision trees randomly, hence the name as random forest. It is a popular ensemble technique, and majorly uses bagging method.

Here, in this classification, the number of estimators is taken as 10 (i.e., 10 trees are taken to build for maximum and accurate prediction) and the random state is taken as 0. The accuracy of the training dataset in random forest classifier is given as 99.53%.

19.6 Results and Discussion

19.6.1 Confusion Matrix

The confusion matrix in Figure 19.6 shows the accuracy for all the model classifiers based on the test data. The number of cancer patients who were mistakenly diagnosed as not having cancer (false negative). The number of patients who were misdiagnosed as having cancer when they didn't have it - false positives. The total number of correct diagnoses (both true positive and true negative).

		Prediction	
		0	1
Actual	0	TN	FP
	1	FN	TP

Figure 19.6 Confusion matrix for test data.

19.6.2 Model Evaluation and Prediction

The model is evaluated for the test dataset and the performance of the model is verified. Here the actual target column (diagnosis) values are compared with the values predicted by the model. The performance criteria used are F1 score and accuracy measure. Table 19.2 demonstrates the performance of all the three models.

From Table 19.2, accuracy and metrics, the classification model that worked best with the test data was the random forest classifier with an accuracy score of 98.0%. To make predictions/classifications on the test data and display both the random forest classifier model's classifications/ predictions and the actual values of the patient indicating whether they have cancer or not.

The best model is used to predict the diagnosis value for a new patient data which is not there in the training dataset. The new patient record with row number 571 and its last 6 attributes has been shown in Figure 19.7.

Table 19.2 Model evaluation.

Algorithm model	Logistic regression	Decision tree classifier	Random forest
F1 score	0.97	0.93	0.98
Testing Accuracy	0.98	1.0	0.99

559	925236	B	9.423	27.88	59.26	271.3	0.08123	0.04971	0	0	0.1742	0.06059	0.5375
560	925277	B	14.59	22.68	96.39	657.1	0.08473	0.133	0.1029	0.03736	0.1454	0.06147	0.2254
561	925291	B	11.51	23.93	74.52	403.5	0.09261	0.1021	0.1112	0.04105	0.1388	0.0657	0.2388
562	925292	B	14.05	27.15	91.38	600.4	0.09929	0.1126	0.04462	0.04304	0.1537	0.06171	0.3645
563	925311	B	11.2	29.37	70.67	386	0.07449	0.03558	0	0	0.106	0.05502	0.3141
564	925622	M	15.22	30.62	103.4	716.9	0.1048	0.2087	0.255	0.09429	0.2128	0.07152	0.2602
565	926125	M	20.92	25.09	143	1347	0.1099	0.2236	0.3174	0.1474	0.2149	0.06879	0.9622
566	926424	M	21.56	22.39	142	1479	0.111	0.1159	0.2439	0.1389	0.1726	0.05623	1.176
567	926682	M	20.13	28.25	131.2	1261	0.0978	0.1034	0.144	0.09791	0.1752	0.05533	0.7655
568	926954	M	16.6	28.08	108.3	858.1	0.08455	0.1023	0.09251	0.05302	0.159	0.05648	0.4564
569	927241	M	20.6	29.33	140.1	1265	0.1178	0.277	0.3514	0.152	0.2397	0.07016	0.726
570	92751	B	7.76	24.54	47.92	181	0.05263	0.04362	0	0	0.1587	0.05884	0.3857
571	924934	M	10.29	27.61	65.67	321.4	0.0903	0.07658	0.05999	0.02738	0.1593	0.06127	0.2199

Figure 19.7 Prediction for new data.

These values are given as input to the model and it predicts a value of 1 in the target column (diagnosis). A value of 1 in the target column means the tumour is Malignant.

19.7 Conclusion

In this work the Wisconsin dataset was taken to compare the performance of various classification algorithms to predict breast cancer in patients. Logistic regression, decision tree classifier and random forest classifier are implemented in Python to detect breast cancer. The performance of the study is measured by its testing accuracy and F1 score. The results showed that random forest has the highest testing accuracy and F1 score of 0.98 and 0.99. To further assure this prediction, a new patient's data was added to the Wisconsin dataset and tested with a random forest model. The predicted value obtained was in parallel with the actual value. The optimization of the model also should be done in future. This further optimization requires more standardization and validation.

Future work may involve methods to further improve the performance of the classifier.

References

1. Sasikala, S, Appavu, S. and Geetha, S. "A novel feature selection technique for improved survivability diagnosis of breast cancer", *Procedia Computer Science*, Vol. 50, 2015.
2. "On Breast Cancer Detection: An Application of Machine Learning Algorithms on the Wisconsin Diagnostic Dataset" by Abien Fred M. Agarap, 7 February 2019.
3. Wang, D. Zhang and Y. H. Huang "Breast Cancer Prediction Using Machine Learning," Vol. 66, No. 7,2018.
4. Westerdijk L. Predicting Malignant Tumor Cells in Breasts. Vrije Universiteit Amsterdam; Amsterdam, The Netherlands, 2018.
5. Kasim, A., Shkedy, Z., Lin, D., Van Sanden, S., Abrahantes, J.C., Göhlmann, H.W., *et al.*, "Translation of disease associated gene signatures across tissues", *International Journal of Data Mining and Bioinformatics*, Vol. 11, 2015.
6. Vikas Chaurasia and S. Pal, "Using Machine Learning Algorithms for Breast Cancer Risk Prediction and Diagnosis," 2016.
7. Ch. Shravya, K. Pravalika, Shaik Subhani, "Prediction of Breast Cancer Using Supervised Machine Learning Techniques", *International Journal of Innovative Technology and Exploring Engineering (IJITEE)*, Vol. 8, Issue 6, April 2019.

8. B. Akbugday, "Classification of Breast Cancer Data Using Machine Learning Algorithms," 2019 Medical Technologies Congress (TIPTEKNO), Izmir, Turkey, 2019.
9. M. Amrane, S. Oukid, I. Gagaoua and T. Ensarİ, "Breast cancer classification using machine learning," 2018 Electric Electronics, Computer Science, Biomedical Engineerings' Meeting (EBBT), Istanbul, 2018.
10. Radhakrishnan S. & Priyaa S. A Critical Study on Data Mining Techniques in Health-care Dataset. International Research Journal of Engineering and Technology, 2(5), 2015.
11. Wenbin Yue, Zidong Wang, "Machine Learning with Applications in Breast Cancer Diagnosis and Prognosis," 9 May 2018.
12. Kaur, S. & Singh, R. A Review of Data Mining Based Breast Cancer Detection and Risk Assessment Techniques. *Int J ComputSci and Inform Secur*, 2016.
13. Fuller, M. S., Lee, C. I., & Elmore, J. G. Breast Cancer Screening: An Evidence-Based Update. The Medical Clinics of North America, 2015.
14. P. Hamsa Gayatri, P. Sampath, Performance Analysis of breast cancer Classification using Decision tree Classifier, *International Journal of Current Pharmaceutical Research*, Vol. 9, Issue 2, 2017.
15. Keles, M. Kaya, "Breast Cancer Prediction and Detection Using Data Mining Classification Algorithms: A Comparative Study." Tehnicki Vjesnik - Technical Gazette, Vol. 26, no. 1, 2019.
16. Y.-S. Sun, Z. Zhao, Z.-N. Yang, F. Xu, H.-J. Lu, Z.-Y. Zhu, W. Shi, J. Jiang, P.-P. Yao, and H.-P. Zhu, "Risk factors and preventions of breast cancer," *Int. J. Biol. Sci.*, Vol. 13, no. 11, p. 1387, 2017.
17. M. D. Ganggayah, N. A. Taib, Y. C. Har, P. Lio, and S. K. Dhillon, "Predicting factors for survival of breast cancer patients using machine learning techniques," *BMC Med. Inform. Decis. Making*, vol. 19, no. 1, 2019.
18. M. Togacar and B. Ergen, "Deep learning approach for classification of breast cancer," in *Proc. Int. Conf. Artif. Intell. Data Process. (IDAP)*, Sep. 2018.
19. M. Tiwari, R. Bharuka, P. Shah, and R. Lokare, "Breast cancer prediction using deep learning and machine learning techniques".
20. D. A. Omondiagbe, S. Veeramani, and A. S. Sidhu, "Machine learning classification techniques for breast cancer diagnosis," *IOP Conf. Ser., Mater. Sci. Eng.*, Vol. 495, Jun. 2019.

20

T-Count Optimized Quantum Comparator Circuit

Gayathri S. S., R. Kumar* and Samiappan Dhanalakshmi

Department of Electronics and Communication Engineering, SRM Institute of Science and Technology, Kattankulathur, Chengalpet, Tamil Nadu, India

Abstract
Quantum circuits that solve arithmetic problems find immense application in many quantum algorithms like Grover's search algorithm, quantum cryptography, and quantum image processing applications. In this paper, a T-count optimized quantum comparator structure for single-qubit and two-qubit input is proposed. When compared to previous works, the proposed quantum structures show a better reduction in the number of T gates count and offer less T-depth.

Keywords: Quantum circuit, T-gates, quantum comparator, quantum gates, fault-tolerant, quantum computing

20.1 Introduction

Qubits are units of data that operate on quantum gates that can take $|0\rangle$ state, $|1\rangle$ state and superposition of both states represented as $|\psi\rangle = \alpha\,|0\rangle+\beta\,|1\rangle$ where, α and β are satisfying $||\alpha||^2 + ||\beta||^2 = 1$. The output states can be measured by finding the values of α and β. Upon measurement the output state collapses to one of its measurement in standard basis states 0 and 1 [1]. To maintain unique mapping between input and output vectors, quantum circuits consume ancillary qubits and garbage outputs in the input and output section, respectively [2–4]. To employ the quantum circuits completely on quantum algorithms the garbage generated by the circuit has to be erased by running the circuit backwards after copying the

**Corresponding author*: kumarr@srmist.edu.in

A. Chitra, V. Indragandhi and W. Razia Sultana (eds.) Intelligent and Soft Computing Systems for Green Energy, (265–272) © 2023 Scrivener Publishing LLC

required output. The quantum computing hardware are highly susceptible to noise errors [5]. Noise resisting quantum circuits can be created utilizing fault-tolerant quantum gates as shown in Table 20.1. Quantum CCNOT or Toffoli gate is widely used in quantum algorithms but it does not fall into the fault-tolerant gates set family.

Hence, fault-to Fault-tolerant implementation of CCNOT gate has received much attention in research and several notable decompositions were proposed. Among the handful of decomposition schemes, the Clifford+T decomposition of CCNOT gate proposed by Amy *et al.* [6] is shown in Figure 20.1 which decomposes to seven T-gates and with three

Table 20.1 Universal Clifford+T quantum gates.

S. no	Gate	Matrix	Symbol
1	Pauli-X	$\begin{bmatrix} 0 & 1 \\ 1 & 0 \end{bmatrix}$	X- gate (or) $\oplus$
2	Hadamard	$\frac{1}{\sqrt{2}}\begin{bmatrix} 1 & 1 \\ 1 & -1 \end{bmatrix}$	H
3	CNOT gate	$\begin{bmatrix} 1 & 0 & 0 & 0 \\ 0 & 1 & 0 & 0 \\ 0 & 0 & 0 & 1 \\ 0 & 0 & 1 & 0 \end{bmatrix}$	
4	T-Gate	$\begin{bmatrix} 1 & 0 \\ 0 & e^{i\pi/4} \end{bmatrix}$	T
5	Phase Gate	$\begin{bmatrix} 1 & 0 \\ 0 & i \end{bmatrix}$	S
6	T-Gate Hermitian	$\begin{bmatrix} 1 & 0 \\ 0 & e^{-i\pi/4} \end{bmatrix}$	T^{-1} (or) $T^{\dagger}$
7	Phase Gate Hermitian	$\begin{bmatrix} 1 & 0 \\ 0 & -i \end{bmatrix}$	S^{-1} (or) $S^{\dagger}$

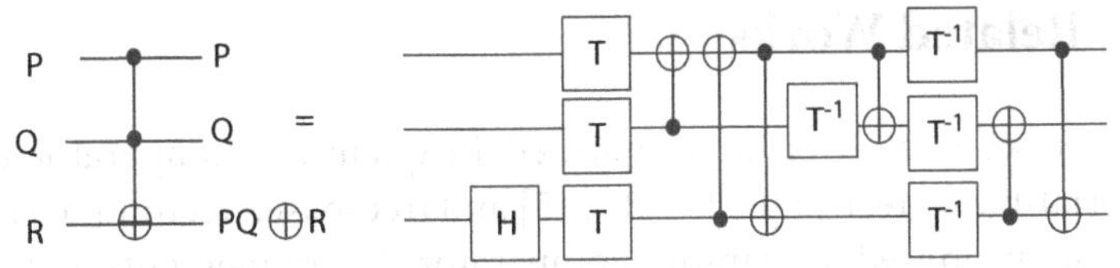

Figure 20.1 Toffoli gate decomposition using Clifford+T gates [6].

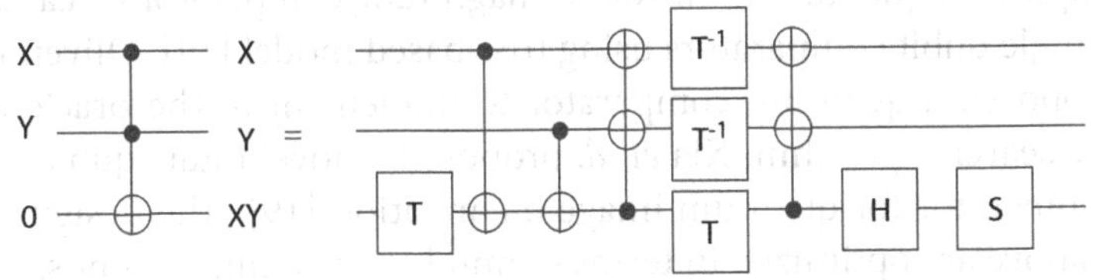

Figure 20.2 Temporary logic and decomposition of Toffoli gate using Clifford+T gates [7].

layers of T-gates. This decomposition shows resource efficiency by using a smaller number of T-gates and can be used for quantum circuit design that can implement functions other than classical logic AND operation. To realize a logic AND function using CCNOT gate a decomposition by Gidney can be utilized [5]. Figure 20.2 shows the logic AND realization of a Toffoli gate decomposition using Clifford+T gates that utilizes only four T-gates with two layers of T-gates. Fault-tolerant quantum algorithms can be implemented in two ways, namely (i) deploying error correction circuits, and (ii) designing the circuit using fault-tolerant gates. The former may result in high time complexity by increasing the overhead job for the processor and the latter is preferable in many algorithms [8–11].

Fault-tolerant quantum circuits can be analyzed using the following parameters,

1. T-Count: T-count is defined as the cumulative number of T-gates utilized the entire quantum circuit.
2. T-depth: T-depth is defined as the number of layers of T-gates that have the ability to perform parallel quantum information processing.
3. Qubits: It is defined as the total number of input qubits in the quantum circuit including ancillary qubits.

20.2 Related Works

Several works have been accomplished in quantum comparator design using serial-based technique [12, 13, 15] and tree-based technique [15, 16]. Wang *et al.* proposed a quantum comparator that implements a quantum comparator using serial-based design with *n* qubits [12]. Al-Rabadi proposed a quantum comparator design by cascading a group of single-qubit magnitude comparator using tree-based technique [17]. Thapliyal *et al.* also proposed a quantum multi-qubit magnitude comparator by cascading many single qubit comparators using tree-based model [14]. Oliveira *et al.* [18] proposed a quantum comparator to implement in the oracle of the Grover's search algorithm. Xia *et al.* proposed a novel multi-qubit magnitude comparator for quantum image binarization [19]. The designs mentioned above are optimized in terms of number of auxiliary qubits.

20.3 Proposed Quantum Comparator

The proposed single qubit magnitude comparator is inspired from the classical Boolean model and utilizes two CCNOT gates with two ancillary qubits that is producing no garbage output as shown in Figure 20.3. The proposed single qubit magnitude comparator is compared with existing counterparts and it is presented in Table 20.2.

20.3.1 Multi-Qubit Magnitude Comparator

Any quantum circuit proposed must be uniformly scaled to perform comparison operation on a string of qubits. In this section a serial-based comparator design is proposed using serial-based technique [17]. The proposed multi-qubit magnitude comparator computes the greater, lesser, and equal output between strings by employing two stages: In stage 1, single qubit comparator cells are cascaded to compute the outputs of each qubit

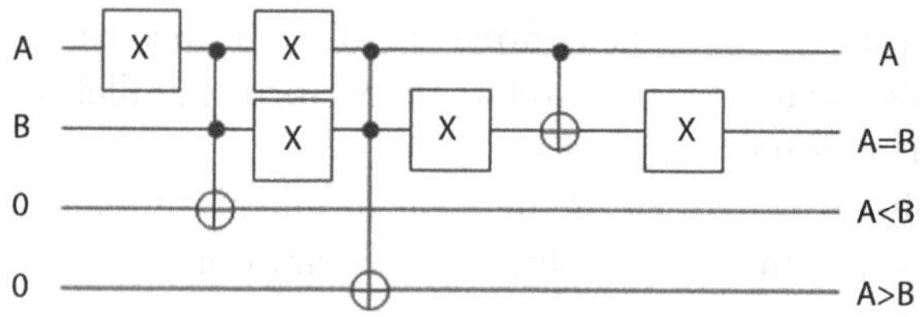

Figure 20.3 Proposed single qubit magnitude comparator.

Table 20.2 Comparison of the proposed single qubit magnitude comparator with existing works.

Designs	Parameters		
	T-count	T-depth	Ancilla
[15]	24	12	6
[18]	8	4	2
[19]	11	7	1
[20]	14	6	0
Proposed	8	4	2

in the string. In the second stage the outputs of the single qubit comparator cells are processed as per the Boolean expressions provided below,

$$G(p, q) = G_q + E_q.G_p \tag{20.1}$$

$$E(p, q) = E_p.E_{(}q) \tag{20.2}$$

$$L(p, q) = [G(p, q) \oplus E(p, q)]' \tag{20.3}$$

Here p and q denote the operand bits of the given string where $p > q$. Figure 20.4 shows the quantum circuit of two qubit magnitude comparator.

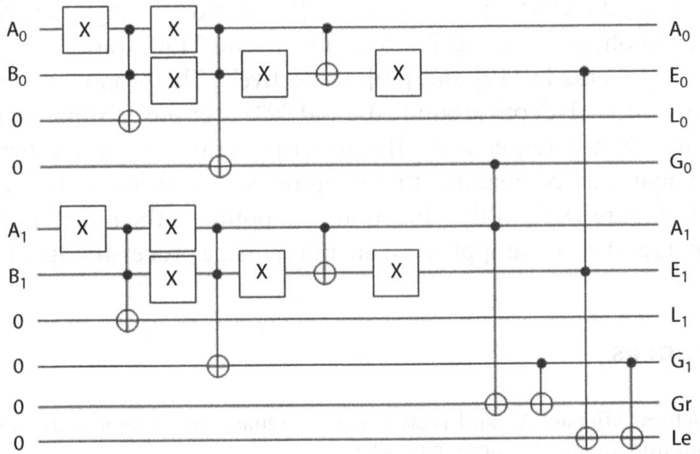

Figure 20.4 Proposed two qubit magnitude comparator.

Table 20.3 Comparison of the proposed two qubit magnitude comparator with existing works.

Designs	Parameters		
	T-count	T-depth	Ancilla
[19]	35	15	2
[20]	56	24	2
Proposed	24	12	6
% of imp. with ref.to [19]	31.42 %	20 %	-
% of imp. with ref.to [20]	57.14%	50 %	-

The proposed two qubit magnitude comparator offers reduced T-count and T-depth when compared with the existing counterparts, and the comparison is shown in Table 20.3.

The proposed magnitude for two qubit string offers a T-count savings of 31.42% and 57.14% over the existing works by [18, 19], respectively, and shows a T-depth savings of around 20% and 50% by [18, 19] over the existing works, respectively.

20.4 Conclusion

The proposed single-qubit magnitude comparator shows a better T-count savings than the existing counterparts. The proposed magnitude for two qubit string offers a reduced T-gate count around 31.42% and 57.14% over the existing works in [19] and [20], respectively. The proposed work also shows a reduced T-depth around 20% and 50% over the existing works presented in [19] and respectively. This paper presents a single and two qubit string magnitude comparator that is optimized in terms of T-count and T-depth. Future work in this direction is to optimize the number of ancillary qubits and it can be applied on quantum image processing applications.

References

1. Nielsen, Michael A., and Isaac Chuang. "Quantum computation and quantum information." (2002): 558–559.

2. Bennett, Charles H. "Logical reversibility of computation." *IBM journal of Research and Development* 17.6 (1973): 525–532.
3. Haghparast, Majid, and Keivan Navi. "Design of a novel fault tolerant reversible full adder for nanotechnology-based systems." *World Applied Sciences Journal* 3.1 (2008): 114–118.
4. Asadi, Mohammad-Ali, Mohammad Mosleh, and Majid Haghparast. "Toward novel designs of reversible ternary 6: 2 Compressor using efficient reversible ternary full-adders." *Journal of Supercomputing* 77.5 (2021): 5176–5197.
5. Bravyi, Sergey, and Alexei Kitaev "Universal quantum computation with ideal Clifford gates and noisy ancillas." *Physical Review A* 71.2 (2005): 022316.
6. Amy, Matthew, *et al.* "A meet-in-the-middle algorithm for fast synthesis of depth- optimal quantum circuits." *IEEE Transactions on Computer-Aided Design of Integrated Circuits and Systems* 32.6 (2013): 818–830.
7. Gidney, Craig. "Halving the cost of quantum addition." *Quantum* 2 (2018): 74.
8. Thapliyal, Himanshu, Edgard Mun˜oz-Coreas, and Vladislav Khalus. "T-count and qubit optimized quantum circuit designs of carry lookahead adder." arXiv preprint arXiv:2004.01826 (2020).
9. Muñoz-Coreas, Edgard, and Himanshu Thapliyal. "Quantum circuit design of a t-count optimized integer multiplier." *IEEE Transactions on Computers* 68.5 (2018): 729–739.
10. Thapliyal, Himanshu, *et al.* "Quantum circuit designs of integer division optimizing t-count and t-depth." *2017 IEEE International Symposium on Nanoelectronic and Information Systems (iNIS)*. IEEE, 2017.
11. Li, Hai-Sheng, *et al.* "Quantum implementation circuits of quantum signal rep- resentation and type conversion." *IEEE Transactions on Circuits and Systems I: Regular Papers* 66.1 (2018): 341-354.
12. Wang, D., Liu, Z., Zhu, W., Li, S. Design of quantum comparator based on extended general Toffoli gates with multiple targets. *Comput. Sci.* 39(9), 302–306 (2012).
13. Thapliyal, H., Ranganathan, N. Design of reversible sequential circuits optimizing quantum cost, delay, and garbage outputs. *ACM J. Emerg. Technol. Comput. Syst.* 6(4), 14–45 (2010).
14. Gayathri, S. S., *et al.* "T-Count Optimized Quantum Circuit Designs for Single- Precision Floating-Point Division." *Electronics* 10.6 (2021): 703.
15. Thapliyal, H., Ranganathan, N., Ferreira, R. Design of a comparator tree based on reversible logic. In *2010 10th IEEE Conference on Nanotechnology (IEEE-NANO)*, pp. 1113–1116 (2010).
16. Vudadha, C., Phaneendra, P.S., Sreehari, V., Ahmed, S.E., Muthukrishnan, N.M., Srinivas, M.B. Design of Prefix-Based Optimal Reversible Comparator. In *2012 IEEE Computer Society Annual Symposium on VLSI (ISVLSI)*, pp. 201–206 (2012).

17. Al-Rabadi, A.N. Closed-system quantum logic network implementation of the viterbi algorithm. Facta universitatis-Ser.: *Elec .Energ.* 22(1), 1–33 (2009).
18. Oliveira, David Sena, and Rubens Viana Ramos. "Quantum bit string comparator: circuits and applications." *Quantum Computers and Computing* 7.1 (2007): 17–26.
19. Xia, Haiying, *et al.* "Novel multi-bit quantum comparators and their application in image binarization." *Quantum Information Processing* 18.7 (2019): 1–17.
20. Xia, Haiying, *et al.* "An efficient design of reversible multi-bit quantum comparator via only a single ancillary bit." *International Journal of Theoretical Physics* 57.12 (2018): 3727–3744.

21

IoT-Based Heart Rate Monitoring System for Smart Healthcare Applications

Jaba Deva Krupa Abel, Samiappan Dhanalakshmi*, Sanjana N.L. and R. Kumar

Department of ECE, College of Engineering and Technology, Faculty of Engineering and Technology, SRM Institute of Science and Technology, SRM Nagar, Kancheepuram, Chengapattu Dt., Tamil Nadu, India

Abstract

Non-invasive and continuous monitoring of the heart is necessary to ensure the good health of every human being. The underlying challenge faced is the distance between the doctor and the patient, followed by lack of transportation, poverty, and poor-quality services. The other constraints are obtaining the proper morphology of the QRS complex of the Electrocardiogram (ECG) signal followed by R peak detection and calculation of heart rate. Various techniques were proposed to resolve this problem but obtaining the accurate morphology remains one of the biggest challenges since the raw ECG signal is corrupted by noises like power line interference and motion artifacts. This paper aims at developing an IoT-based heart rate monitoring system. Fractional Fourier transform is used to estimate the QRS complex of the ECG signals and thereby the heart rate. The proposed algorithm is validated using ECG signals from the Physionet MIT arrhythmia database. The proposed algorithm ensures accurate heart rate detection and is implemented on the Amazon Web Services (AWS) cloud on the internet. The proposed robust algorithm for heart rate detection combined with secure and flexible cloud servers like AWS provides an efficient heart rate monitoring unit.

Keywords: Fractional Fourier Transform, Internet of Things (IoT), ECG monitoring, R-peak detection

**Corresponding author*: dhanalas@srmist.edu.in

A. Chitra, V. Indragandhi and W. Razia Sultana (eds.) Intelligent and Soft Computing Systems for Green Energy, (273–286) © 2023 Scrivener Publishing LLC

21.1 Introduction

Unexpected deaths due to cardiovascular diseases are one of the significant causes of mortality these days. According to a recent survey, the number of people affected by this disease has increased from 25.7 million in 1990 to 54.5 million in 2016 [1]. In India, the number of deaths due to cardiovascular diseases has increased from 1.3 million in 1990 to 2.8 million in 2016. Continuous monitoring of cardiac activity is necessary to ensure good health conditions and prevent fatality due to heart diseases. The challenges underlying the availability of timely and quality medical intervention include the distance between the doctor and the patient and accurate detection of heart rate. Internet of things (IoT) based real-time monitoring of patients can reduce the mortality rates as the continuous information on health status is made available to the physicians.

The primary functionalities of IoT-based heart rate monitoring include: (1) Acquiring a real-time ECG signal from the patient using a sensor and transferring it to the cloud. The following two steps are performed in the IoT cloud: (2) Denoising acquired signal to remove noises like powerline interference and baseline wander and storing the data to have a historical log. (3) Extracting the vital parameters like heart rate from the processed signal. (4) Transferring the extracted information to the physician. The quality of the information provided to the physician has a significant, life-saving impact on the patients. Hence, a robust algorithm for the accurate detection of vital parameters like heart rate is essential. The proposed approach uses Fractional Fourier transform (FrFT) to accurately detect the R-peaks and provide precise heart rate [2]. FrFT is a linear operator that rotates the signal by an angle $\alpha = a\pi/2$ in the time-frequency plane, where is the fractional parameter. FrFT transforms a signal to an intermediary domain between time and frequency, i.e., fractional frequency plane. The proper choice of can well distinguish the QRS complex from other ECG components and ensure accurate detection of R-peaks. The availability of such a powerful tool in connection with IoT-based platforms provides efficient real-time heart rate monitoring systems.

The authors propose a prototype for monitoring the heart rate of a patient using a cloud environment. The rest of the paper is organized as follows: Section 21.2 presents the related works. Section 21.3 discusses the methodology proposed. Section 21.4 provides details on obtained results, and the conclusion is presented in section 21.5.

21.2 Related Work

Several algorithms were presented in the literature related to IoT-based ECG monitoring. In [3], the authors propose an IoT-based ECG monitoring system with a local liquid crystal display (LCD) to display the ECG waveform through Web Interface/Mobile-Application. The acquisition of real-time ECG signals is made using three electrodes. Data are sent to IoT Cloud stored. The ECG data can be visualized and stored in the IoT cloud servers. In [4], the authors present real-time heart rate monitoring. The IoT system is connected through a wireless cloud for long-distance transmission. The sensor module for data acquisition uses an embedded device, myRIO 1900, that can be interfaced with LabView.

In [5], the authors present a technique to acquire the ECG remotely. The algorithm utilizes single-lead ECG, which is appropriate for the IoT environment. The novelty of the method is with selecting authenticated ECG using template matching by the sum of squared difference. The availability of such a classification procedure improves the efficiency of the unit. A real-time ECG monitoring system is implemented using AD8232 in [6]. The signals acquired are pre-processed using digital filters to remove primary noises and interfaced with Arduino Uno. The data is transferred to the cloud using the communication module, and the results are accessed using the graphical user interface (GUI).

A prototype for monitoring multiple parameters like SpO2, pulse rate, ECG, and blood glucose is developed, and the data is communicated to the cloud in [7]. The authors investigated four different transmission modes for efficiently utilizing the computing resources. The multi-parameter monitoring enables complete supervision of a patient's health status. In [8], the authors presented Cypress Wireless Internet Connectivity for Embedded Devices (WICED) IoT module for effective ECG monitoring. Several communications and device management protocols, including Message Queuing Telemetry Transport (MQTT), User Datagram Protocol (UDP), and Hypertext Transfer Protocol (HTTP), etc., are investigated. The wireless data transmission is enabled between the sensor and the cloud using a Wi-Fi module.

ECG signals acquired by the AD8232 sensor are processed to extract features such as Mean RR, root mean square of successive differences (RMSSD), Ratio SR, and Standard deviation of NN intervals (SDNN) is proposed in [9]. The complete framework consists of a sensor unit integrated with a Wi-Fi module and the feature extraction algorithm running on the IoT cloud. A similar combination of hardware modules for implementation for quality-assisted ECG monitoring is proposed in [10]. The additional ability of the proposed approach underlies the availability of an intelligent technique to accept only the quality signals for processing. The existence of such signal quality assessment techniques further enhances the efficiency of IoT-based monitoring systems. The integrated unit with ECG acquisition sensor, Arduino, Bluetooth communication module, and cloud servers is proposed in [11] to provide continuous ECG monitoring. The novelty of the contribution lies with enabling secure data transmission using Lightweight Secure IoT and Lightweight Access Control. After careful study of existing techniques, it is observed that real-time ECG monitoring using cloud servers relies on similar hardware architecture. However, their performance can vary based on other aspects like the type of communication protocol adapted, the ability to promote secure transmission, and the nature of the feature extraction algorithm. The primary focus of the proposed technique is on the efficiency of the algorithm in extracting accurate features like heart rate. The robust method Fractional Fourier transform is adopted for detecting the heart rate. The integration of such powerful algorithms in IoT cloud like AWS can perform efficient real-time monitoring.

21.3 Methodology

The objective of the paper is to implement an efficient real-time heart rate monitoring system. The complete framework involves two primary folds: (1) Robust algorithm to accurately detect heart rate from the ECG signals, and (2) Secure and flexible AWS platform for implementing real-time monitoring. There are three primary modules in the generic IoT-based monitoring systems:

1. Sensing network where the ECG signals are acquired from the patient using a hardware module. The hardware unit consists of electrodes as transducers followed by digital filters like bandpass filter and notch filter to remove noise as power line interference and baseline wander. The sensing network is usually integrated with a communication module like Wi-Fi, Zigbee, or Bluetooth for data transmission from the patient's locality to the IoT cloud.
2. IoT Cloud is where three primary functionalities are taken care of: data cleaning, data analysis, and data storage. Data cleaning is responsible for pre-processing the ECG signals

to eliminate any noises added during transmission. The signal quality assessment is part of data cleaning to ensure valid signals are processed. Data analysis involves processing the ECG signals to extract relevant features like heart rate. Efficient algorithms with minimum error detections are implemented for such purposes. Data storage enables the availability of historical data of the patient to assist the physician.

3. Graphical User Interface is the end terminal application implemented on smartphones or it can be available as a webpage. This is made available so that physicians can track the health status of their patients, and the patients can, too.

Figure 21.1 depicts the block diagram of the proposed IoT-based heart rate monitoring system. The ECG signals acquired from the sensor are sent to the IoT cloud using transmission protocols like Wi-Fi or Zigbee, or Bluetooth. However, the paper is evaluated using the real-time data from the Physionet database, and the details are presented in the results section. Figure 21.1 depicts the practical framework involved in acquiring the signal and transmitting it to the cloud. The proposed method uses AWS cloud due to its more secure and beneficial storage and other flexibilities. On the cloud, the algorithm is run to denoise the ECG signal and estimate the heart rate. A robust algorithm based on Fractional Fourier transform (FrFT) is proposed. The ECG signal is initially pre-processed using digital filters to suppress the noises. Later, the FrFT is applied to the pre-processed ECG to transform it to the fractional-frequency domain. The purpose of converting the signal to fractional frequency is to distinguish the QRS complex from other components of ECG waves. The application of FrFT well enhances the QRS peaks and then improves the detection of R-peaks

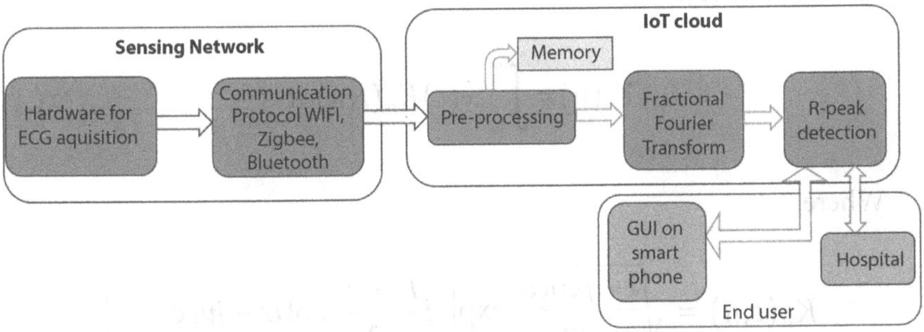

Figure 21.1 Block diagram of proposed IoT-based heart rate monitoring unit.

at higher accuracy. The final step is to run an R-peak detection algorithm to identify the peaks. Then the heart rate can be computed based on the R-R interval detected.

The algorithm's output is the heart rate that can be communicated to the end user or the physician using a mobile application or web page. Further, the processed ECG and the detected heart rate can be logged in the cloud storage and the time-stamp to have a history of patient data. The theoretical details of FrFT and AWS cloud servers are presented in the following sections.

21.3.1 Fractional Fourier Transform

The Fractional Fourier transform is capable of transforming the signal into the fractional frequency domain. It was initially proposed by Namias and Ozaktas [12] and is an improvement to the already existing Fourier Transform. This method has become more prevalent and widely applied in the signal processing domain due to its exciting characteristics and features. The Fractional Fourier transform has proven its importance and significance in the Separation of Multicomponent Chirp signals [13], radar, noise removal in the frequency domain, quantum physics, and analysis of optical signals [14].

The objective of the paper is to detect the R-peaks of an ECG signal and compute the heart rate. FrFT is a linear operator capable of rotating the signal in the time-frequency plane by a subjective angle $\alpha = a\pi/2$ where a is the fractional parameter. Since FrFT represents the signal in a fractional frequency plane or transforms the signal to an intermediary domain between the time and frequency, it becomes hugely easier to differentiate the QRS complex from other components of the ECG signal. The range of α is usually between $0 < \alpha < \pi/2$. When a is substituted as 1 the angle becomes $\pi/2$ and this results in the ordinary Fourier transform of this signal. The FrFT of the continuous abdominal ECG signal (t) is presented in equation (21.1) [2].

$$Y_x(u) = \int_{-\infty}^{\infty} \hat{x}(t) K_a(t,u) \mathrm{d}t \tag{21.1}$$

Where

$$K_a(t,u) = \sqrt{\frac{1 - j\cot\alpha}{2\pi}} \exp\left(j\frac{t^2 + u^2}{2}\cot\alpha - jut\csc x \right)$$

when

$$\alpha \neq p\pi \tag{21.2a}$$

$$K_a(t,u) = \delta(u-t)$$

when

$$\alpha = 2p\pi \tag{21.2b}$$

$$K_a(t,u) = \delta(u+t)$$

when

$$\alpha = (2p+1)\pi \tag{21.2c}$$

Here $K_a(t, u)$ is called the fractional Fourier kernel.

For a discrete-time abdominal signal $x(n)$, the expression for discrete Fractional Fourier transform and its kernel is as defined in equations (21.3) and (21.4).

$$Y_x(u) = \int_{-\infty}^{\infty} \hat{x}(n) K_a(n,u)\, dt \tag{21.3}$$

Where

$$K_a(n,u) = \sqrt{\frac{1 - j\cot\alpha}{2\pi}} \exp\left(j\frac{n^2+u^2}{2}\cot\alpha - jut\csc x \right)$$

when

$$\alpha \neq p\pi \tag{21.4a}$$

$$K_a(n,u) = \delta(u-n)$$

when

$$\alpha = 2p\pi \tag{21.4b}$$

$$K_a(n,u)=\delta(u+n)$$

when

$$\alpha=(2p+1)\pi \tag{21.4c}$$

The FrFT of the signal is estimated and is subjected to thresholding for estimating the r-peaks of the ECG signal. The r-peak locations are configured and heart rate is computed.

21.3.2 Amazon Web Services

In addition to the algorithm proposed, the availability of cloud infrastructure makes it easy and affordable to determine the heart rate. Amazon Web Services (AWS) provides virtual servers on the internet, called the public cloud. The benefits of AWS include providing additional processing, memory, and storage to process the algorithm. AWS allows the user to increase the capacity with ease and on a pay per use model, making it economical and straightforward to use AWS to provide processor, memory, and storage on a pay per use model. We can allocate and use the system as required and pay for the time we use. It is connected to the internet on a high-speed network.

The steps for loading the system on the cloud are as follows: 1) Launching an AWS windows instance; 2) connecting to the instance; 3) Installing MATLAB on AWS Windows Instance Elastic Compute Cloud. It allows an AWS subscriber to request and provision a computer server. This instance is chosen because it provides cloud-based amendable computing power. It is a high-performance, stable, safe, and reliable environment for deploying Windows-based applications. AMI is the acronym for Amazon Machine Image. It is a model for an Operating System and serves as a foundation in the creation of the desired instance. The type of instance selected is t2. medium which is a 2vCPU and 4GB memory server. The Remote Desktop Protocol is used to establish the connection to the instance. This is a Microsoft protocol that lets users connect to another device or system over an internet connection. This cloud infrastructure provides a robust method to deal with storing and processing of data. The data is secured and stored in the cloud and can be continuously monitored and used. This architecture ensures Instant enhancement of processing, storage, and memory capacity for additional requirements. One of the best and affordable cloud infrastructures is the Amazon Web Services (AWS). AWS is flexible as we can allocate and use the system as required and pay for the time we use. In addition to the flexibility and good security, it also provides processor, Memory, and Storage on a pay per use model.

21.4 Results and Discussion

The presented paper implements IoT-based heart rate monitoring using an AWS cloud server. An algorithm combining Fractional Fourier transform and R-peak detection is proposed for estimating the heart rate from the ECG signal. The proposed algorithm's performance is validated using MATLAB 2020b on a 64 bit 8bit RAM processor. The efficiency of the proposed approach is tested using ECG signals available in the Physionet MIT arrhythmia database [15, 16]. There are two prominent channels available in the database: a modified limb II (MLII) and each modified lead, otherwise called precordial leads, M1, M2, M4, or M5. Each record is sampled at 360Hz with an 11-bit resolution. The details of results obtained on these data with their inferences are presented in this section.

Figure 21.2(a) presents the ECG signal acquired from the MIT arrhythmia dataset, and its pre-processed version is shown in Figure 21.2(b). The bandpass filtering having cut-off frequencies followed by a notch filter of resonant frequency 50Hz is done to remove baseline wander and power line interferences. The pre-processed ECG is then transformed to the fractional frequency domain using FrFT. The FrFT for various values of the fractional parameter are presented in Figure 21.3. Figure 21.3(a) illustrates the pre-processed ECG signal, which is transformed to fractional frequency domain for various rotation angles as in Figures 21.3(b)-(e). As observed in the Figure 21.3(b), for the value of the fractional parameter $a = 0$, the signal is not rotated

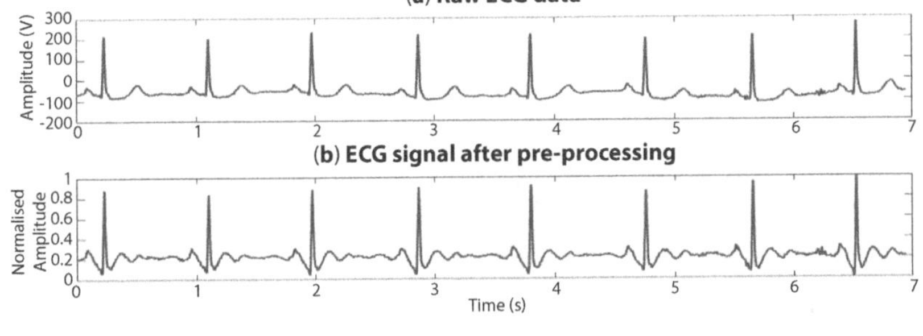

Figure 21.2 (a) Raw ECG signal (b) Filtered ECG using digital filters.

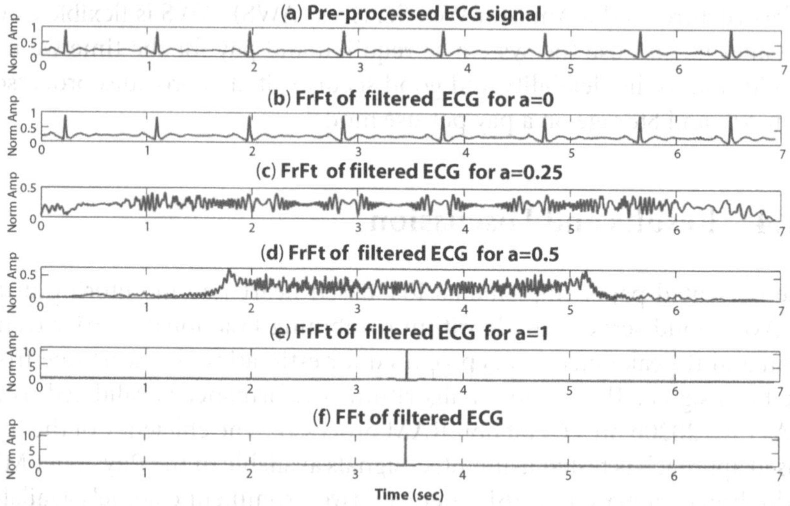

Figure 21.3 FrFT of filtered ECG for various values of fractional parameter.

and the FrFT of signal is the signal itself. For the value of $a = 1$, the signal is rotated by 90 degrees, and the result is the same as the Fourier transform of the signal as shown in Figures 21.3(e) and 21.3(f), respectively. Figures 21.3(c) and 21.3(d) represents the signal in the intermediate time-frequency domain. For the intermediate values of between 0 and 1, the signal is mapped to the fractional-frequency domain. The proper choice of will help in analysing the signal in fractional frequency domain.

For the current problem of interest, the role of FrFT is with the identification of R-peaks. Hence, the choice of is such that the QRS complex of the ECG wave is well distinguished from the other components. After careful analysis, the value of a is chosen as 0.01. Figure 21.4 presents the FrFT of the ECG signal for the value of $a = 0.01$. From the figure, it is observed that

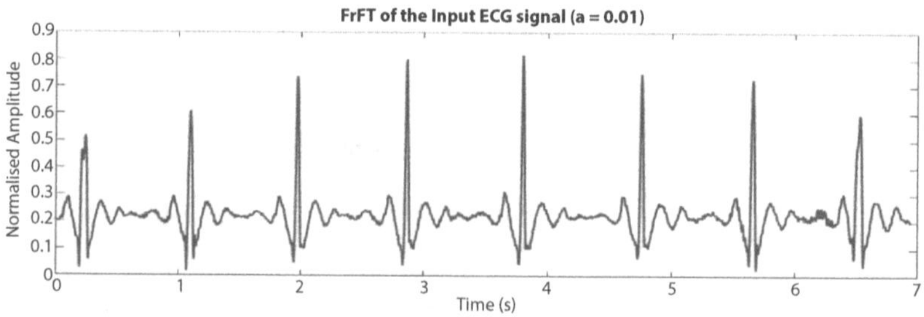

Figure 21.4 Fractional Fourier transform of ECG signal for a = 0.01.

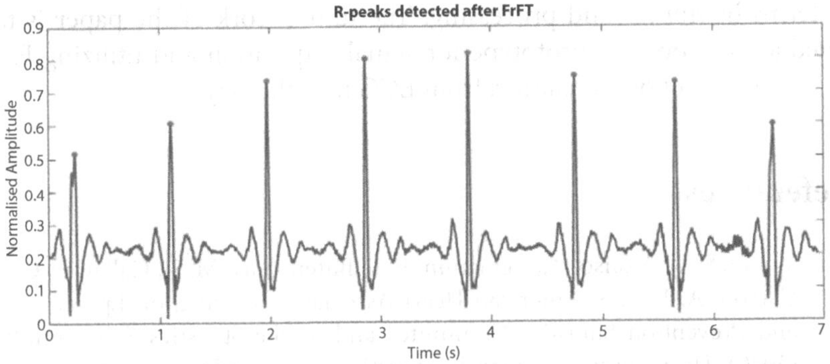

Figure 21.5 FrFT of ECG with detected R-peaks.

the application of FrFT with the right choice of a can enhance the QRS complex and make them distinguishable from other components of ECG signals like P and T waves. This behavior is the ability of FrFT to transform the signals into the fractional-frequency domain and separate the signal components that overlap in time and frequency domains.

The final step in the proposed algorithm is to find the R-peaks for which threshold is applied to the FrFT signal. Figure 21.5 presents the output of the peak detection algorithm. The identified peaks are marked on the FrFT signal, and it is observed that the R-peaks are detected at higher accuracy. Any false detection can be eliminated by considering the higher peak over a window of 50ms. The estimated heart rate is found to be 76 bpm. The proposed algorithm is implemented on the AWS server by following the steps explained in section 21.3.2. The results show that real-time ECG monitoring can be effectively implemented on IoT cloud and extended for remote monitoring applications.

21.5 Conclusion

This paper implements an IoT-based heart rate monitoring system. The proposed method is validated using the signals from the Physionet MIT arrhythmia database. The proposed algorithm based on Fractional Fourier transform is implemented on AWS cloud, and heart rate is computed. The potential benefits of FrFT are utilized to distinguish the QRS complex from other ECG components. The proposed system can efficiently provide a solution for real-time monitoring of heart rate and can be a lifesaver. Further, the algorithm implemented on AWS cloud offers flexibility in terms of storage and processing. The future work of the paper is targeted at developing a prototype for signal acquisition and utilizing FrFT for extraction of other features from ECG morphology.

References

1. Virani, S. S., Alonso, A., Benjamin, E. J., Bittencourt, M. S., Callaway, C. W., Carson, A. P., ... & American Heart Association Council on Epidemiology and Prevention Statistics Committee and Stroke Statistics Subcommittee. (2020). Heart disease and stroke statistics—2020 update: a report from the American Heart Association. *Circulation, 141*(9), e139–e596.
2. Gupta, V., & Mittal, M. (2019). A comparison of ECG signal pre-processing using FrFT, FrWT and IPCA for improved analysis. *IRBM, 40*(3), 145–156.
3. Kamble, P., & Birajdar, A. (2019, March). IoT Based Portable ECG Monitoring Device for Smart Healthcare. In *2019 Fifth International Conference on Science Technology Engineering and Mathematics (ICONSTEM)* (Vol. 1, pp. 471–474). IEEE.
4. Arun, U., Natarajan, S., & Rajanna, R. R. (2018, October). A Novel IoT Cloud-based Real-Time Cardiac Monitoring Approach using NI myRIO-1900 for Telemedicine Applications. In *2018 3rd International Conference on Circuits, Control, Communication and Computing (I4C)* (pp. 1–4). IEEE.
5. Rehman, A., Saqib, N. A., Danial, S. M., & Ahmed, S. H. (2017, November). ECG based authentication for remote patient monitoring in IoT by wavelets and template matching. In *2017 8th IEEE International Conference on Software Engineering and Service Science (ICSESS)* (pp. 91–94). IEEE.
6. Basu, S., Sengupta, A., Das, A., Ghosh, M., & Barman, S. (2019, December). IoT-Based Real-Time Remote ECG Monitoring System. In *International Conference on Computers and Devices for Communication* (pp. 23–28). Springer, Singapore.
7. Li, C., Hu, X., & Zhang, L. (2017). The IoT-based heart disease monitoring system for pervasive healthcare service. *Procedia Computer Science, 112*, 2328–2334.
8. Deshpande, U. U., & Kulkarni, M. A. (2017). Iot based real time ECG monitoring system using cypress WICED. *International Journal of Advanced Research in Electrical, Electronics and Instrumentation Engineering, 6*(2).
9. Sobya, D., Muruganandham, S. K., Nallusamy, S., & Chakraborty, P. S. (2018). Wireless ECG monitoring system using IoT based signal conditioning module for real time signal acquisition. *Indian J. Public Health Res. Dev, 9*, 294–299.
10. Satija, U., Ramkumar, B., & Manikandan, M. S. (2017). Real-time signal quality-aware ECG telemetry system for IoT-based health care monitoring. *IEEE Internet of Things Journal, 4*(3), 815–823.

11. Xu, G. (2020). IoT-assisted ECG monitoring framework with secure data transmission for health care applications. *IEEE Access*, *8*, 74586–74594.
12. McBride, A. C., & Kerr, F. H. (1987). On Namias's fractional Fourier transforms. *IMA Journal of Applied Mathematics*, *39*(2), 159–175.
13. Shao, Z., He, J., & Feng, S. (2019). Separation of Multicomponent Chirp Signals Using Morphological Component Analysis and Fractional Fourier Transform. *IEEE Geoscience and Remote Sensing Letters*, *17*(8), 1343–1347.
14. Almeida, L. B. (1994). The fractional Fourier transform and time-frequency representations. *IEEE Transactions on Signal Processing*, *42*(11), 3084–3091.
15. Moody, G. B., & Mark, R. G. (2001). The impact of the MIT-BIH arrhythmia database. *IEEE Engineering in Medicine and Biology Magazine*, *20*(3), 45–50.
16. Goldberger, A. L., Amaral, L. A. N., Glass, L., Hausdorff, J. M., Ivanov, P. C., Mark, R. G., ... & Stanley, H. E. (2000). Components of a new research resource for complex physiologic signals. *PhysioBank, PhysioToolkit, and Physionet.*
17. Pan, J., & Tompkins, W. J. (1985). A real-time QRS detection algorithm. *IEEE Transactions on Biomedical Engineering*, (3), 230–236.

22

Neural Collaborative Filtering-Based Hybrid Recommender System for Online Movies Recommendation

S. Priyanka[1], P. Saravanan[1*], V. Indragandhi[2] and V. Subramaniyaswamy[1]

[1]School of Computing, SASTRA Deemed University, Thanjavur, India
[2]School of Electrical Engineering, Vellore Institute of Technology, Vellore, India

Abstract

A recommendation process is usually the most commonly used form of commercial website. The custom recommender method is of vital importance in modeling users' choice of movies based on their previous interest. In today's internet apps, recommender services play a crucial role. However, the existing Deep Learning (DL) approaches have some limitations, which have a negative impact on the efficiency of the suggestion models. This paper provides an innovative deep learning architecture to improve filtering results in recommender systems. This methodology proposes new movies for users based on individual and related tastes. By studying user trends of film watching, we offer a technique of anticipating and suggesting a film. The similarity between any set of users is estimated using movie rating information and review data. Further we classified the user with similar film preferences and analysed the user group's consumption behavior to increase forecast accuracy by factoring the change in preferences over time. As films are an important source of entertainment, we have proposed a Recommender System (RS) in this work. Collaborative filtering and content-based filtering methods provide a conventional approach to recommendation systems with sentiment restrictions. We use a Hybrid filtering system based on simple Recurrent Neural Network (RNN) to demonstrate the efficiency of the suggested technique. The proposed system outperforms in terms of Root Mean Square Error (RMSE) and Mean Absoulte Error.

**Corresponding author*: sharan.doit@gmail.com

A. Chitra, V. Indragandhi and W. Razia Sultana (eds.) Intelligent and Soft Computing Systems for Green Energy, (287–302) © 2023 Scrivener Publishing LLC

Keywords: Neural network, multi-layer perceptron, collaborative filtering, recommender systems, content-based filtering, hybrid rnn, sentiment analysis

22.1 Introduction

Recommender mechanisms have become ever more important in today's unbelievably busy environment. In the limited 24 hours, the various tasks they need to do are still out of time. The consulting programs are often vital since they help people decide correctly without using their emotional resources. A Recommendation System goal is primarily to find content that is valuable to an individual. It also provides a range of explanations for making customized user-specific and interesting content lists. RS are artificial intelligence-orientated algorithms that cover all possible options and have a personalized, individual and fascinating list of items. Those results are based on their profile; search/browsing backgrounds, what other people are looking for, and how many films you can view with similar characteristics and demographics. The details and heuristics are used to do this using mathematical analyses [1, 2]. Usually, the recommenders include content-based filters, knowledge-based filters, and shared methods of filters [3].

For content-driven filtering, articles based on user profile similarities and item specifications are recommended. It is a widely used way to filter texts and articles and has explicit material on artifacts, user preferences, and knowledge-based filtering approaches guidelines [4]. The most frequent use of the independently chosen approach is Collaborative Filtering (CF). Users receive intuitive predictions in this filtering by assembling statistics from different users [5]. Memory-based and model-based is the main classification of collaborative filtering. The rating information and suggestions during the mutual memory filtering process are used to calculate the similarities for users or artifacts. An autoencoder-based recommender system for movie recommendation was proposed in [6]; the deep learning model with the multi-layer concept has been utilized to improve the performance of RS.

Different computer and mathematical approaches are cast off to understand in model-based CF. The model-based CF method of the Netflix Prize [7] is the most popular model-based CF method. In the interior product, the user's relationship with objects is modeled on latent space utilizing Matrix Factorization Netflix Prize, which is the greatest widely held model-based CF method. The matrix factorization procedure is effective, but only by interacting as a user's dot and a latent vector can hinder success.

The paper is structured as follows. Section 22.2 defines the comprehensive portrayal of the corresponding work and the place of the article. Section 22.3 provides a proposed methodology. In Section 22.4 the results are discussed, and Section 22.5 summarizes the conclusion with future work.

22.2 Related Works

Another CF-based algorithm is proposed in a deep Neural Network (NN) by continuously increasing the number of times problems are caused with fitting in standard algorithms [8]. Using the batch normalization technology on each layer of NN, the proposed algorithm overcomes the overall problem for the continuous rise in the number of eras. The NN layer receives the data as a user classification vector and a standard object classification vector. The classifier is constructed by a stochastic gradient descent system with back propagation. The forecast was made based on conditional likelihood. The algorithm was tested with 100k and 1M dataset film lenses.

Katarya *et al.* have suggested a model of a data clustering method of recommendations [9]. The method was proposed for a cuckoo search-based K-means clustering model. This model focused on collective filters. Li *et al.* suggested a technique of data sparsity handling for the recommendation [10]. This model bases on two main problems in the recommendation system; data sparsity management and the identification and management of the drifting principle. Deep neural networks were used for computer vision testing and Natural Language Processing (NLP) problems many years ago. Writers are the first to extend profound learning in predictive analytics to the greatest of our experience and have also recommended Neural Collaborative Filters (NCF). The NCF approach addresses the problem of low-dimension representation of Matrix Factorization (MF).

In [11], the authors suggested a recommendation framework method of Deep Matrix Factorization (DMF). The characteristics were collected from a user-item matrix and integrated into the NN by the authors, with this profound learning approach. They also utilize DMF feedback results, explicit and implicit. Applying Bi-clustering or Bi-max algorithm for similarity measures and further for item recommendation [12] is another well-known technique in RS.

In [13], the authors used the CNN model with the Method of probability factorization and suggested the Conv. MF quality service guideline. In [14], Collaborative Neural Social Recommendation (CNSR) presented a neural architecture that combines social networking information and the interaction between users and social media in an online social networking approach. In [15], the writers explored a novel technique focused on customized Long-Short Term Memory techniques (LSTM) and matrix factorization. This could include the tacit representation of many different users and services; the predictive procedure may be redesigned to process new knowledge.

In most cases, the user ID and object ID are only used to achieve excellent outcomes for the Recommender System domain. In [16], a Deep Hybrid Web Service Recommendation (DHSR) method has been proposed, which used the Multi-sectoral Perceptron to integrate textual knowledge similarity to learn the non-linear connection between services and the construction of mashups. The Stacked Denoising Auto-Encoder (SDAE) approach [17] has been proposed to construct a Deep Learning-based Long-Tail Service Recommendation (DLTSR) architecture to report the long-tail problem of the web services recommendation mechanism.

The writers used the picture-based methodology for the suggestion of resources in [18]. The picture functionality is obtained using the JPEG coefficient algorithm, and by using the NN convolution, the Random Forests set technique is often used as a basis for learning.

While the new user dilemma is well debated, the new group and the new item issues are equally critical when building a good recommender system. A new user challenge is when the user has not submitted sufficient ratings to receive customized reviews. New customers also feel ignored, and they may even abandon the application.

22.3 Proposed Methodology

The ultimate objective is to suggest products that are more closely related to the user's interests. To achieve user recommendation, this project proposed a recommendation system with an approach based on user behavior-oriented. We use the Douban data set to compare different classification algorithms. A hybrid method is used for this recommendation. We have to do some steps before we reach our recommendation process. We consider both the ratings and reviews which the user gives to the movies [19].

Initially, sentiment analysis has been implemented to filter the negative feedback. Only positive feedbacks are given to the users. The similar user prediction will help the user to predict exactly what they need to them. We are applying RNN for the data for both ratings and reviews. The proposed system uses a collaborative filtering-based movie services recommendation framework as shown in Figure 22.1.

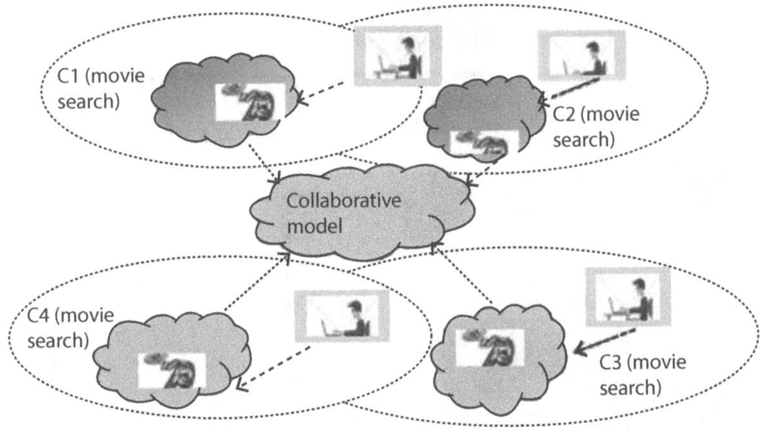

Figure 22.1 CF movie services recommendation scenario.

It works on four online cloud servers: cloud C1, cloud C2, cloud C3, and cloud C4. The dotted lines indicate bags of wireless paths. The complete movie service is the collaborative works of user1, user2, user3 and user4. The system may recommend the movie for the user who searches for it based on the rating or other parameters provided by other users.

22.3.1 Dataset Used for the Proposed System

The Douban dataset is a Chinese dataset that has a huge collection of both movies and books data. It has both ratings and reviews given by the user and has different forms of interaction. The dataset contains 14,864 users' information, 17,008 item information, 638,393 user-item interactions, 10,816 number of lists, and 775,877 user–list interaction details.

22.3.2 Architecture Diagram

The workflow of the proposed Recommendation system is given in Figure 22.2. Initially, the data is split into review features and rating features. Stop word analysis, Tokenization, and Stemming word analysis have been implemented as preprocessing steps. As we need only positive feedback for the recommendation process, the sentimental analysis has been implemented. To extract the Key phrase natural language processing algorithm has been used. RNN classification is applied on the features to classify them into positive, neutral, and negative.

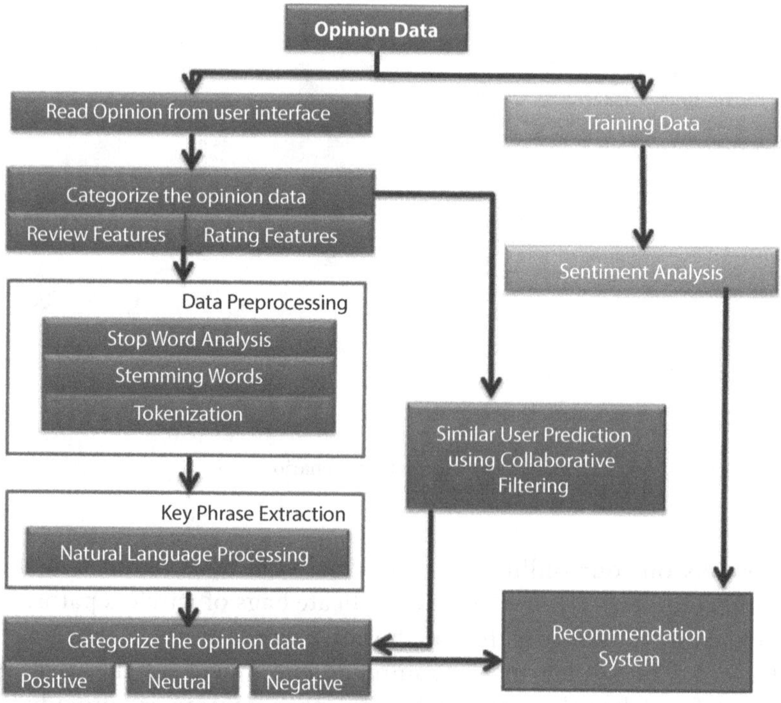

Figure 22.2 Process flow of the proposed system for movie recommendation.

22.3.3 Sentiment Analysis

Sentiment analysis is a prediction activity where the model is taught to predict the polarity of textual information or feelings such as Positive, Neutral, and Negative. It becomes quite difficult to interpret the enormous volume of textual data, and we are already overburdened with many unstructured data. However, the study of sentiments can be quite valuable to label such messages by companies. Feedback, film reviews, etc., may be calculated with a sentimental analysis. Even the identification of emotions is part of the sentiment analysis, which can influence how good the film is, how average and terrible.

In the realm of machine learning, sentiment analysis has been a hot issue. It's mostly used with data that has been tagged with a sale, like product reviews from product datasets. A scalar score is included with the review text a user submits, and it allows for accurate text polarity identification. When it comes to social data, this capacity to detect positive or negative emotion behind a piece of text is even more intriguing. Every second, new user data is added to reviews. We'll be able to identify the most current user attitude toward a range of issues if our algorithm can predict sentiment labels for incoming product reviews.

Recurrent Neural Networks are made up of neurons with weights and biases that may be learned. Each neuron gets some inputs, does a dot product, and optionally adds non-linearity to the result. From the raw picture pixels on one end to class scores on the other, the entire network still represents a single differentiable scoring function. On the fully connected last layer, they still have a loss function, and all methods we established for learning conventional neural networks still apply. We use three main types of layers to build LSTM, and Fully Connected Layers. Based on RNN, reviews are classified and labeled as "Positive", Negative", and "Neutral" with improved accuracy.

22.3.4 Hybrid Recommendation

Collaborative filters imitate the user-to-user suggestions. It forecasts the preferences of the users as a weighted and linear mixture of other preferences.

22.3.4.1 Filtering Based on Content

It is based on similarities between the products' characteristics. It suggests things to a client based on the highest goods that the same client previously evaluated. It is necessary to construct a list of these elements. A profile of the object is included with each item. These qualities are organised into a table. Comparison function on ratings has been used to best items. The highest-ranked item is recommended. The user will be given the best scoring match. This technique depends solely on items and not on user choices.

22.3.4.2 Collaborative Filtering

It depends upon how other people have reacted to the identical products. It does not depend on the element's features but on other users' choices. Figure 22.3 shows how the NCF part will work. There needs to be a similar user survey. Users have a table with several classified goods they like or prefer. Based on similarities, what the user likes may be predicted based on what comparable users have done. The list is sorted and adapted to users who have used the same goods to compare and advise. All are summarized and suggested for the highest score. Code is based on an algorithm and given a user x, an item that x may enjoy is recommended. The highest scoring item is preferable. The problem is that you need data to make suggestions.

From highly rated movies to the lowly rated movie for our understanding. Next, we are moving to the significant part which is user and item interaction. This user-item interaction matrix will give details of every movie. The particular movie has how many users provide the rating for the particular movie and also has a total number of ratings. The User item interaction matrix has two more important steps they are user-based interaction and item-based interaction. These give the individual contribution of user and item details. The matrix also contains NAN values which may affect the results also. The Pearson correlation method will help to fill the NAN values in the matrix.

22.3.5 Neural Collaborative Filtering (NCF)

NCF depicts the user/item as a vector of hidden features projected onto a common feature space. In this feature space, the inner product of user-item latent vectors might be utilised to represent user-item interactions. This basic implementation of Matrix Factorization (MF) can be improved by merging it with neighbor-based models, combining it with item-content topic models, and extending it to factorization machines for broader feature modeling. In Neural Collaborative Filtering, the user-item inner product is replaced by a neural architecture. NCF hoped to accomplish the following by doing so:

1. Under this framework, NCF attempts to express and generalize MF: MF is a specific instance of NCF, generalized and expressed. Because MF is so effective in the recommendation arena, doing so will boost NCF's credibility.
2. NCF uses a multi-layer perceptron to learn user-item interactions: Neural network design is used to model the interaction between user and object features. It learns user-item interactions using a Multi-Layer Perceptron (MLP). This is an improvement over MF since MLP can learn any continuous function (theoretically) and has a high level of nonlinearities (due to numerous layers), making it ideally suited to learn user-item interaction functions.

NCF joins two subnetworks together. The GMF is a Generalized Neural Network variant of matrix factorization, with the element-wise product of the user and item latent factors as the input. It is made up of two layers.

One is the Hadamard product of vectors in user matrix P and item latent matrix Q, and the prediction score of the user u on item i. They are given by equations 22.1 and 22.2.

$$\text{Hadmad product } = p_u \odot q_i \tag{22.1}$$

$$\text{Prediction Score } \hat{y}ui = \alpha\left(h^T x\right), \tag{22.2}$$

Where, pu is the uth row of P and qi is i^{th} row of Q.

MLP is just another element of this architecture. The MLP subnetwork does not exchange user and item embeddings with GMF to increase model flexibility. As input, it takes the combination of user and item embeddings. It can estimate the various interactions between users and items thanks to the sophisticated linkages and nonlinear transformations. The MLP subnetwork is more clearly described in equation 22.3.

$$\hat{y}ui = \alpha\left(h^T \phi^L\left(z^{(L-1)}\right)\right) \tag{22.3}$$

Where ϕ is the function of the concerned layer, α is the activation function and z is the output of that layer.

The pictorial representation of the NCF is shown in Figure 22.3. To collect the sequential/time-series data, RNN was built. In RNN, we multiply

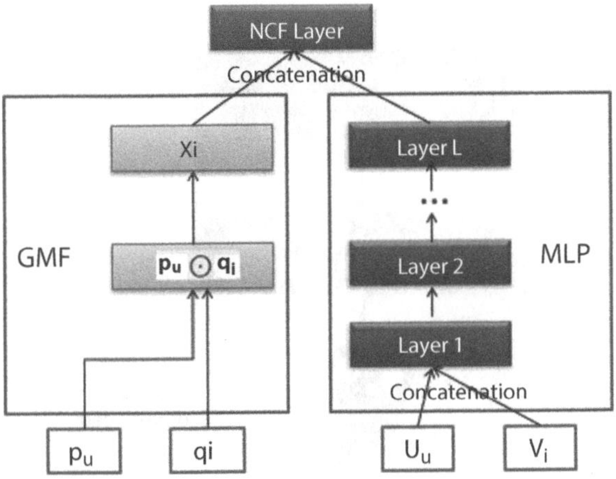

Figure 22.3 Flow of neural collaborative filtering.

the weight of the input from the preceding state by the result from the preceding state. Then we transfer them to the new status Tanh function. Now we multiply the new status with a Tanh function output to achieve the output vector.

22.3.6 User-Based Recurrent Neural Networks (RNN)

In the preceding section on time-dependent suggestions, we discussed several developing techniques for formalising a sequential problem. We'll go through how we employ DL approaches in RS to account for temporal dynamics, and we'll show a novel gated architecture for RNNs with a unique gated recurrent unit that allows user-related information to be easily integrated into the model. Other DL methods for RS, which usually focus on consumption sessions without particularly reflecting on the user, are distinguished by the distinct idea of user information in our study.

The process of recurrent neural networks for this hybrid recommender system is shown in Figure 22.4. Initially, it must make the generalized,

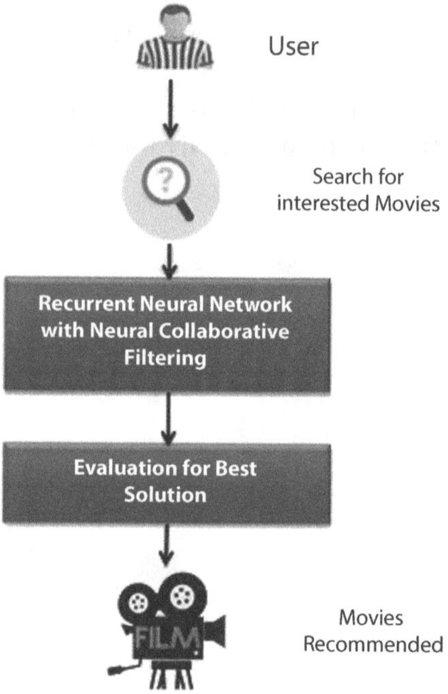

Figure 22.4 Process of recurrent neural network.

domain-independent formulation of RNNs concrete and move to RS in order to make sequential recommendations.

RNNs are effective tools for modelling any type of sequence, especially when combined with grading layers. We depend on a variant of RNNs that generates the $o^{(t)} \in R^{I}$ at each phase of the time by smooth processing of secret current state to $h(t) \in \mathbb{R}n$.

- o(t) Indicates the output of the model at the index sequence t. X(t) is input of the training sample sequence
- h(t) representing the model's hidden state at the index sequence t. h(t) determine by (Z(t)), x(t), h(t-1), b
- b is bias and σ is activation function and D,E,V are linear parameters
- y(t) indicates output of the training sample sequence

$$o(t) = Vh(t)+c \tag{22.4}$$

Let ϒ be the set of users, then the one-user is given by the variable υ (t) ∈ {0, 1} |ϒ|. It is used to integrate the actual set of characteristics into the proposed model. We want the data to consist of user-item tuples that may be interpreted as a point of the user-uncontrollable item'i' at a specific time step t in the structure. We extend the creative concept of the unseen state based on this:

$$h(t) = \sigma(Z(t)) = \sigma(Dx(t) + Eh(t-1)+b) \tag{22.5}$$

$$y(t) = \sigma(o(t)) \tag{22.6}$$

Explicitly remembering ideas about the user into the recurrent cells may improve the network's prediction capability. We'll go over a few different implementations of (22.5) that structurally change the basic recurrent unit. We may extensively incorporate user-related information into the network by integrating a user variable.

22.4 Results and Discussion

Keras and sklearn packages have been used to incorporate the proposed model. For prediction success measurement, this model uses Movies data collection. The dataset comprises 1 million recommendations and is a

Regular Recommendation System Benchmark. Information on movies, consumers, genres and reviews are given. This dataset is separated by 70:30 ratios of training and testing results. The performance of the proposed Mean Absolute Error (MAE) models was calculated by,

$$\text{MAE} = \frac{1}{N}\sum\left|P_{(z,i)} - R_{(z,i)}\right| \tag{22.7}$$

By equations 22.7 & 22.8, the user rating for P(z,i) on item I, is R(z,i) and N is anticipated as the overall rating on the item set. As the MAE falls, the predictor of user raters is increasingly precise. The low values of MAE indicate the high power of the predictive models. When compared to the other models, our model has the best MAE values.

Figure 22.5 shows MAE loss dependency on the rating density and fixed the item and User count at 6,000 and 3,000. They mainly focused on the Hybrid method in specific category, i.e., Content-based approach and collaborative filtering method. The density and the item counts are fixed at range 3,000 and 4%, respectively. the density and the user counts are fixed at range 6,000 and 4%, respectively.

Root Mean Square Error (RMSE) emphasizes more on greater absolute error, and if RMSE is little, it is better than the accuracy of a recommendation system

$$\text{Root Mean Square Error} = \sqrt{\frac{\sum_{i=1}^{n}\left(P_{(z,i)} - R_{(z,i)}\right)^2}{N}} \tag{22.8}$$

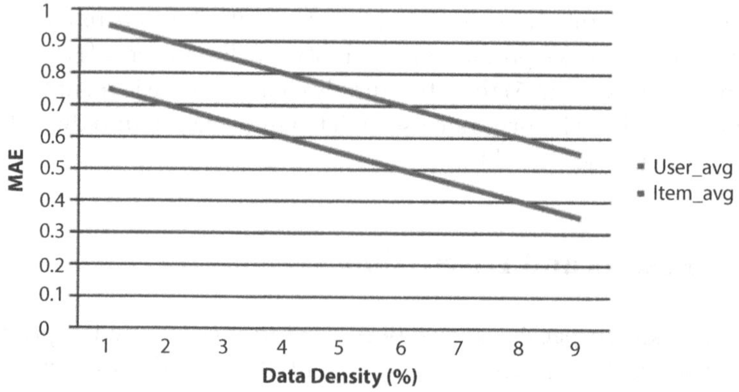

Figure 22.5 MAE by the proposed system for varying density value.

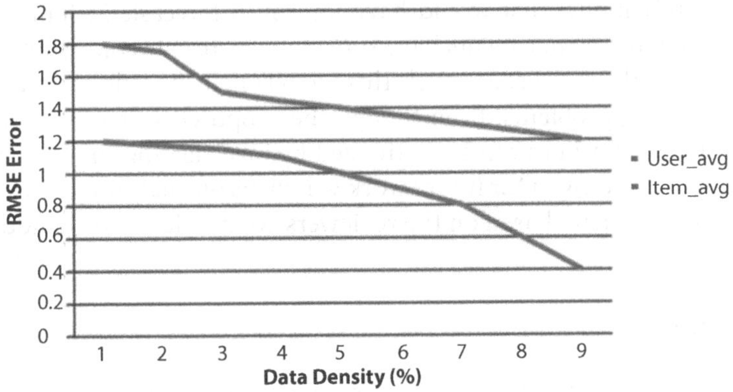

Figure 22.6 RMSE by the proposed system for varying Density value.

Figure 22.6 shows the RMSE comparisons of the recommender algorithms at different levels of data density. Our results are compared with the previous literature and its similar matches, and the recommendation algorithms also demonstrate the highly predicted accuracy of the Douband datasets. The important one is the right side of the graph in Figure 22.6, which shows the rating data's density level, Here the notable impact on this work is neighborhood-based collaborative filtering techniques where the rating data become sparser by RMSE. In our sparsest samples, however, these approaches are marginally more sensitive to add new ratings as the data grows sparser. While considering the rating density with 0.351%, the RMSE values of, both CF_Item and CF_User are significant on the scale of rating 1 to 5 and positive reviews similar to movies. The baseline and hybrid methods, on the other hand, are significantly more stable and compatible across density levels. In particular, the experimental results demonstrate that the Hybrid method (User_Avg, Item_Avg, User_Item_Avg) is very resistant to changes in data density. The RMSE analysis yields a result that supports this approach. We ignore the persistent prediction rule's basic baseline.

Our proposed system improves the performance of RNN with regularization with a dataset split of 80%-10%-10%. The loss of training is significantly as much as the affirmation at first loss.

22.5 Conclusion and Future Work

The system of recommendations allows one to avoid the information overload that the modern world gives us. This article suggested a profound

Hybrid RNN model that would have improved forecasts in film recommender systems. Experiments have shown better results concerning MAE and RMSE in the proposed model. This model's main benefit is to tackle the cold beginning problems. Even though the proposed system still outperforms, the MAE values can be optimized by investigating the methods of avoiding fake reviews. Our future work will focus on analyzing movie websites for fake reviews based on the reviewers' MAC address or IP address.

References

1. Goodfellow, I., Bengio, Y., & Courville, A. (2016). *Deep Learning*. MIT press.
2. Zhang, S., Yao, L., Sun, A., & Tay, Y. (2019). Deep learning based recommender system: A survey and new perspectives. *ACM Computing Surveys (CSUR)*, 52(1), 1–38.
3. Sivaramakrishnan, N., Subramaniyaswamy, V., Viloria, A., Vijayakumar, V., & Senthilselvan, N. (2020). A deep learning-based hybrid model for recommendation generation and ranking. *Neural Computing and Applications*, 1–18.
4. Nassar, N., Jafar, A., & Rahhal, Y. (2020). Multi-criteria collaborative filtering recommender by fusing deep neural network and matrix factorization. *Journal of Big Data*, 7, 1–12.
5. Nilashi, M., Ibrahim, O., & Bagherifard, K. (2018). A recommender system based on collaborative filtering using ontology and dimensionality reduction techniques. *Expert Systems with Applications*, 92, 507–520.
6. Yi, B., Shen, X., Zhang, Z., Shu, J., & Liu, H. (2016, December). Expanded autoencoder recommendation framework and its application in movie recommendation. In *2016 10th International Conference on Software, Knowledge, Information Management & Applications (SKIMA)* (pp. 298–303). IEEE.
7. Zhuang, F., Zhang, Z., Qian, M., Shi, C., Xie, X., & He, Q. (2017). Representation learning via dual-autoencoder for recommendation. *Neural Networks*, 90, 83–89.
8. Frémal, S., & Lecron, F. (2017). Weighting strategies for a recommender system using item clustering based on genres. *Expert Systems with Applications*, 77, 105–113.
9. Katarya, R., & Verma, O. P. (2017). An effective collaborative movie recommender system with cuckoo search. *Egyptian Informatics Journal*, 18(2), 105-112.
10. Li, J., Xu, W., Wan, W., & Sun, J. (2018). Movie recommendation based on bridging movie feature and user interest. *Journal of Computational Science*, 26, 128–134.

11. Xue, H. J., Dai, X., Zhang, J., Huang, S., & Chen, J. (2017, August). Deep Matrix Factorization Models for Recommender Systems. In *IJCAI* (Vol. 17, pp. 3203–3209).
12. Akshaya, B., Akshaya, S. K., Gayathri, S., & Saravanan, P. (2016). Investigation of Bi-Max Algorithm for On-Line Purchase Recommender System using Social Networks. *Indian Journal of Science and Technology*, 9(44).
13. Kim, D., Park, C., Oh, J., Lee, S., & Yu, H. (2016, September). Convolutional matrix factorization for document context-aware recommendation. In *Proceedings of the 10th ACM Conference on Recommender Systems* (pp. 233–240).
14. Zhang, S., Yao, L., Sun, A., & Tay, Y. (2019). Deep learning based recommender system: A survey and new perspectives. *ACM Computing Surveys (CSUR)*, 52(1), 1–38.
15. Xiong, R., Wang, J., Li, Z., Li, B., & Hung, P. C. (2018, July). Personalized LSTM based matrix factorization for online QoS prediction. In *2018 IEEE International Conference on Web Services (ICWS)* (pp. 34–41). IEEE.
16. Xiong, R., Wang, J., Zhang, N., & Ma, Y. (2018). Deep hybrid collaborative filtering for web service recommendation. *Expert systems with Applications*, 110, 191–205.
17. Bai, B., Fan, Y., Tan, W., & Zhang, J. (2017). Dltsr: A deep learning framework for recommendations of long-tail web services. *IEEE Transactions on Services Computing*, 13(1), 73–85.
18. Ullah, F., Zhang, B., & Khan, R. U. (2019). Image-based service recommendation system: A JPEG-coefficient RFs approach. *IEEE Access*, 8, 3308–3318.
19. Ullah, F., Zhang, B., Khan, R. U., Chung, T. S., Attique, M., Khan, K., ... & Jan, S. (2020). Deep Edu: A deep neural collaborative filtering for educational services recommendation. *IEEE Access*, 8, 110915–110928.

23

Farmer's Eye Using CNN

Elam Cheren S., Yuvan Raj Kumar M.*, Vivek G., Udhayakumar N. and Saravanakumar M. V.

Department of EEE, Sri Krishna College of Engineering and Technology, Coimbatore, India

Abstract

This work is based on identifying disease in plants by using Neural network (Convolutional Neural Network) that suggests pesticide or insecticide for that disease and also providing the overview of the disease. In addition we need to gain environmental details around the plant using different types of sensors. This whole process is done by rover, which captures images and feeds them to the processor which does image classification and then sends results to the user via wireless technology.

***Keywords*:** Processor, rover, convolutional neural network, image classification, neural network

23.1 Introduction

The Indian economy majorly depends on agriculture and farm production. India is the major exporter of farm products around the globe, amounting to about 943.53 lakhs tons [2]. The GDP share of agriculture has grown 19.9% in recent years. Due to sudden climatic changes and increased pollution plants are more prone to different diseases which farmers nowadays are not able to detect [1, 8]. The major challenge in agriculture is identifying disease, giving the remedy for it, and stopping the spread of disease in the field [4, 5]. To achieve this we have to combine traditional methodology with new, modern technologies which create a revolution in the field of agriculture and enable farmers to not only identify and classify diseased plants but also make it easier to choose medicine for the diseased plant [3, 7]. The process is controlled

**Corresponding author*: yrk1599@gmail.com

A. Chitra, V. Indragandhi and W. Razia Sultana (eds.) Intelligent and Soft Computing Systems for Green Energy, (303–314) © 2023 Scrivener Publishing LLC

by wireless technology which makes modernized agriculture for the future [6, 11]. To make this scenario possible we designed, analyzed and tested a rover that is controlled by a microprocessor [9]. The rover can be monitored and controlled by the user from a remote area [10]. The rover contains a camera which captures images then feeds that captured image into the neural network. This proposed idea is focused on solving problems like identifying animal intervention in field, delivering real-time notification to the farmer and processing without human intervention [13, 14]. In this rover there are sensors and electronic devices and they are combined using Python scripts. Based on attempted test cases, we were able to achieve a success rate of 85% while using Convolutional Neural Network [12]. The Convolutional Neural network uses image classification algorithm to classify images based on a binary classification method to predict diseases in plants. After image classification the predicted details will be sent to the farmer via wireless communication technology.

23.2 Related Works

A. Image classification

In recent years, image classification has boomed. Majorly, Image classification using neural network has become a new trend of the era; nowadays, a special neural network called convolutional neural network is used for segmenting and classifying images. From 2012 the competition on Image Net the Alex net was proposed, then in 2015 VGG was proposed. After this, NasNet was proposed by Google in 2018 and it is trained in GPU. In Image Net competition it is noted that the accuracy of image classification increased up to 92.85%.

B. Convolutional Neural Network

The Artificial Neural Network is simply called as Neural Networks. The Neural Network is a combination of neurons of circuit or network or algorithm which are used to identify the relationship between the series of data through a process of mimicking how the human neuron works. A neural network is an interconnectional layer of previous one or previous function. The Convolutional neural network is a special type of Neural Network that is used to work with image data.

23.3 PV Module

The proposed idea is based on identifying diseased plants using deep learning technique by the concept of image classification algorithm by using convolutional neural network. The rover is controlled by the farmer from a remote area. The control signal for the rover is sent by the farmer/user by wireless transmission using ZigBee transmitter. In the rover there is a ZigBee receiver that receives the signal and controls the rover according to the signal sent by the user. The Pi camera in the rover captures the image of a leaf and sends it to the Microprocessor (Raspberry Pi); in the processor the captured image is fed into the Neural Network. Here, image classification takes place, the Deep Neural Net finds the plant is diseased or healthy. If the Neural Net finds the leaf is healthy it sends a message to the user to that effect; if the leaf is diseased the Neural Net describes the disease of the plant and offers some recommendation of pesticides or insecticides. Sensors present in the rover send information about the environmental conditions in the plant location to the user; the environmental data of temperature, moisture and humidity is measured often and updates in the data base storage. The proposed block diagram is shown in Figure 23.1.

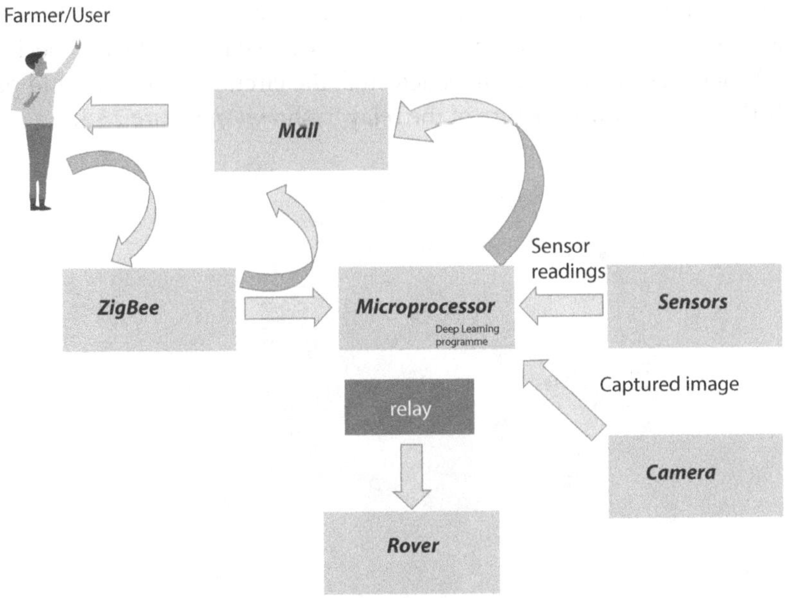

Figure 23.1 Block diagram of proposed system.

23.4 Hardware Description

A. Microprocessor

It is a central controlling unit minicomputer which is fabricated on a small chip which is capable of doing some operations like a computer, like arithmetic, logical and controlling operation. The proposed system uses Broadcom BCM2837 which is a 64-bit Microprocessor with a clock frequency of 1.25 GHz and consisting of 1GB RAM. It is 40 GPIO pins. The chosen microprocessor also has one port for connecting the camera and also for connecting display. It supports Python programming language to code in it. The Processor uses ARM A35 architecture, which is faster for running Deep Learning program. It has 4 USB 2 ports Micro SD card support to load Operation System and to store data. It also supports wireless connectivity and Bluetooth. The microprocessor used in the proposed system is shown in Figure 23.2.

B. Relay

Relay is just a switch but it is operated electrically, which is why it is called as electrically operated switch. It contains a coil; inside the coil the flux is created. This flux induces a magnetic field and cuts the contact between the switches. It consists of DPDT. It only allows one circuit to work at a time and keeps the second not functioning until the relay allows for the current flow to the second circuit. There won't be any electrical connections in between those two circuits. The link can be done only via magnetic field and mechanical. Whenever the overcurrent is detected, the circuit disconnects and protects the system. The structure of the relay in shown in Figure 23.3.

Figure 23.2 Broadcom BCM2837 Microprocessor.

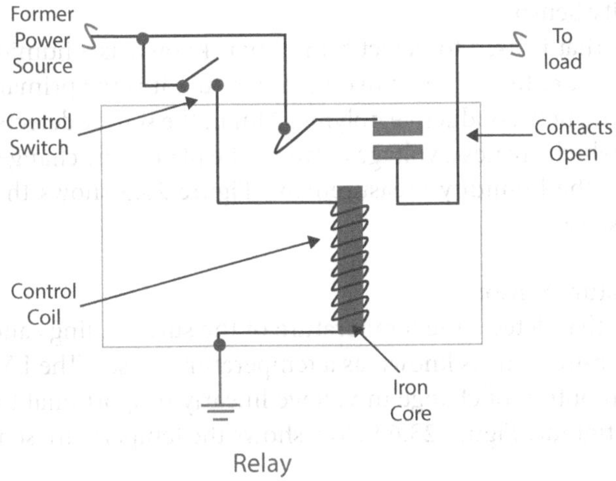

Figure 23.3 Relay circuit.

C. ZigBee

ZigBee is a wireless data transmission device which operates at a frequency of 2.4Ghz having a maximum data transfer rate of 250Kbps. It transfers data over a short distance range of 10-100 meters. The ZigBee image in shown in Figure 23.4.

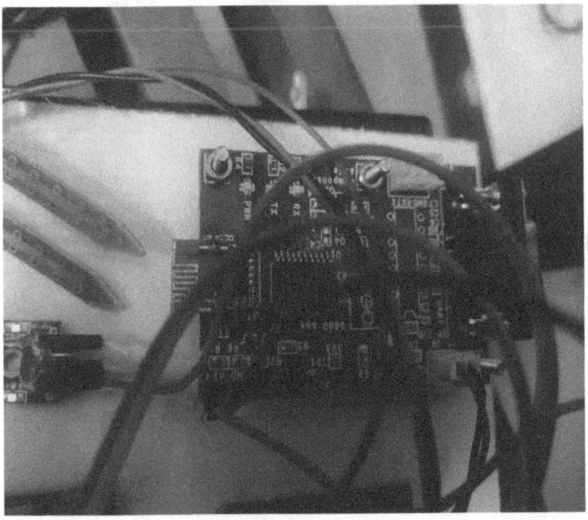

Figure 23.4 ZigBee.

D. Humidity Sensor

The sensor that is used to detect humidity is known as a humidity sensor. Since agriculture highly dependson temperature it is the primary detail to be noted. The non-conductive polymer film in the sensor detects the moisture in air which creates a voltage between the plates. The change in voltage level shows the humidity measurement. Figure 23.5 shows the image of Humidity sensor.

E. Temperature Sensor

The sensor that detects the temperature of the surroundings and provides the temperature value is known as a temperature sensor. The LM35 sensor provides an output of change in voltage linearly proportional to temperature in centigrade. Figure 23.6 below shows the temperature sensor.

F. Moisture Sensor

This type of sensor is majorly used to find the moisture level in soil. It provides high output if the moisture level is good and low output when there is low moisture level. Figure 23.7 shows a moisture sensor.

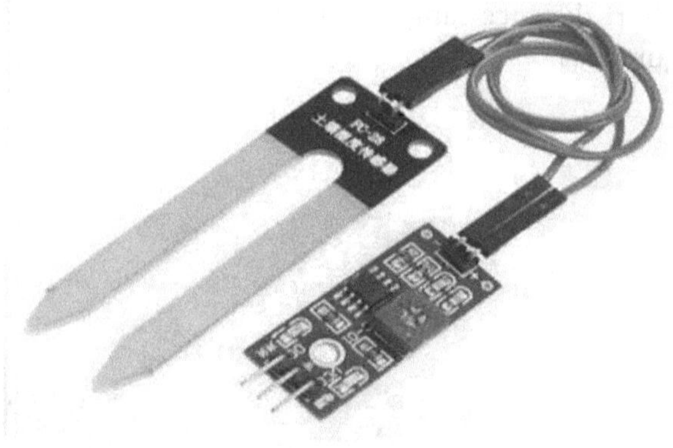

Figure 23.5 Humidity sensor.

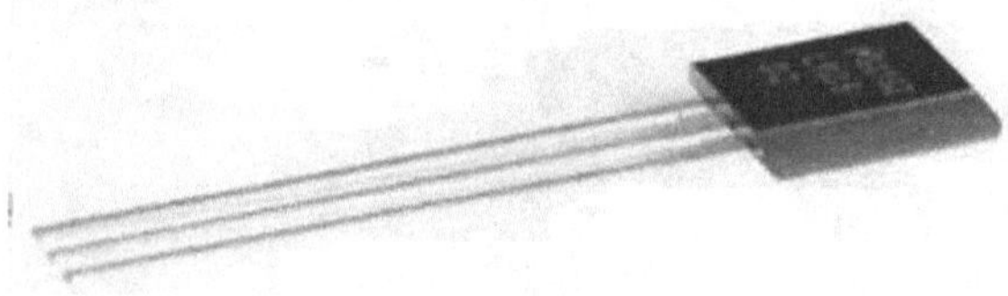

Figure 23.6 Temperature sensor.

Figure 23.7 Moisture sensor.

G. Pi Camera
This device is used to capture an image of the plant. After the capture, the image is saved in microprocessor for the next step. It consists of 1-Megapixel high-qualitycamera with IR filter removal. Figure 23.8 below shows the Pi camera.

23.5 Software Implementation

A. Deep Learning
Deep learning is the subfield of Machine Learning that functions like the human brain using Neural Network by pre-processing data and creating a

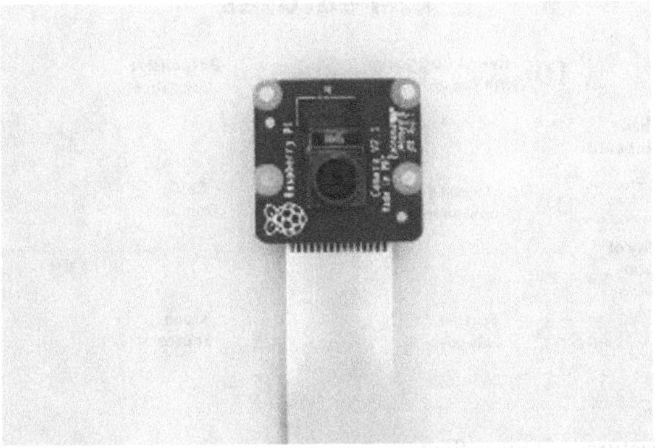

Figure 23.8 Pi Camera.

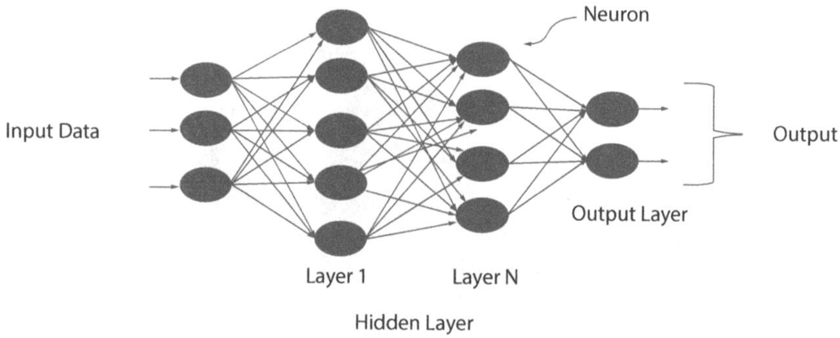

Figure 23.9 Deep neural network.

pattern to make a decision. This deep learning is used in the field of computer vision, speech recognition, machine translation, bio-informatics and in the medical field. Figure 23.9 shows the outline of Deep Neural Network.

B. Tensor Flow

It was created by Google Brain team. Tensor is a free open-source tool for machine learning. It consists of a variety of tools and libraries. It has a wide variety of uses but specifically focuses on deep neural networks. It can be accessed with Python that makes tensor flow easier to code and faster in solving problems. Figure 23.10 below shows the features of tensor flow.

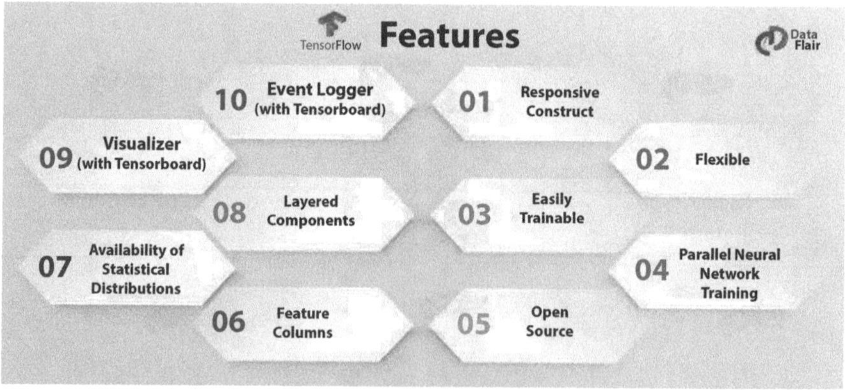

Figure 23.10 Features of tensor flow.

23.6 Hardware Implementation

The experiment was conducted in standard environment where we achieved the accuracy of 89.75%. Here we used image augmentation technique to crop the correct shape of the image and to avoid unwanted background learning. The prototype developed is shown in the Figure 23.11.

Figure 23.12 below shows the relay module which is used to control the movement of the rover and speed of the rover.

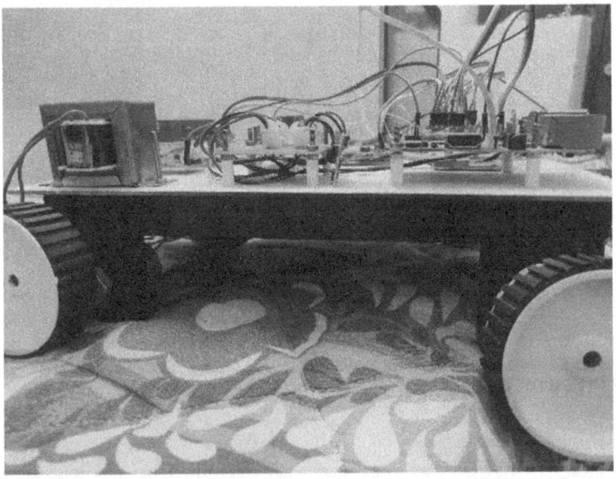

Figure 23.11 Rover prototype.

Figure 23.12 Relay module.

Figure 23.13 shows the ZigBee module is divided into two parts, namely data receiving area and data transmission area. The transmission ZigBee is used to transmit signal from user regarding the motor direction and the ZigBee receiver receives the signal from the user. Figure 23.14 shows the experimental output and model prediction.

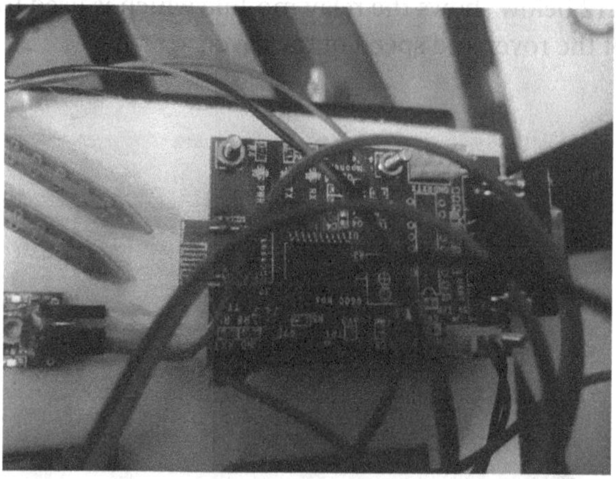

Figure 23.13 ZigBee module.

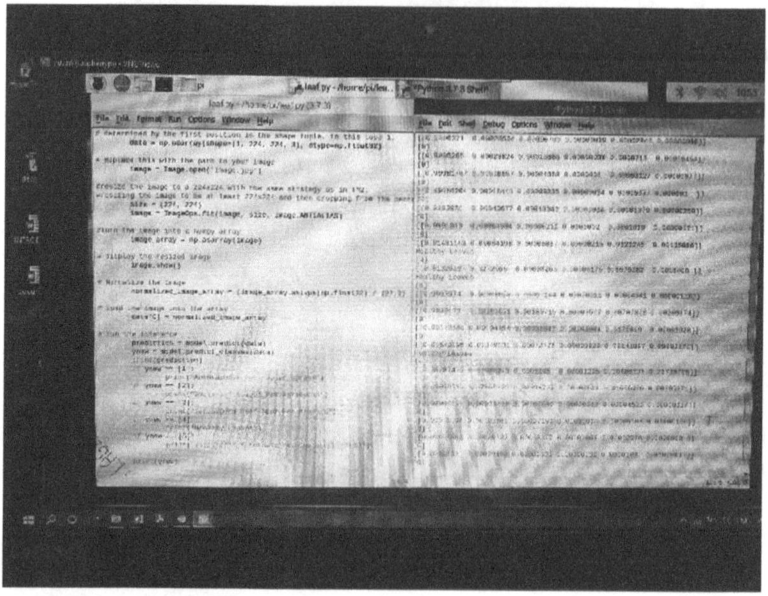

Figure 23.14 Experimental output.

23.7 Conclusion

The proposed Farmer's Eye was done using pretrained leaf models and is effectively used for detecting disease in plants with an accuracy of 85%. The proposed system was implemented and tested successfully and effectively.

References

1. K. A. Garrett, S. P. Dendy, E. E. Frank, M. N. Rouse, and S. E. Travers, "Climate change effects on plant disease: genomes to ecosystems," *Annual Review of Phytopathology*, vol. 44, pp. 489–509, 2006.
2. S. M. Coakley, H. Scherm, and S. Chakraborty, "Climate change and plant disease management," *Annual Review of Phytopathology*, vol. 37, no. 1, pp. 399–426, 1999.
3. S. Chakraborty, A. V. Tiedemann, and P. S. Teng, "Climate change: potential impact on plant diseases," *Environmental Pollution*, vol. 108, no. 3, pp. 317–326, 2000.
4. A. J. Tatem, D. J. Rogers, and S. I. Hay, "Global transport networks and infectious disease spread," *Advances in Parasitology*, vol. 62, pp. 293–343, 2006.
5. J. R. Rohr, T. R. Raffel, J. M. Romansic, H. McCallum, and P. J. Hudson, "Evaluating the links between climate, disease spread, and amphibian declines," *Proceedings of the National Academy of Sciences of the United States of America*, vol. 105, no. 45, pp. 17436–17441, 2008.
6. T. Van der Zwet, "Present worldwide distribution of fire blight", in *Proceedings of the 9th International Workshop on Fire Blight*, vol. 590, Napier, New Zealand, October 2001.
7. S. A. Miller, F. D. Beed, and C. L. Harmon, "Plant disease diagnostic capabilities and networks," *Annual Review of Phytopathology*, vol. 47, pp. 15–38, 2009.
8. L. Semler and J. Furst, "Wavelet-based texture classification of tissues in computed tomography," in *18th IEEE Symposium on Computer-Based Medical Systems (CBMS'05)*, pp. 265–270, Dublin, Ireland, 2005.
9. G. Paschos, "Perceptually uniform color spaces for color texture analysis: an empirical evaluation," *IEEE Transactions on Image Processing*, vol. 10, no. 6, pp.932–937, 2001.
10. X. Liu and D. Wang, "Texture classification using spectral histograms," *IEEE Transactions on Image Processing*, vol. 12, no. 6, pp. 661–670, 2003.
11. M. Pietikäinen, T. Mäenpää, and J. Viertola, *Color Texture Classification with Color Histograms and Local Binary Patterns*, IWTAS, New York, NY, USA, 2002.
12. S. Zenker, E. E. Aksoy, and D. Goldschmidt, "Visual terrain classification for selecting energy efficient gaits of a hexapod robot," in *2013 IEEE/ASME*

International Conference on Advanced Intelligent Mechatronics, pp. 577–584, Wollongong, NSW, Australia, July 2013.
13. Bir Bhanu, Sungkee Lee, John Ming, "Adaptive image segmentation using a genetic algorithm," *IEEE Transactions on Systems, Man and* Cybernetics, vol. 25, no.12, pp. 1543–1567, 1995.
14. ArtiN. Rathod, Bhavesh Tanawal, Vatsal Shah, "Imageprocessing techniques for detection of leaf disease," in *International Journal of Advanced Research in Computer Science and Engineering*, vol. 3, no.11, pp. 397–399, 2013.

24

Solar Powered Density and Emergency-Based Traffic Control System Using NI LabVIEW

M. Devika Rani*, **G. Bhavani, K. Kartheek, A. Sindhura and D. Nikhila**

Department of Electrical and Electronics Engineering, Prasad V Potluri Siddhartha Institute of Technology, Kanuru, Vijayawada, Andhra Pradesh, India

Abstract

The undertaking is aimed at planning a response-based traffic signal system where the exigency of signal timing will change consequently on identifying traffic strength at any intersection. Gridlock is an issue, in which continuous movement of vehicles along an entire network of intersecting streets in all directions comes to a complete halt. Present-day traffic mark out system is time limit-based, where the chance is wasteful for the off traffic paths. To overcome this, we have made a new traffic signal control structure. The heavy density side of the intersection needs longer green time than the standard time. We proposed the controlled operation of red light and green light based on the thickness of traffic present around that surrounded area. This task is achieved by photoelectric proximity Infrared sensors. Based on proper determination of traffic thickness; green light is assisted by the processor. The specified sensors are placed on roadsides. These will identify the existence of vehicles and send the data to the microcontroller, where it will choose a flank to alter the indication lights.

Keywords: Signal timing, traffic strength, traffic signal, infrared sensors, microcontroller

24.1 Introduction

As of late, because of fast expansion in the number of vehicles, gridlock has become a critical issue in numerous places in the world. Because of this issue,

**Corresponding author*: devikamothukuri@gmail.com

A. Chitra, V. Indragandhi and W. Razia Sultana (eds.) Intelligent and Soft Computing Systems for Green Energy, (315–334) © 2023 Scrivener Publishing LLC

there has been a decline in the normal speed of the vehicles. Individuals lose time, miss openings, and get diverted. Gridlock directly impacts organizations.

Because of this gridlock there is a deficiency of cash, profitability from labour, exchange openings are lost, conveyance gets postponed, and accordingly expenses continue to rise. To settle these blockage issues, we need to create new offices and also build up most recent foundation and yet make it accurate. Development of streets and paths isn't conceivable in every situation, yet incorporating knowledge into the streets and paths with trend-setting innovation is surely conceivable. Thus, there is need for a superior and proficient traffic signal framework. The clocks and electrical sensors will decide the timing and operation of programmed traffic signal. In previous designed systems, each of the conversion stage is accommodated with mathematical model in the clock. Based on this, lights are getting ON & OFF upon the clock changes. When appropriately utilized, traffic light signals are significant gadgets for the control of vehicular traffic in a street. They allocate the right-of-way to a decision of traffic developments and thereby profoundly impact traffic stream. Traffic light signals that are appropriately planned, found, worked, and kept up will have at least one of the accompanying benefits:

a) Provide systematic development of traffic
b) Minimize finishing development
c) Coordinated for nonstop development
d) Provide driver certainty by showing the right way.

In order to have a feasible plan for the system and to overcome day-to-day issues such as gridlock, Graphical programming language Lab View is used. NI Lab View is generally graphical code environment which allows system level designers to perform fast designing and testing. In client controlled applications for hardware interface, research facility estimations, information perception and examination, it is composed of highly enriched library capacities and programming devices.

24.2 Literature Review

N. Dinesh Kumar, G. Bharagava Sai, and K. Shiva Kumar [1] revealed that with increased metropolitan traffic, there is a need for new innovations and models for improving traffic control. Due to advanced technology, the number of vehicles has grown and the expansion is required based on the time. So, a reenacting and improved analysis is needed for controlling traffic signals. The simulation model can be improved to control the timespan traffic signal based on traffic analysis. The traffic framework for a metropolitan city is based on thickness of traffic and the effective measures implemented to overcome various issues. This can be even extended to incorporate traffic in the executive framework for a metropolitan city in view of the thickness of traffic. Savvy Approach to Traffic Management utilizing Lab VIEW [2] depicted that as per the World Wellbeing Organization report, India records the most noteworthy number of street mishaps each year. Around 5 lakh (1 lakh = 100,000) mishaps occur on Indian streets, killing about 1.3 lakh individuals and harming about 5.2 lakh every year. These numbers are alarmingly high. Thus, the request of the time is to plan a framework that makes the driver of the vehicle cautious regarding the current street conditions. Aside from it, the execution of the proposed framework is a push to utilize the current innovation sagaciously in our everyday life and thus convey the on-street traffic requirements progressively to the driver.

An Internet of Things (IoT) formed Apt Traffic Management System, SCOPE – Ultrasonic sensor can be utilized to quantify traffic thickness [3–6]. However, Wi-Fi isn't accessible everywhere. Executed in 2019 by Abdul Kadar Muhammad. Minimal effort Traffic Control System for Emergency Vehicles Using ZigBee. In 2019, by M.E Harikumar. Degree – Simple plan yet not cost efficient and burns-through more force. The square outline was attracted in order to give a clarification of how the framework functions initially and the sun-based board gives power which is utilized over dc power supply to control the complete system operation. Sensors provide information to regulator for performing some sensible tasks to control the traffic signals as yield used for controlling traffic at street crossing points. Extension Stolen Vehicles can be recognized; however, consumes more power.

24.3 Methodology

24.3.1 Block Diagram

The way to deal with this plan is acknowledged through the plan and execution of its info subsystem, control unit (control program) and yield subsystem. The info subsystem is made of sensors, modified what's more, carried out utilizing some generally existing standards to accomplish ideal execution. The control unit is acknowledged by a microcontroller-based control program, which deciphers the information and qualifies it to create an ideal yield. The block diagram of the whole framework is shown in Figure 24.1 with the significant parts of the framework.

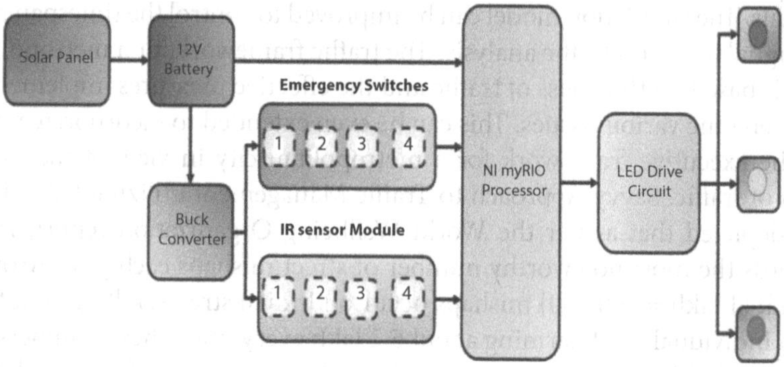

Figure 24.1 Basic block diagram representing various components.

These include:

- Solar Panel
- Power Supply
- Buck Converter
- Emergency Switches
- IR Module
- NI myRIO Processor
- LED Module

The square outline was attracted in order to give a clear clarification of how the framework functions initially, in this manner; the sun-based board gives power which is changed over by the DC power supply utilized to control the sensor clusters, the regulator, and the traffic signals. The sensors give contribution to the regulator which at that point performs some sensible tasks to control the traffic signals as yield utilized for controlling traffic at street crossing points. In picking the sensors, the accompanying highlights were contemplated: precision, range, alignment, goal and reasonableness. Although the infrared (IR) sensors are normally upset by commotion in the encompassing like radiations, surrounding light and so on, they were utilized for this plan since they are modest and promptly accessible on the lookout and are not difficult to interface with.

The block diagram comprises NI myRIO processor, solar panel, battery, buck converter, emergency switches, led drive circuit and led lights which are used for the traffic control.

The emergency switches and IR sensors are interfaced to the processor via port pins. Emergency switches are used to clear the traffic in one particular direction in case of emergency condition; the densities of the vehicles are sensed by the IR sensors. The signals which are coming from the emergency switches and IR sensors are fed to the processor.

Here the solar panel is used for the power supply and it is connected to a chargeable battery; the rating of the battery comes with 12V. Most of the digital circuits, control processors need a 5V supply. To provide regulated 5V power source, we used a buck converter to step down the voltage for the required level.

The LED drive circuit controls the traffic signals based upon the commands given by the processor.

24.3.2 LabVIEW

LabVIEW is a system plan policy and improvement environment for visual programming language from National Instruments.

LabVIEW is a flexible programming language that utilizes graphical user interface (GUI). Due to its generally basic UI, it permits the developer to compose practical projects in no time. Specifically, LabVIEW is appropriate for controlling different equipment gadgets and recording information. LabVIEW is very simple to use and, for generally small projects, simple to troubleshoot. The entirety of the writing computer program is done graphically with symbols and doesn't need such a muddled syntax. Note that contrasted with other programming dialects, the graphical portrayal of LabVIEW utilizes more memory: this reduces the speed of projects. A program like LabVIEW is an ideal prologue to programming. The LabVIEW instructional exercise will comprise the following:

- Getting Started
- Basic Mathematical Functions
- "For" Loops, Sequences, and Plotting
- Boolean Operators, Rings, and Case Structures
- Working with Data Files
- Troubleshooting

For measurement and automation, NI LabVIEW consists of Graphical programming software. Also LabVIEW Real-Time Module, LabVIEW FPGA Module, LabVIEW PDA Module, LabVIEW Datalogging and Supervisory Control Module support this software. Block diagram objects include terminals, Sub VIs, functions, constants, structures, and wires, which transfer data among other block diagram objects.

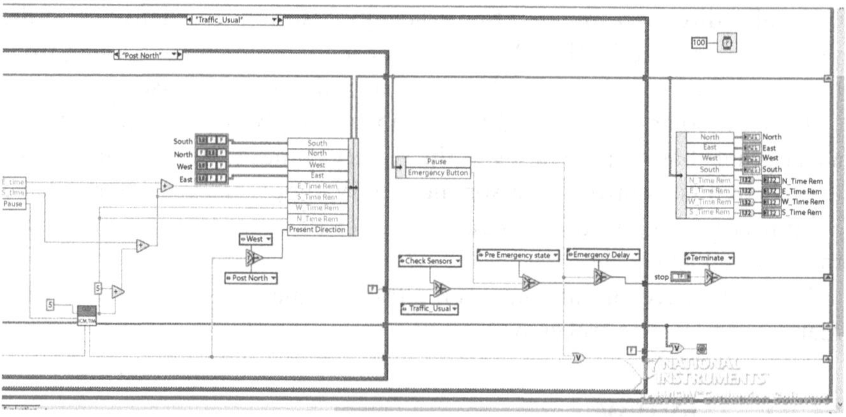

Figure 24.2 Basic block diagram of LabVIEW.

In the working of our project, we have used both software and hardware. In this section, we concentrate on the design of the traffic light using LabVIEW. The blockdiagram shown in Figure 24.2 represents the functionality of the Virtual Instruments (VI) where it contains the graphical source code. The code is termed as graphical source code, G code or block diagram code. For common measurement applications various VIs are available on the LabVIEW template such as sub-VIs, functions, structures, and front panel objects. As the appearance and operation of LabVIEW programs are similar to the physical instruments, these are called as Virtual Instruments or VIs. Every VI uses input from user interface or other sources and displays that information. A VI contains three components:

- Front panel – Serves as the user interface.
- Block diagram – contains the graphical source code that defines the function of the VI.
- Icon and connector pane – identifies the VI so that the VI can be used in another VI. A VIwithin another VI is called a sub-VI. A sub-VI corresponds to a subroutine in text-based programming languages.

LabVIEW is utilized for four primary purposes:

First, Computerized Manufacturing trial of a segment/sub-framework/framework. Second, Robotized Product plan approval of a part/sub-framework/framework. Third, control and additionally checking of a machine/piece of modern hardware/measure. Fourth, condition observing of a machine/piece of mechanical hardware. LabVIEW is utilized across numerous industry verticals inside the assembling domain, including:

- Military/aviation
- Auto
- Semiconductor
- Telecom.
- Energy/power
- Customer gadgets
- PCs and hardware
- Clinical device

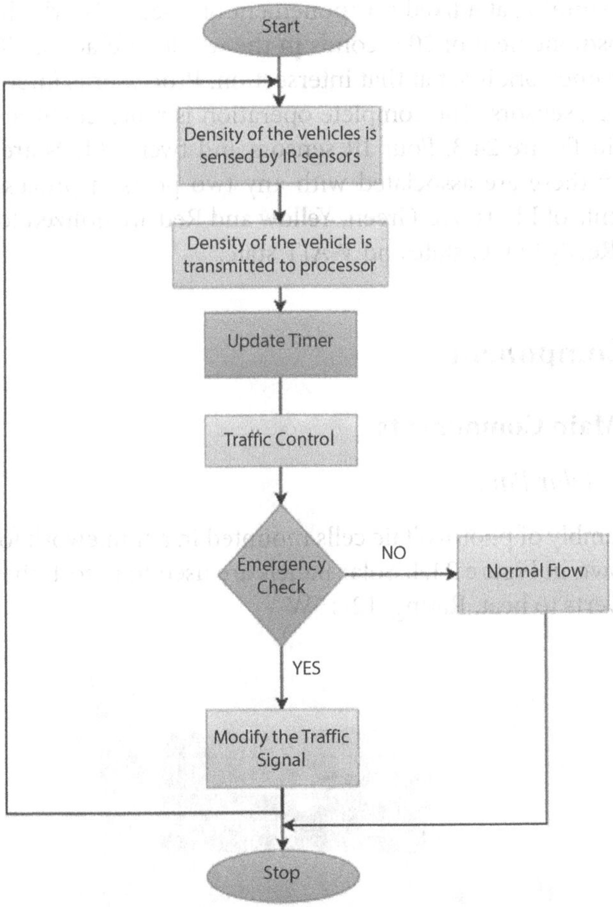

Figure 24.3 Flowchart showing sequence of operation.

Depending on the number of vehicles entering through a portion of the street, the designed system selects the regular measure of changing postponement of traffic lights. Multiple sensors are placed on four sides of four-way street that identify the number of vehicles passing around the region covered by the sensors. To ensure an effective traffic signal system, IR sensors are utilized. An IR sensor contains IR transmitter IR beneficiary (photodiode) in itself. These IR transmitters and IR beneficiaries will be mounted on the same roadsides at a specific distance. As the vehicle goes through these IR sensors, the IR sensor will distinguish the vehicle and will send the data to the microcontroller. The microcontroller will check the quantity of vehicles, and give the shining opportunity to LED by the thickness of vehicles. In the event that the thickness is higher, LED will sparkle for higher time than normal or the other way around. The traffic signals are at first running at a fixed postponement of 5 seconds, which thus produces a postponement of 20 seconds in the whole interaction. This whole installed framework is set at that intersection. Processor is interfaced with LEDs and IR sensors. The complete operation is represented in flowchart as shown in Figure 24.3. Four IR sensors and twelve LEDs are required. In this way these are associated with any two ports of processor. Three arrangements of LEDs, viz. Green, Yellow and Red are utilized to show the GO state, Ready to GO state and WAIT state.

24.4 Components

24.4.1 Main Components

24.4.1.1 Solar Panel

It is an assembly of photovoltaic cells mounted in a framework for installation as shown in Figure 24.4. Solar panels are used to absorb the sun's rays and it converts to heat. Rating: 12/15W

Figure 24.4 Solar panel.

24.4.1.2 *Battery*

Electrochemical cells can be charged electrically to provide a static potential for power or release electrical charge. This combination of cells represents the basic energy source battery shown in Figure 24.5.

Figure 24.5 Battery.

24.4.1.3 *Buck Converter*

To obtain an output voltage less than the input voltage DC-DC buck converter as shown in Figure 24.6 is used with rating 5V, 2A for required output voltage of 5V dc from 12V dc.

Figure 24.6 Buck converter.

24.4.1.4 *IR Sensor*

To sense or identify surrounding happenings an electronic device is used called as Infrared sensor is used as shown in Figure 24.7. Heat measurement and object detection are also possible with this sensor.

Figure 24.7 IR sensor.

24.4.1.5 NI my RIO

myRIO is a portable device as shown in Figure 24.8. It was introduced by National Instruments. It consists of dual core ARM cortex A9 programmable processor and also Xilinx Field Programmable Gate Array. FPGA support in myRIO helps to design complicated systems for solving real-life problems faster than a microcontroller.

Figure 24.8 NI myRIO.

24.4.1.6 NPN Transistor

A p type material is placed between two n type materials which is termed as npn transistor. Used NPN transistor is shown in Figure 24.9. Mainly it is used to amplify weak signals and generates a strong amplifying signal.

Figure 24.9 NPN transistor.

24.4.1.7 Glue Sticks

Glue sticks are little tubes that are used for encapsulation and wire tacking as shown in Figure 24.10.

Figure 24.10 Glue sticks.

24.4.1.8 Heat Sink Slive Tubes

These are shrinkable plastic tubes. Uses of slive tubes are wires insulation, building of cables and protecting the wires from external factors. Used heat sink slive tubes are shown in Figure 24.11.

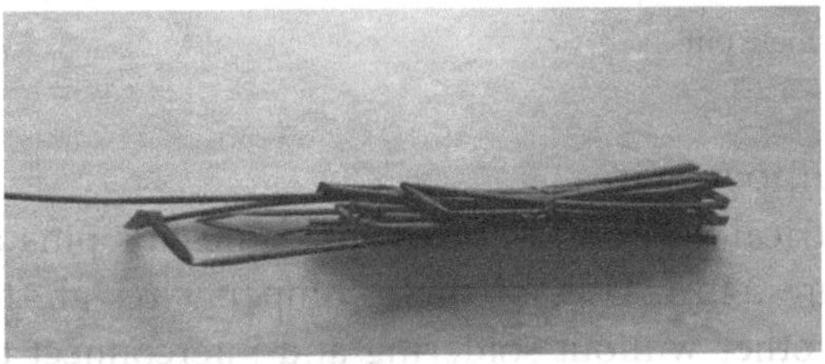

Figure 24.11 Heat sink slive tubes.

24.4.1.9 Toggle Switch

The switch which can be marked by its handle or lever that creates possible environment for control of flow of electric current or signal is the toggle switch as shown in Figure 24.12. It consists of a hinged switch that provides two operating modes, namely ON or OFF.

Figure 24.12 Toggle switch.

24.4.2 Supporting Components

24.4.2.1 Connecting Pins

24.4.2.1.1 DC Male Pin

The pin is mounted in the socket and makes contact with a second internal contact. DC male pin is shown in Figure 24.13. It consists of two parts, the outer part called as barrel and inner part called as tip.

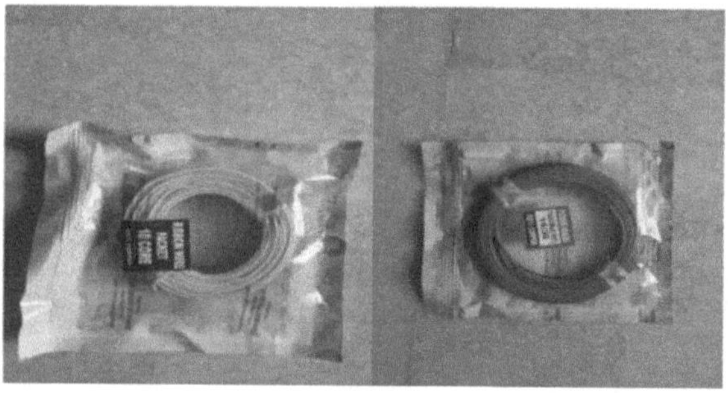

Figure 24.13 DC male pin.

24.4.2.2.2 Jumper Wires

These are electrical wires and consist of connector pins at each end as shown in Figure 24.14. Uses of these jumper wires are to connect two points to each other without soldering and interconnect the breadboard components easily to change the circuit as per need.

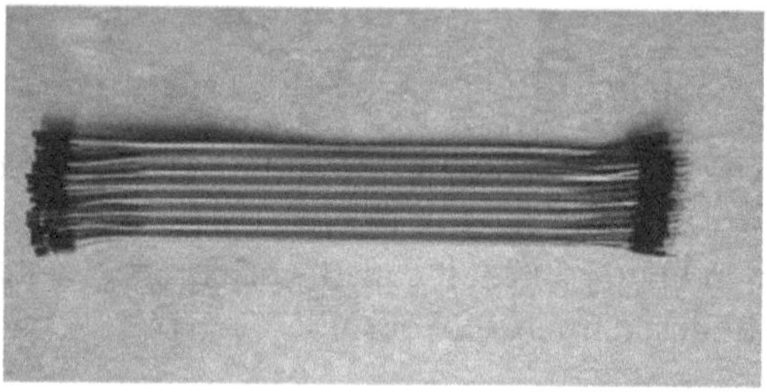

Figure 24.14 Jumper wires.

24.4.2.2 LEDS

24.4.2.2.1 Yellow LED Resistor

Basically, the supply voltage is increased with increasing the resistor value. The forward voltage of this resistor is 1.8V. Yellow LED resistor is shown in

Figure 24.15. These type of LEDs having the resistor values are 75,160,360.510 ohm can be utilized by corresponding voltages are 3.3, 3, 9, 12V respectively.

Figure 24.15 Yellow LED resistor.

24.4.2.2.2 Red LED Resistor

A red LED typically drops around 1.7 to 2.0 volts, but since both voltage drop and light frequency increase with band gap. Red LED resistor is shown in Figure 24.16.

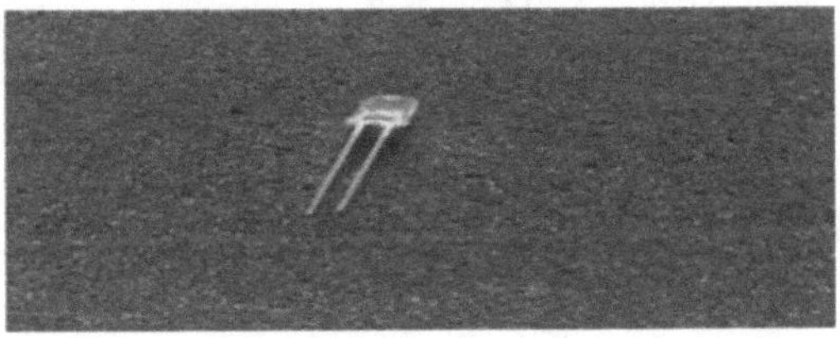

Figure 24.16 Red LED resistor.

24.4.2.2.3 Green LED Resistor

A standard Green LED has forward voltages in the range 1.4V to 2.6V, depending on the desired brightness and the choice of forward current. Green LED resistor is shown in Figure 24.17.

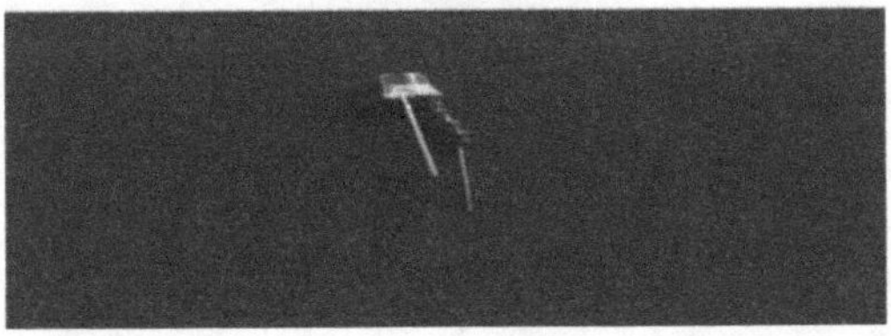

Figure 24.17 Green LED Resistor.

24.4.2.2.4 Resistor

Resistor is an electrical component having two terminals. It can be used for regulating current flow in an electrical circuit. A 1Kohm resistor is used in this electrical circuit.

24.5 Result

24.5.1 SIM View

Complete lab view with all four paths is shown in Figure 24.18. Here the lights change state YELLOW; one is turned ON for 2 seconds and GREEN light is turned on for a period distributed by the "clock subVI". In the clock subVI at first the date is gained utilizing the get date and time in seconds function. Figures 24.19-24.22 shows Green ON in four different directions. When the GREEN light is turned on for the necessary time, after the time gets completed at that point ideal will be the following state to be set off where the GREEN light will be killed and at the same time YELLOW and RED light will be turned on.

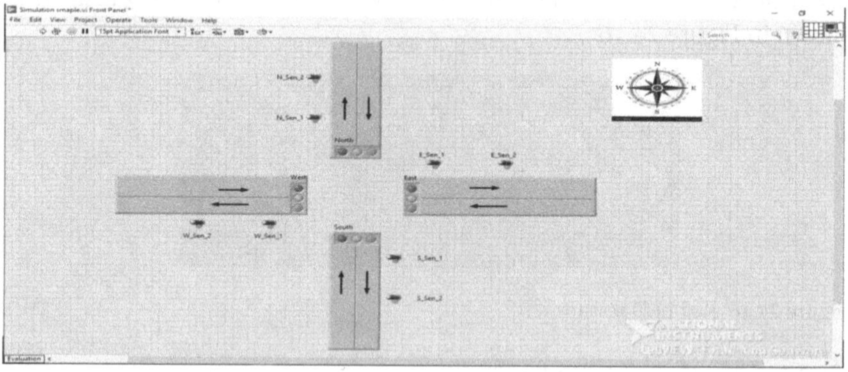

Figure 24.18 Lab view showing all four paths.

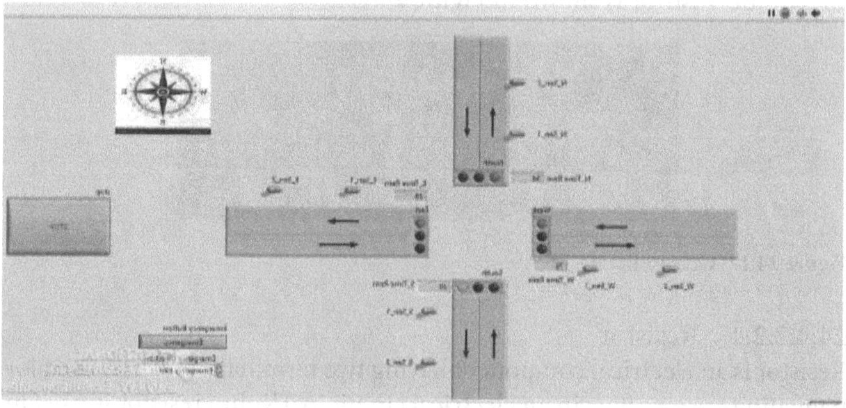

Figure 24.19 Lab view showing Green ON for one direction.

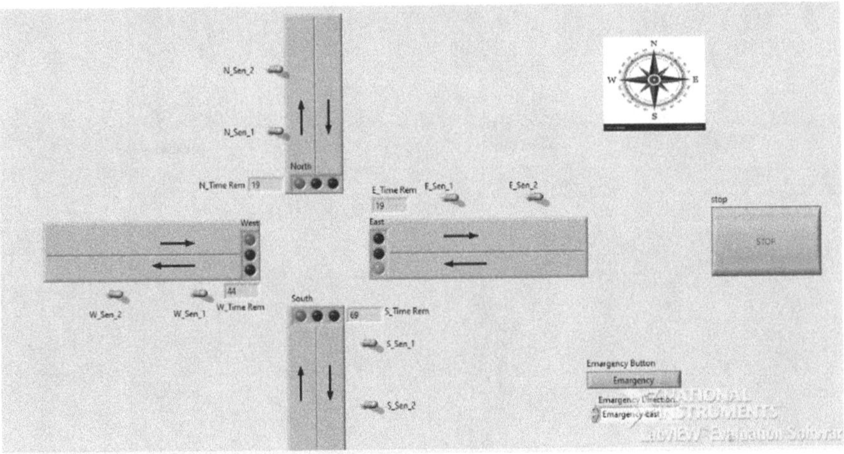

Figure 24.20 Lab view showing Green ON for one direction.

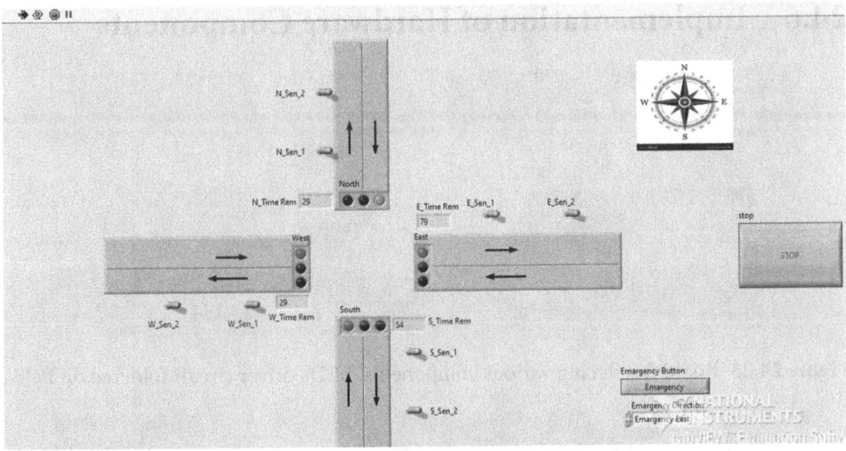

Figure 24.21 Lab view showing Green ON for one direction.

Assuming there an emergency vehicle in one of the paths, priority should be given at that specific side also; GREEN light should be given to that specific side. For that we should take the pictures of the multitude of paths on the double and we need to handle them and find if there is any rescue vehicle and give green light to that specific side. For preparing the application to recognize the rescue vehicle at that specific side we adjust a strategy called as Shading order strategy utilizing NI Vision associate.

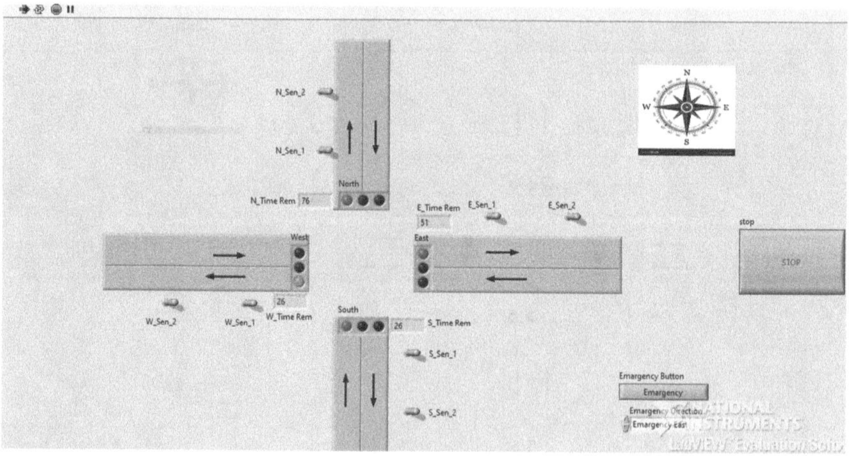

Figure 24.22 Lab view showing Green ON for one direction.

24.6 Implementation of Hardware Components

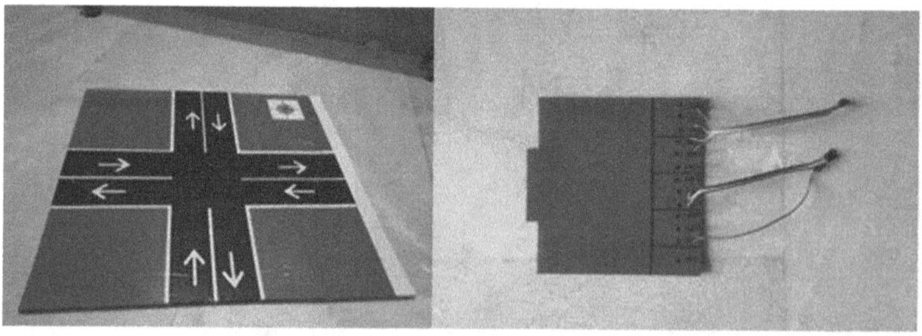

Figure 24.23 Board for placing various components, LED's driver circuit soldered on PCB.

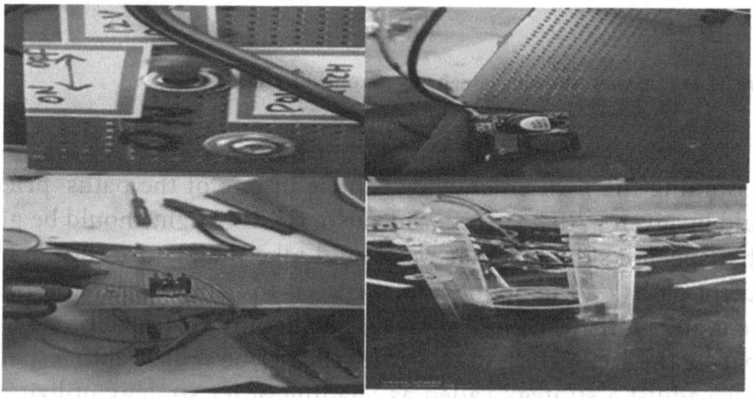

Figure 24.24 Placing various components, soldered on PCB.

The Board used for placing various components is shown in Figure 24.23. LED's driver circuit, other parts are soldered on PCB as shown in Figures 24.24 and 24.25. Initially all the hardware will be in OFF state. After the main switch is turned on, the controller which we are using, that is NI My Rio, will glow after 30 sec. The designed system provides necessary delay of signals based on vehicle count through a particular area or section. Arranged four sensors at four sides count the number of vehicles travelled in that covered sensor area. Also Emergency switch is soldered on PCB as shown in Figure 24.26.

EAST direction is kept as DEFAULT, so after the controller glows, the sensors present in the EAST direction will activate and the GREEN LED will glow in this side. At the 25th second, the YELLOW LED will glow in the EAST DIRECTION; that is an indication that red light is about to be shown in the next immediate step. Similarly, if yellow is shown, start coming to initial condition based on red light indicator. The same happens with the NEXT DIRECTION, i.e., WEST DIRECTION. At the 25th second, the YELLOW LED will glow up in the NORTH DIRECTION. As soon as 30

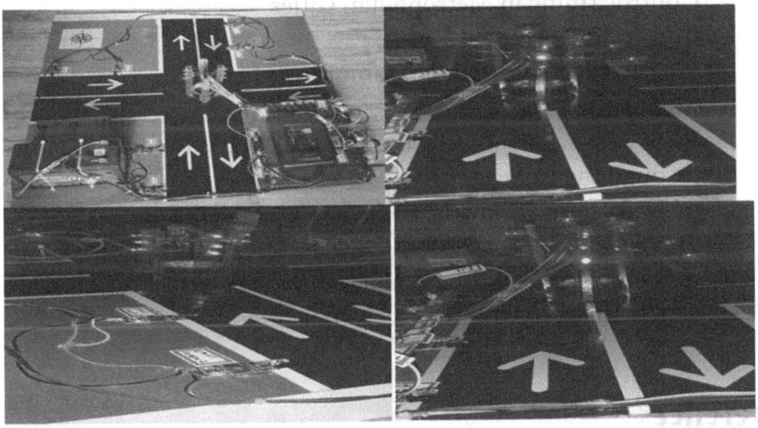

Figure 24.25 Placing various components, soldered on PCB.

Figure 24.26 Emergency switch soldered on PCB, Emergency operation.

seconds completes for NORTH DIRECTION, the GREEN LED glows in the WEST DIRECTION.

24.7 Conclusion

In this design work, to overcome the undesired time wastage and minimize road traffic casualties, a density-based traffic light control system is developed for traffic control at four-road intersections in which the existing conventional traffic light control system has failed to achieve efficiency. The traffic control system we introduced in this project overcame all the difficulties that earlier traffic control systems that were built with timer. So this process is a solution for the previous system.

Applications

- Control Traffic in Metropolitan Cities
- Controlling of Signal Lights.
- We can save a considerable amount of time.
- Emergency.
- With less or zero waiting time, emergency vehicles may pass through the traffic.
- In case of fire accidents, the number of deaths during travel to hospital can be reduced.
- IR sensors are used to evaluate the density of the vehicles which are fixed within a fixed space.

References

1. M. Naveen, S. Raghavendra, D. Imran Basha, & P. Kiran Kumar (2019). Density basedTraffic Signals Controlling Using ARDUINO and IR Sensors. 2019 *International Journal of Electronics Engineering (IJEE)*. ISSN: 0973-7383. Volume 11, Issue 1. pp. 348–351.
2. Saiba P A, Afeefa M U, Aruna T S, Clincy Jose, & Radhika V M (2017). Density BasedTraffic Signal System using PIC Microcontroller. *International Journal of Computer Trends and Technology (IJCTT)*. ISSN: 2231–2803. Volume 47, Number 1.
3. Maheshwari, Pranav, Deepanshu Suneja, Praneet Singh, and Yogeshwar Mutneja. "Smarttraffic optimization using image processing." In *2015 IEEE*

3rd International Conference on MOOCs, Innovation and Technology in Education (MITE), pp. 1–4. IEEE, 2015.
4. Ankit Dhiman, Atul Kumar, Ajay Kumar, Yatin Sokhal, Harpreet Kaur Channi, "Modellingof Traffic Light Control Using LabVIEW" in *International Journal of Scientific Research in Computer Science, Engineering and Information Technology* © 2017 IJSRCSEIT, Vol. 2, Issue 6, ISSN: 2456–3307.
5. Elizabeth Basil, Prof. S. D. Sawant, "IoT based Traffic Light Control System using RaspberryPi" in *International Conference on Energy, Communication, Data Analytics and Soft Computing (ICECDS-2017)*
6. LabVIEW User Manual- www.ni.com/pdf/manuals/320999e.pdf

25

Observation of TCSU: Travel Cold Storage Unit Operated by SPV Technology

Devesh Umesh Sarkar[1]*, Tapan Prakash[1], Madhur Zadegaonkar[2], Ritu Bhimgade[2], Abhijeeta Gupta[2] and Nidhi Ambekar[2]

[1]Electrical Engg., VIT, Vellore, India
[2]Electrical Engg., GHRIET, Nagpur, India

Abstract
The project is aimed at designing a monitoring system for a travel cold storage unit which helps to reduce the wastage of food products that are preserved at frozen temperature due to the lack of detailed observation of the cold storage unit while transporting from manufacturer to dealer. Therefore, it would be advantageous to implement a monitoring system for a cold storage unit based on IoT technology that analyzes the temperature, location and various parameters inside the unit. In this monitoring system, the temperature of the cold storage unit can be observed anytime by the manufacturer and dealer using NodeMCU programmed by Arduino Software. Also, the temperature of the cold storage unit is sensed by a temperature sensor and if the temperature of the unit exceeds the desired temperature then it will give an alert message to both ends by using GSM technology, and the location of the transporting unit is detected through the GPS technology. The system will be operated by solar panel which will provide supply to the cold storage unit. Hence, this system works effectively to avoid unnecessary wastage of food products.

***Keywords*:** NodeMCU, DS18B20-temperature sensor, global positioning system, global system for mobile, peltier plate, SPV, battery

25.1 Introduction

Cold storage is used to preserve something for a longer period of time, as with refrigerators. Cold storage is usually used to keep things like food and agricultural products, and in pharmacies, etc. [1]. It has been the most

**Corresponding author*: devesh.sarkar2020@vitstudent.ac.in

A. Chitra, V. Indragandhi and W. Razia Sultana (eds.) Intelligent and Soft Computing Systems for Green Energy, (335–344) © 2023 Scrivener Publishing LLC

effective and the best way to transport products from one place to another until recently, when news has been spreading of products getting spoiled in transit because no one knows what the temperature is inside the cold storage [1]. In some cases, no one knows if the product inside requires a lower temperature or has any other requirements [2]. Basically, when the product is in transit, we cannot handle it, and that has now become a major concern. We had the opportunity to talk to Shree Vinayaka foods Karanja MIDC about the problem of monitoring cold storage online and the need to give the alert message on mobile to both ends for some errors like high temperature and power failure [2]. So, we decided to work on it by monitoring the cold storage system based on Internet of Things (IoT), where one can access the temperature of the cold storage from anywhere, anytime [3]. Also, our system will give an alert message if anything goes wrong while transporting the products. In this proposal we implement a framework for cold storage management system based on IoT technology by using heterogeneous IoT devices [3]. This is used to preserve the various parameters of yield such as degeneration time, temperature parameter, etc., for longer period [4].

25.2 Working Methodology

A. Description

The solar panel is used to charge the battery and the battery provides supply to the freezer. For monitoring the temperature, DS18b20 (Temperature sensor) temperature sensor is used for this system [5]. It has two dissimilar metals which generate the electrical voltage and is indirectly proportional to change the temperature [6]. If the temperature of the cooling system increases, there is a voltage drop and it is sensed by the sensor. Temperature sensor is connected to NodeMCU. The programming of NodeMCU is done on Arduino IDE Software [6]. The Internet of Things (IoT) is a new research field which connects the physical world objects to the internet in order to monitor them. A temperature sensor records the data which is stored on the cloud server to monitor with respect to date and time as well as in the form of graph, so the temperature of cold storage is continuously monitored through IoT [7]. The cloud server provides wireless communication which is also connected to NodeMCU that sends an alert message if any rise in temperature occurs. By using Global Positioning System (GPS) technique, manufacturer and dealer can get the exact location of a vehicle. Global System for Mobile (GSM) provides wireless communication which is also connected to NodeMCU that sends an alert message if any rise in temperature occurs. By using GPS technique manufacturer and dealer can get the exact location of a vehicle is shown in Figure 25.1.

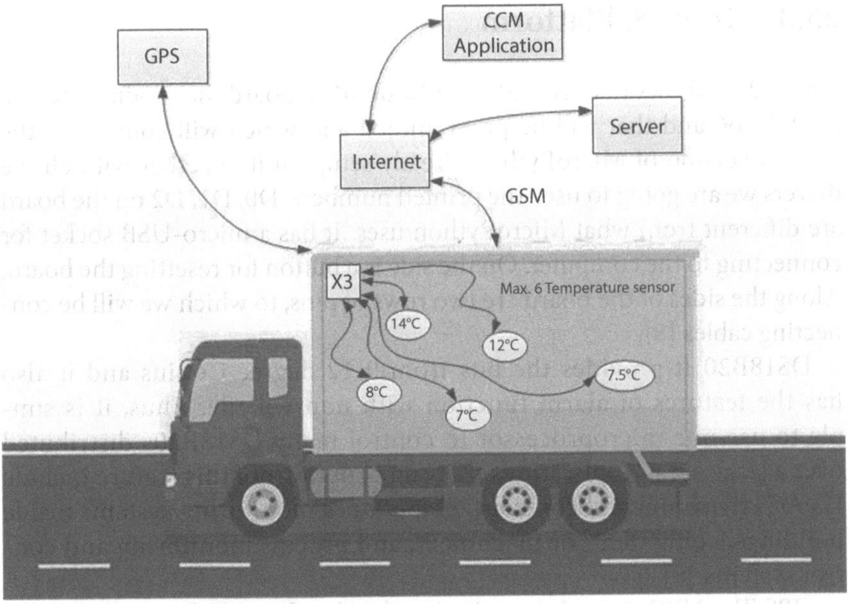

Figure 25.1 Monitoring system diagram.

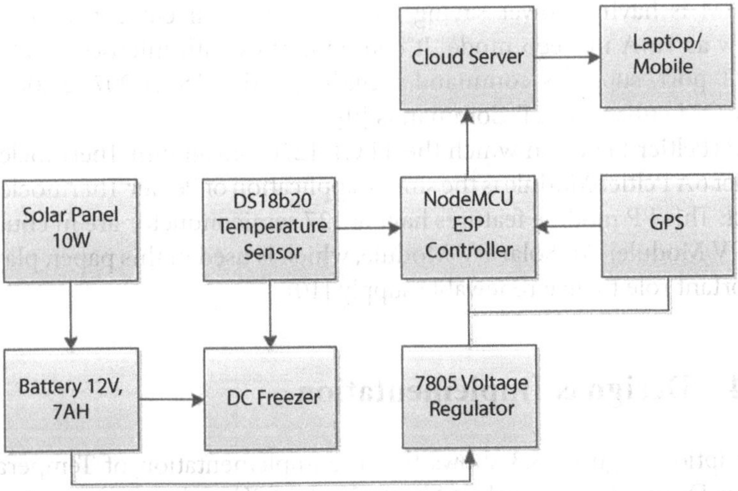

Figure 25.2 Processing diagram.

25.3 Tools & Platform

Figure 25.2 shows that, NodeMCU: NodeMCU board the module include is ESP8266 and that will be programmed and which will come with the current version of MicroPython already setup on it, together with all the drivers we are going to use. The printed numbers D0, D1, D2 on the board are different from what MicroPython uses. It has a micro-USB socket for connecting to the computer. On the side is a button for resetting the board. Along the sides of the board are two rows of pins, to which we will be connecting cables [8].

DS18B20 It provides the bits from 9-12 degree Celsius and it also has the features of alarm function with non-volatile. Thus, it is simple to use one microprocessor to control many DS18B20s distributed over a large area. Applications that can benefit from this feature include HVAC environmental controls, temperature monitoring systems inside buildings, equipment, or machinery, and process monitoring and control systems [8].

GPS The NEO-6 module series is a family of stand-alone GPS receivers featuring the u-blox 6 positioning engines whose performance is very high. So its cost-effective receivers offer numerous connectivity options in a miniature 16 x 12.2 x 2.4 mm package. Their flexible-architecture with their power & memory keys makes NEO-6 modules which is ideal for mobile devices operated by battery with space constraints [9].

GSM is having power saving technique, the current consumption is as low as 1mA in sleep mode. It communicates with microcontroller via UART port, supports command including 3GPP TS 27.007, 27.005 and SIMCOM enhanced AT Commands [9].

PP (Peltier Plate): In which the TEC1-12706 40x40mm Thermoelectric Cooler 6A Peltier Module is the simple application of Peltier Thermoelectric Effect. This PP module features having 127 semiconductor are in coupled.

SPV Module: An Solar PV Module, which is used in this paper, plays an important role to give renewable supply [10].

25.4 Design & Implementation

Description: Figure 25.3 shows that the implementation of Temperature Sensor DS18B20, GPS and GSM into NodeMCU is shown above through the pin diagram of NodeMCU ESP8266 [10]. The supply of 5V is provided to the VCC of NodeMCU. Temperature sensor DS18B20 is connected to

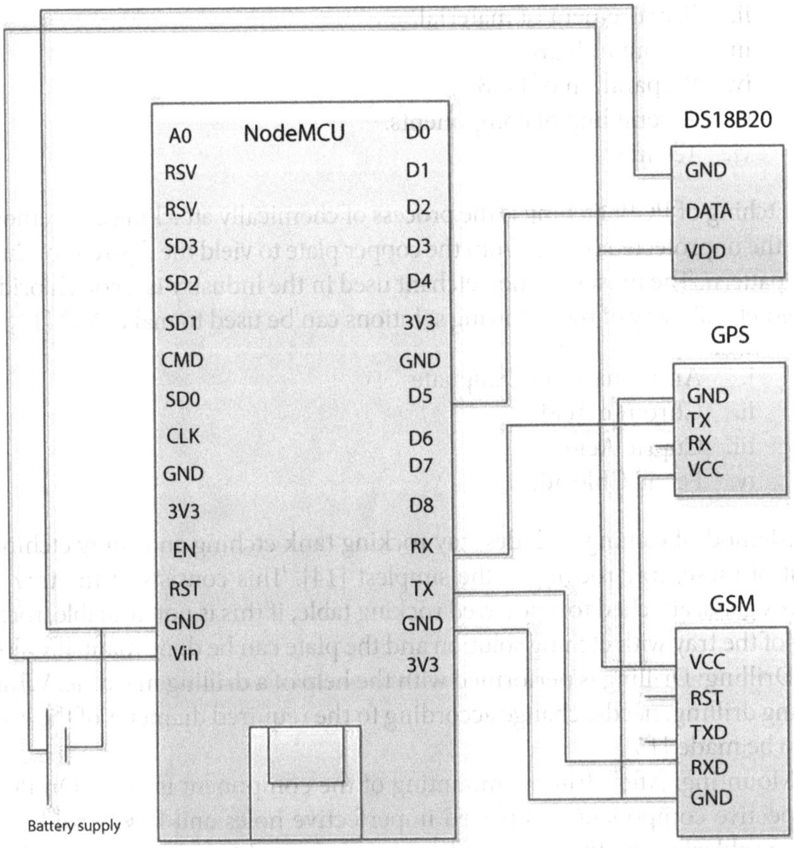

Figure 25.3 Connection diagram.

D5 of NodeMCU to sense theongoing temperature inside the cold storage unit and transmit the data through the GSM and the ground of temperature sensor is connected at the GND of NodeMCU [11]. The Transmission pin TX of GPS is connected to the Receiver pin RX of NodeMCU so that the GPS will transmit the data and the NodeMCU will receive the data through the receiver pin RX and the ground of GSM is connected at the GND of NodeMCU. Also, the Transmission pin TX of NodeMCU is connected to the receiver pin RX of GSM so that the GSM will receive the data from NodeMCU and operate its function by transmitting an alert and the ground of GSM is connected at the GNDof NodeMCU [12].

PCB Designing: Following are the steps for PCB designing:

i. Designing of actual material.

ii. Procurement of material.
iii. Layout of PCB.
iv. Preparation of PCB.
v. Assembling of components.
vi. Testing.

Etching of PCB: Etching is the process of chemically attacking and removing the unprotected copper from the copper plate to yield the desired conductor pattern. The most common etchant used in the industry is ferric chloride. Theoretically, any of the following solutions can be used to make PCB [13]:

i. Ammonium Per Sulphate
ii. Chromic Acid
iii. Cupric Acid
iv. Ferric Chloride

Method of etching includes tray rocking tank etching and spray etching. Out of these, tray rocking is the simplest [14]. This consists of the tray of Pyrex glass, attached to a powered rocking table; if this is not available, rocking of the tray with etching solution and the plate can be done manually also.

Drilling: Drilling is performed with the help of a drilling machine. While doing drilling, needle change according to the required diameter of the hole is to be made [15].

Mounting: After drilling, mounting of the component is done. On PCB respective component was placed imperfective holes and finally soldered. After soldering the PCB was ready to be connected to the respective relays and supply. Before then wiring diagram areas are drawn which decide the external wire connection to the PCB [15, 16].

Testing: The main objective of the testing is to check the output performance as per our assumption.

Advantages of PCB: Advantages of PCB over normal wiring are as follows [16]:

i. PCB is necessary for interconnection of a large number of electronics component in a very small area with minimum parasitic wiring effects.
ii. PCB is stable for mass production with less chance of wiring error.
iii. Small component can be easily mounted on PCB.
iv. Construction is neat, small and truly a work of art.
v. By using PCB, the electronic equipment becomes more reliable in size and less costly.

25.5 Advantages & Application

Advantages:

i. Reduces wastage of food products while transporting the food products from manufacturer to dealer.
ii. Proper information of the product's temperature and location is provided to the manufacturer and dealer.
iii. Automatic temperature is being monitored in the cold storage unit.
iv. This Cold Storage Monitoring System is a simple and affordable system.

Application:

i. For industrial application.
ii. For pharmaceutical application.
iii. For logistics industries.

25.6 Conclusion

The cold storage unit has been monitored online and gave the alert message on mobile via GSM to the manufacturer and dealer when errors occurred like high temperature and power failure. Using IoT, detection of different possibilities of problems occurring while transportation resulting in wastage of food products has been avoided. Fuel consumption has been reduced by using the solar operated cold storage unit.

By using this monitoring system, wastage of food products has been avoided and monitoring of travel cold storage unit has been done by using IoT. The manufacturer and the dealer has received the proper information about the respective cold storage unit regarding its temperature through an SMS with the help of GSM and also, they have received the exact location of the vehicle with the help of GPS.

Future scope of proposed work is as follows:

i. Solar panels can be used to charge the battery that operates the cold storage unit so this can save the charging of the vehicle battery.

ii. Manufacturer and dealer can detect the live temperature inside the cold storage unit and location of the vehicle anytime and anywhere through a website.
iii. Cold storage unit can be operated with the battery which can be charged with the help of solar panels instead of operating with the help of vehicle battery.

References

1. Mario Frustaci, "WSN based Online Parameter Monitoring in Cold Storage Warehouses in Cloud Using IoT Concepts", *IEEE Internet of Things Journal*, e-ISSN: 2327-4662, Volume 05, Issue 04, 27 October 2017.
2. V. C. Chandanashree, "Cold Storage Traceability System", *IEEE Transactions on Cold Storage Traceability System*, e-ISSN: 1558-2515, Volume 21, Issue 3, 30 December 2018.
3. Y.B. Kim, "Paper for the realization of humidity and temperature sensor", *IEEE Sensors Journal*, e-ISSN: 1558-1748 Volume 15, Issue 5, 19 December 2017.
4. Matthew Clayton, "Battery energy storage for power generated by solar pane", *IEEE Transactions on Smart Grid*, e-ISSN: 1949-3061, Volume 3, Issue 2, 11 May 2017.
5. Wai-Ru Lai, "Analysis and modeling of dual-band GSM networks", *Journal of Communications and Networks*, e-ISSN: 1976-5541, Volume 1, Issue 3, Sept. 2018,
6. Pierre Jacques Verlinden, "Performance of Rooftop Solar PV System with Crystalline Solar Cells", *IEEE Journal of Photovoltaics*, e-ISSN: 2156-3403, Volume 6, Issue 1, 26 October 2017.
7. Zhao X, "The Design of the Internet of Things Solution for Food Supply Chain", *IEEE Journal Transactions*, e-ISSN: 5386-7266, Volume 5, Issue 2, May 2017, pp. 30-33.
8. Abel Avitesh Chandra & Seong Lee, "Advanced Monitoring of Cold Chain using wireless sensor network and sensor cloud", *Electronic Journals on Sensors and Applications*, e-ISSN: 5386-3852, Volume 10, Issue 1, 19 March 2018, pp. 228-234.
9. Manoj Kumar Tyagi and Raghu Ram, "Warehouse monitoring and management with the help of Solar energy", *International Journal of Engineering Science and Advanced Technology*, e-ISSN: 2395-0056, Volume 12, Issue 3, 23 July 2017, pp. 503-507.
10. Amrita Srivastava, Ankita Gulati, "iTrack: IoT framework for Smart Food Monitoring System", *International Journal of Computer Applications*, 2016.

11. Qinghua Zhang, Yi Wang, "Warehouse environment monitoring system based on WSN", *IEEE 9th Conference on Industrial Electronics and Applications (ICIEA)*, 2014.
12. Abel Avitesh Chandra, Seong-Ro Lee, "Advanced Monitoring of cold chain using WSN", *International electronic conference on sensors and applications*, 2014.
13. Yi-ping Chen, Jian Wang, "A Corn Warehouse Monitoring System Based on IOT", *International Conference on Materials, Manufacturing and Mechanical Engineering*, 2016.
14. A.Sathish Kumar, S.Sudha, "Design of Wireless Sensor Network based Fuzzy Logic Controller for a Cold Storage System", *IEEE 7th power India International Conference (PIICON)*, 2016.
15. Manoj Kumar Tyagi, Balanagu Raviteja, Rammohanarao Koduri and Gangi Raghu Ram, "Warehouse monitoring and management with the help of WSN and Solar energy", *International Journal of Engineering Science and Advanced Technology*, Volume 2, Issue 3, pp 503–507, 2012.
16. Sandeep Kaushik, Charanjeet Singh, "Monitoring and Controlling in Food Storage System Using Wireless Sensor Networks Based on Zigbee & Bluetooth Modules", *International Journal of Multidisciplinary in Cryptology and Information Security*,Vol. 2, no. 3, 2013.

Index

Also of Interest

Check out these other related titles from Scrivener Publishing

SMART GRIDS AND INTERNET OF THINGS, Edited by Sanjeevikumar Padmanaban, Jens Bo Holm-Nielsen, Rajesh Kumar Dhanaraj, Malathy Sathyamoorthy, and Balamurugan Balusamy, ISBN: 9781119812449. Written and edited by a team of international professionals, this groundbreaking new volume covers the latest technologies in automation, tracking, energy distribution and consumption of Internet of Things (IoT) devices with smart grids.

SMART GRIDS AND GREEN ENERGY SYSTEMS, Edited by A. Chitra, V. Indragandhi and W. Razia Sultana, ISBN: 9781119872030. Presenting the concepts and advances of smart grids within the context of "green" energy systems, this volume, written and edited by a global team of experts, goes into the practical applications that can be utilized across multiple disciplines and industries, for both the engineer and the student.

SMART GRIDS AND MICROGRIDS: Concepts and Applications, Edited by P. Prajof, S. Mohan Krishna, J. L. Febin Daya, Umashankar Subramaniam, and P. V. Brijesh, ISBN: 9781119760559. Written and edited by a team of experts in the field, this is the most comprehensive and up to date study of smart grids and microgrids for engineers, scientists, students, and other professionals.

MICROGRID TECHNOLOGIES, Edited by C. Sharmeela, P. Sivaraman, P. Sanjeevikumar, and Jens Bo Holm-Nielsen, ISBN: 9781119710790. Covering the concepts and fundamentals of microgrid technologies, this volume, written and edited by a global team of experts, also goes into the practical applications that can be utilized across multiple industries, for both the engineer and the student.

INTEGRATION OF RENEWABLE ENERGY SOURCES WITH SMART GRIDS, Edited by A. Mahaboob Subahani, M. Kathiresh and G. R. Kanagachidambaresan, ISBN: 9781119750420. Provides comprehensive coverage of renewable energy and its integration with smart grid technologies.

INTEGRATED GREEN ENERGY SOLUTIONS VOLUME 2, Edited by Milind Shrinivas Dangate, W. S. Sampath, O. V. Gnana Swathika, and Sanjeevikumar Padmanaban, ISBN: 9781394193660. This second volume in a two-volume set continues to present the state of the art for the concepts, practical applications, and future of renewable energy and how to move closer to true sustainability.

INTEGRATED GREEN ENERGY SOLUTIONS VOLUME 1, Edited by Milind Shrinivas Dangate, W. S. Sampath, O. V. Gnana Swathika, and Sanjeevikumar Padmanaban, ISBN: 9781394193660. This first volume in a two-volume set presents the state of the art for the concepts, practical applications, and future of renewable energy and how to move closer to true sustainability.

RENEWABLE ENERGY SYSTEMS: Modeling, Optimization, and Applications, Edited by Sanjay Sharma, Nikita Gupta, Sandeep Kumar, and Subho Upadhyay, ISBN: 9781119803515. Providing updated and state-of-the-art coverage of a rapidly changing science, this groundbreaking new volume presents the latest technologies, processes, and equipment in renewable energy systems for practical applications.

Encyclopedia of Renewable Energy, by James G. Speight, ISBN: 9781119363675. Written by a highly respected engineer and prolific author in the energy sector, this is the single most comprehensive, thorough, and up to date reference work on renewable energy.

INTELLIGENT RENEWABLE ENERGY SYSTEMS: Integrating Artificial Intelligence Techniques and Optimization Algorithms, Edited by Neeraj Priyadarshi, Akash Kumar Bhoi, Sanjeevikumar Padmanaban, S. Balamurugan, and Jens Bo Holm-Nielsen, ISBN: 9781119786276. This collection of papers on artificial intelligence and other methods for improving renewable energy systems, written by industry experts, is a reflection of the state of the art, a must-have for engineers, maintenance personnel, students, and anyone else wanting to stay abreast with current energy systems concepts and technology.

POWER ELECTRONICS FOR GREEN ENERGY CONVERSION, Edited by Mahajan Sagar Bhaskar, Nikita Gupta, Sanjeevikumar Padmanaban, Jens Bo Holm-Nielsen, and Umashankar Subramaniam, ISBN: 9781119786481. Written and edited by a team of renowned experts, this exciting new volume explores the concepts and practical applications of power electronics for green energy conversion, going into great detail with ample examples, for the engineer, scientist, or student.

INTEGRATION OF RENEWABLE ENERGY SOURCES WITH SMART GRIDS, Edited by A. Mahaboob Subahani, M. Kathiresh and G. R. Kanagachidambaresan, ISBN: 9781119750420. Provides comprehensive coverage of renewable energy and its integration with smart grid technologies.

Green Energy: Solar Energy, Photovoltaics, and Smart Cities, Edited by Suman Lata Tripathi and Sanjeevikumar Padmanaban, ISBN: 9781119760764. Covering the concepts and fundamentals of green energy, this volume, written and edited by a global team of experts, also goes into the practical applications that can be utilized across multiple industries, for both the engineer and the student.

Energy Storage, Edited by Umakanta Sahoo, ISBN: 9781119555513. Written and edited by a team of well-known and respected experts in the field, this new volume on energy storage presents the state-of-the-art developments and challenges in the field of renewable energy systems for sustainability and scalability for engineers, researchers, academicians, industry professionals, consultants, and designers.

Energy Storage 2nd Edition, by Ralph Zito and Haleh Ardibili, ISBN: 9781119083597. A revision of the groundbreaking study of methods for storing energy on a massive scale to be used in wind, solar, and other renewable energy systems.

Hybrid Renewable Energy Systems, Edited by Umakanta Sahoo, ISBN: 9781119555575. Edited and written by some of the world's top experts in renewable energy, this is the most comprehensive and in-depth volume on hybrid renewable energy systems available, a must-have for any engineer, scientist, or student.

Progress in Solar Energy Technology and Applications, edited by Umakanta Sahoo, ISBN: 9781119555605. This first volume in the new groundbreaking series, Advances in Renewable Energy, covers the latest concepts, trends, techniques, processes, and materials in solar energy, focusing on the state-of-the-art for the field and written by a group of world-renowned experts.

A Polygeneration Process Concept for Hybrid Solar and Biomass Power Plants: Simulation, Modeling, and Optimization, by Umakanta Sahoo, ISBN: 9781119536093. This is the most comprehensive and in-depth study of the theory and practical applications of a new and groundbreaking method for the energy industry to "go green" with renewable and alternative energy sources.

Nuclear Power: Policies, Practices, and the Future, by Darryl Siemer, ISBN: 9781119657781. Written from an engineer's perspective, this is a treatise on the state of nuclear power today, its benefits, and its future, focusing on both policy and technological issues.

Zero-Waste Engineering 2nd Edition: A New Era of Sustainable Technology Development, by M. M. Kahn and M. R. Islam, ISBN: 9781119184898. This book outlines how to develop zero-waste engineering following natural pathways that are truly sustainable using methods that have been developed for sustainability, such as solar air conditioning, natural desalination, green building, chemical-free biofuel, fuel cells, scientifically renewable energy, and new mathematical and economic models.

Sustainable Energy Pricing, by Gary Zatzman, ISBN: 9780470901632. In this controversial new volume, the author explores a new science of energy pricing and how it can be done in a way that is sustainable for the world's economy and environment.

Sustainable Resource Development, by Gary Zatzman, ISBN: 9781118290392. Taking a new, fresh look at how the energy industry and we, as a planet, are developing our energy resources, this book looks at what is right and wrong about energy resource development. This book aids engineers and scientists in achieving a true sustainability in this field, both from an economic and environmental perspective.

The *Greening of Petroleum Operations*, by M. R. Islam *et al.*, ISBN: 9780470625903. The state of the art in petroleum operations, from a "green" perspective.

Printed in the USA/Agawam, MA
September 16, 2026

910661.105